THE COMPLETE **HOW TO**
FIGURE IT

Also by Darrell Huff

How to Lie with Statistics
How to Take a Chance
Cycles in Your Life
Score: The Strategy of Taking Tests
The Complete Book of Home Improvement
How to Work with Concrete and Masonry
Twenty Careers of Tomorrow (with Francis Huff)
The Dog That Came True

THE COMPLETE HOW TO FIGURE IT

DARRELL HUFF

ILLUSTRATED BY CAROLYN R. KINSEY · DESIGNED BY KRISTY MARIA HUFF

W. W. NORTON & COMPANY

New York   London

First published as a Norton paperback 1999

For information about permission to reproduce selections of this book, write to
Permissions, W. W. Norton & Company, Inc., 500 Fifth Avenue, New York, NY 10110.

Library of Congress Cataloging-in-Publication Data

Huff, Darrell.
The complete how to figure it / Darrell Huff with Kristy Maria Huff :
illustrated by Carolyn R. Kinsey.
p.   cm.
Includes index.
ISBN 0-393-03600-6
1. Mathematics—Popular works.   2. Finance, Personal.
I. Huff, Kristy Maria.   II. Title.
QA93.H785   1996

640'.151—dc20   95–46480
                                                       CIP

ISBN 0-393-31924-5 pbk.

W. W. Norton & Company, Inc., 500 Fifth Avenue, New York, N.Y. 10110
www.wwnorton.com

W. W. Norton & Company Ltd., 10 Coptic Street, London WC1A 1PU

3 4 5 6 7 8 9 0

Just a word...

Do you believe with Socrates (who seems to have said it first) and me that the unexamined life is not worth living?

I think you could say the same of the unquantified life. HOW MUCH can be as vital as WHAT.

That, I guess, was the point of a TV show, "Math... Who Needs It?!." With people like Bill Cosby and celebrated teacher Jaime Escolante performing, you knew the answer was going to be "just about everybody."

There lies my justification for offering what may look suspiciously like Everything You Always Wanted to Know About Math But Had the Native Good Sense Not to Ask.

Let me say that I don't think what you hold in your hands is a math book, anyway. I'd rather call it an enabler, and I hope you came upon this copy on the self-help shelves rather than in an obscure corner marked math.

Any mathematical manipulations you learn, or relearn, as you go can be a substantial bonus. But with that miraculous instant-math machine the pocket calculator, a great little enabler itself, you can make use of just about anything in this volume with only cursory recollection of the stuff Miss Fiditch tried to pound into your head in high school.

She didn't have much luck pounding me either. But when I reached a time of hoarding or borrowing or exchanging dollars and choosing timbers and figuring floor loads, the math side of things suddenly came alive for me.

Since so much of what's in this book comes out of my daily life as a family person, it may be relevant that so does the production of the tome itself.

Over the years of collection and creation, my wife Frances made fair copy from foul (when she might have preferred to turn her fingers to her own writing) and thus she performed a very special service in those pre-computer days of typing and retyping. Manually typed copy soon gave way to what emerged from an imposing Selectric—until it in its turn surrendered to the marvels of a 26-pound portable computer called Kaypro. In the end, everything made its way through computer programs variously called Perfect

Writer, Word Star, Word Perfect, and finally to just plain Word for Macintosh. The tool used was Pagemaker. The typewritten part was scanned electronically from paper to computer, acquiring weird errors along the way. These had to be detected and quashed by our granddaughter Lisa Schulman.

Her mother, Kathryn Allan, who is often sharper than I am in matters of math, sedulously read not only proof but raw copy, saving both you and me from many a sad ambiguity or worse. My old college room-mate Jake Coniglio and his wife Faith pitched in here, too.

Kristy Maria Huff, though immersed in computer-oriented graduate study in architecture at the U of Oregon, somehow wrestled a couple of hundred drawings into place within twice that many subchapters to design and create a coherent book out of a mishmash—complete with formatting, subheads, legible type, understandable charts, well-placed illustrations, and other good stuff. The body type and heads are Palatino. Chapter summaries are Architect, knocked out (says the designer Hank Gillette) to simulate lettering on a blueprint.

A friend, David Shellabarger, and our grandsons Will Allan (who did many of the charts and tables) and Jesse Huff-Christensen and grandaughter Maia Huff-Owen provided computer savvy and flying keyboard fingers, while keeping some home fires burning with awesome patience.

The major contribution of another daughter, Carolyn Kinsey, speaks for itself all through the book—more than 200 light-hearted drawings.

These, of course, are also the folks responsible for any and all deficiencies you may find on these pages, so it's quite acceptable to bring errors to my attention.

After saying a good word for every math teacher everywhere, I need only add the name of an inordinately patient editor, Hilary Hinzmann, to complete the list of those to whom this volume is hereby dedicated.

I hope you'll put to use the blank pages you find here and there. Recycle them with supplementary notes or computations of your own so the book will know it is being used.

And don't feel mercenary if you make frequent reference to the financial calculations. We have Woody Allen's word that "Money is better than poverty, if only for financial reasons."

CONTENTS

THE COMPLETE HOW TO FIGURE IT

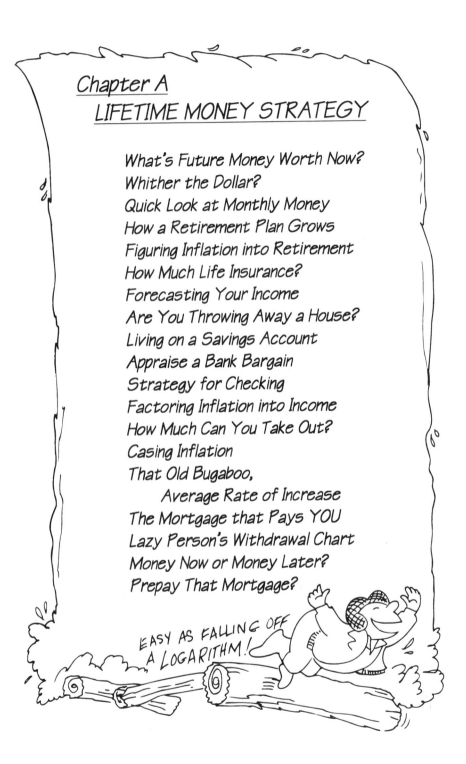

Chapter A
LIFETIME MONEY STRATEGY

EASY AS FALLING OFF A LOGARITHM!

What's Future Money Worth Now?

W<small>HICH</small> is worth more, a $5,000 cash gift today or a guaranteed $15,000 twenty years from now?

Even putting aside the unpredictable personal aspects, is it really possible to give a sensible answer to this—when you have no way to predict the inflation rate to come?

Surprisingly, it is.

To turn your answer from a wild guess into a good working approximation, you need only accept one reasonable probability. This is an assumption that the net (meaning after-taxes) interest rate you can expect to earn will exceed inflation by 6 percent.

If the long-range outlook at the moment of decision seems to favor 5, or 4, or 7 percent, go with that instead. (Wherever you're placing your investment, there should be someone about who can give you current figures.)

Then you won't have to think about the actual rates of either inflation or interest. You're reducing your gamble by noting the historic tendency for those two things to move up and down in harmony.

That said, let's try the computation for $5000 to be tucked away for 20 years at 6 percent interest compounded twice a year.

Begin by expressing the chosen rate of interest as a decimal, by dividing that 6 percent by 100 to get 0.06. That's the rate for a year, so it must be divided by 2 to get the rate for each half year, our chosen compounding period.

To this result, add 1.

That sum, 1.03, must now be raised to the power of the number of compounding periods, which is 40 in 20 years. The 40th power of 1.03 is 3.262 (found in 20 seconds on a simple pocket calculator, by entering 1.03, pressing the times key once and the equals key 39 times. Or times once, equals four times, then times once, equals seven times).

Dividing this result into 1 (finding the reciprocal, to use mathspeak) and multiplying by the specified $15,000 produces the answer you're looking for. If our 6 percent premise proves psychic, it would take only $4598.35 invested now to become $15,000 in 20 years.

If that's all it takes to grow a $15,000 crop in 20 years, I'll take that offer of a full $5000 now—and produce a bigger crop than the offered $15,000.

How big? You can wait a couple of decades and then look at your statement from the money fund where you're invested. Or, considerably more quickly and almost as easily, you can apply the tricks just described to this slightly different recipe.

This time all you have to do is find the 40th power of the periodic interest rate and then multiply that by the number of dollars you'd be starting with.

Once again you arrive at 3.262 for the 40th power of 1.03. Multiplying that by the $5000 gift you started with, you find that 20 years from now you could expect to have $16,310.19 if you haven't had any truck with the tax folks in the meantime.

Whither the Dollar?

FOR AS LONG as we've had inflation—and that's practically forever and everywhere and most of the time—people have made disparaging remarks about our "fifty-cent dollar." Those of us who have been around quite a while are likely to make that "ten-cent."

Actually, though, how long does it take for inflation to cut the value of a currency in half?

The formula is a short one. It says divide the logarithm of 2 by the log of the inflation rate (as a percentage) plus 1.

Given the 6 percent rate of inflation that has been fairly typical of some recent periods, it goes like this. From a table of logarithms, the log of 2 is .30103 and the log of 1 plus .06 (1.06) is 0.025306. Dividing the first of these by the second produces 11.9, which says that the dollar is losing half its value in just under 12 years.

To borrow a term from the physical sciences, its half life is 11.9 years.

It's a happy fact that this kind of problem can be worked backwards. That way you don't have to bother with logarithms at all.

Approaching the previous problem from a slightly different angle, suppose you want to know by what factor the value of your buck is going to be reduced in, say, 12 years.

Just add 1 to the annual rate of depreciation (in decimal form, so that 6 percent is 0.06).

Raise this to the 12th power. Remember that on most simple pocket calculators, this can be done by striking the times key once and the equals key 11 times.

If you've never used this trick, you might test it first by seeing if hitting 2, then times, then equals produces 4. If not, hit times twice instead of once before equals.

It turns out that 1.06 to the 12th power is 2.012, confirming the previous conclusion that in 12 years a dollar declining 6 percent a year will have slipped by a factor of about 2—thus keeping only half its present value.

Now here's how to use this method as an alternative to consulting a log table. We're sometimes told, by apologists for the economic mess we happen to be in at the time, that what we really must worry about is double-digit inflation, meaning 10 percent or more. If 10 percent inflation becomes the norm, how many years will it take to hack a dollar down to half its value?

In terms of your pocket calculator, 1 plus 10 percent of 1 (1.1) must be raised to what power to equal 2?

Enter 1.1 in your calculator. Touch the times key once for the first power (which is the 1.1 you just entered, so the display doesn't change), then the equals for the second power, the equals again for the third, and so on. Keep this up till the displayed amount is about 2. You'll find that on the seventh keystroke (the sixth time you touched equals), the display reads 1.949. That's much closer to 2 than you'll produce with an eighth touch, so you realize that a trifle more than seven years of inflation will cut that buck to half its present worth.

A sobering thought.

And if, by any chance, you doubt the earlier remark about inflation's having been with us just about always, consider the case of Chaucer. Literary historians have sometimes found it amusing that when Chaucer was captured in battle in the mid-1300s, he was ransomed for 16 pounds sterling, which we think of today as maybe twenty-five or thirty dollars. Finally a biographer, John Gardner, has calculated that what this actually amounts to in modern terms is $3840. There's 600 years of off-and-on inflation for you.

Quick Look at Monthly Money

Suppose you had a wad of money under your mattress. Each year, you spent 4 percent, or one-twenty-fifth of it. It would last 25 years, right?

But suppose that same wad were put in a savings account or money fund where it would bear interest. Then how long would it take to use it all up?

You can figure it out from the table for any withdrawal rate from $4 to $10 a year for each $100 invested. Just look opposite the amount chosen and read the figure under the percentage rate of interest that the account bears.

Dollars per $100 invested withdrawn each year	Rate of interest your money is drawing						
	3%	4%	5%	6%	7%	8%	9%
$4	46						
$5	30	41		This is how many years			
$6	23	28	36	you can keep on doing it.			
$7	18	21	25	33			
$8	15	17	20	23	30		
$9	13	14	16	18	22	28	
$10	12	13	14	15	17	20	26

So if you put $15,000 into a 6 percent account and draw out $8 a year for each $100 of that initial amount ($1200 a year or $100 a month) the table says you'll be able to keep it up for just 23 years. And then your money will be all gone.

You probably can see why this table is in a triangular shape instead of the usual rectangle. The missing half represents the situation in which the rate of withdrawal is slower than the rate of increase. If you were to fill it in, it would have to be with infinity signs.

In short, if you were to arrange your affairs that way, you could draw out the chosen amount of money forever and then some. You'd be living on the interest while the principal kept on growing.

How a Retirement Plan Grows

SINCE you don't have to figure in the effects of taxes year by year, it's rather simple to find out how much a retirement fund of the IRA or Keogh type—untaxed till withdrawal—will grow to over the years.

If you have even one of the simplest of business or scientific calculators, its roots-and-powers key will make this computation a cinch.

And it's only moderately more tedious with the ordinary calculator, which will give you the 20th power of the number in its register if you press times once and equals 19 times.

It's more fun—or less agony—to put money aside if you know what it's doing.

To get a handle on this, begin with the constant, or estimated average, rate of return. Express this as a decimal (9 percent becomes 0.09) and divide by the number of compounding periods per year (4 if quarterly).

Add 1.

Raise the result to the power of the total number of compounding periods over the years. Multiply by the number of dollars in the fund.

Using this in our example for 20 years of growth means raising 1.0225 to the 80th power (for the quarterly periods in 20 years). Result: $5.93 for each initial dollar. Encouraging, isn't it?

Figuring Inflation into Retirement

IT'S ONE thing to arrive at an estimate of how much money, or income, you'd need to retire comfortably this year. But if your time to retire is far away, how in the world can you know how much you'll need then—after several more decades of the inflation that has been with us off and on for centuries?

You can educate your guess with even the simplest of pocket calculators, quickly arriving at the inflation factors for 6 percent or any other level you care to postulate. Just enter the percentage, divide by 100, add 1, and multiply the result by itself once for each year after the first.

If you're making a reasonable estimate (it could also be called a wild guess) that there will be 4 percent inflation in the average year from now until your retirement in, let's say, 20 years, this means discovering the 20th power of 1.04.

Found on a typical pocket calculator by entering 1.04 (or entering 4, dividing by 100, and adding 1), then pressing the times key once and equals 19 times, the inflation factor proves to be 2.19.

This tells you that if your guess was good, it will take $2.19 twenty years from now to do the work a dollar does today.

How Much Life Insurance?

THERE'S no brief formula to tell you how much life insurance you should carry. But people who deal with family economic planning think that a family should figure its insurance plan by adding together these elements:

1. Quick cash. This should be enough to cover funeral expenses, current bills and income-tax obligations, possibly estate or inheritance taxes, costs run up during a last illness.

2. Interim. This should be an amount sufficient to support the surviving family for half a year—probably close to half your annual after-tax earnings in recent years.

3. Family support. This should be sufficient to cover the family's needs indefinitely, after deducting earnings by the survivors. For a young widow or widower with children to care for, this can be quite large, even for an employed survivor who can continue employment with good earning prospects.

4. Education. Remember that college costs have been inflating steadily in recent years, at two or three times the rate of living costs generally.

5. Emergency fund. If a couple of thousand dollars is already held in savings (preferably at the maximum interest rate that permits ready access), no additional amount need be provided through insurance.

Forecasting Your Income

Sᴏᴍᴇ of us live on irregular incomes from fees or commissions or from sales of our work. This can be hard on the old nervous system. It doesn't love incomes that are irregular and seem unpredictable. Indeed, many a promising career has been abandoned for no better reason than an illusion of financial insecurity.

In fact, most of the terrors of even a highly irregular income vanish if it can be found to be both sufficient and predictable.

Some years ago, as a free-lance writer with four children to shoe and feed, I finally evolved out of desperation a device for predicting my income for each coming month.

First, I listed all sources from which money might possibly come. Then I estimated the most likely amount of each bit of income and multiplied it by the probability that I would indeed receive that payment and receive it within the period in question.

Thus a check for $75 that had been promised by a source known to be both reliable and prompt would go on my list at full face value or very nearly.

A magazine article written on assignment might go down for $200 if a likely fee for it was $400 and I felt there was an even chance it would fetch its fee within the month.

A purely speculative contribution to an editor I'd never dealt with before might impress me as no better than a 10-percent gamble. So even if the magazine was listed as paying in the thousand-dollar range, I'd put it into the hopper for a mere hundred. Furthermore, if I figured that even if the manuscript did sell with those ten-to-one odds against it, there was an even chance I wouldn't see the check within the month, I'd whack even that modest hundred down to fifty.

In my busy years, when I often had a couple of dozen irons in each month's fire, my forecasts usually proved anywhere from amazingly to uncannily reliable. Many a month they came within a few dollars, for a precision of 95 to 99 percent—even though the individual items on the forecast were never correct.

There's something else this kind of forecast has going for it. Since time is of the essence, as formal agreements sometimes say, my occasional poor forecasts were usually balanced very quickly by comparable differences in the other direction.

I don't think this method can be very reliable unless it is based on enough projects to give the averages a chance to work. In recent years, when I tend to do larger but far fewer projects, I mostly use annual forecasts. These still function well, and they have the additional virtue of persuading me to plan my work more systematically—by the year instead of by the month—now that I can afford that luxury.

Besides, these are just the thing any self-employed laborer needs if he is to make a fair stab at filling out that work sheet each April 15 for pre-payment of estimated federal income tax.

Are You Throwing Away a House?

Having a need at one time to produce a hard-headed view of the relative merits of renting and home ownership, I used some case histories I knew about to make an instructive story. The moral of it, as you will see, is that some folks should rent and some should buy— and the secret of success is knowing which you are and when.

I called my people Bob and Bea Harte and their Uncle Jim. I hope you will find them useful in figuring out moves of your own.

Several decades ago, Uncle Jim and his wife, newly married, concluded that renters have it best. Jim has continued in the same job. They have moved only twice in all those years (one landlord decided to sell; the other was too incredibly stingy with paint). Jim and his wife are living today in essentially the same house as when they started their married life.

After an evening of talk with Uncle Jim, followed by some paper-and-pencil figuring, Bob Harte arrived at a startling conclusion. Before stating that conclusion, let's see how Bob reached it.

When Uncle Jim and his wife moved into their house 23 years earlier, they could have bought it, or something like it, for $36,000. Using his veteran's privileges, Jim would have had no down payment to make, and closing costs would have been minor.

Even at 6 percent for 20 years, Jim's monthly payments would have been $258.12. Taxes and insurance would have added about $70 a month in the early years, possibly $110 by the end (as both rates and the value of the house increased) for an average of $90 a month over the 20 years of the mortgage.

By straining his memory of what his landlord must have laid out, Jim had arrived at a reasonable estimate of maintenance costs for such a house for 20 years. At $500 a year—for small repairs plus an occasional major one, such as exterior painting or a new roof—these came to an average of about $42 a month.

Adding these figures of Jim's together, his nephew Bob arrived at a monthly cost-to-own of something less than $370 for the early years, rising to $410 or so by the twentieth year. Average: $390. But Jim had chosen to rent instead. His landlord, hewing fairly closely to the old rule of thumb that gross monthly rentals should approximate 1 percent of value, had charged him $350 a month when he moved in. As values (and taxes and maintenance) increased over the years, so did the rent Jim paid for his initial rented house and its successors. Jim has no reason to complain about the $550 he currently pays; actually, it represents well under 1 percent of the market value of the house he's in, although this house is similar to one he might have bought years ago for $36,000.

Since Jim's rent increases have been gradual but consistent, following the pattern in most localities, his average rent has been about $450.

And that's how Bob Harte arrived at a startling conclusion: Uncle Jim could have owned the house he's now living in and it wouldn't have cost him an additional cent. There would be no more rent to pay in the future. But now Jim had, as the saying goes, nothing to show for all those years of rent checks, except a bundle of receipts. The difference between Jim's actual rental outlay ($450 a month) and his estimated payments if he had bought ($390) comes to a dazzling $14,400.

On top of that, Uncle Jim could have—depending upon his income and his other deductible expenses—gained by lowering his state and federal income taxes.

Bob Harte had only to multiply Jim's $258 a month by the 240 months the mortgage ran to discover that the total outlay for principal and interest had been $61,920. Taking away the $36,000 that was repayment of principal, $25,920 must have gone for interest. Much or all of this, and property taxes as well, might have been income-tax deductible.

"Gosh," concluded Bob Harte. "Uncle Jim has thrown away a house."

His wife Bea saw it the same way, and the young couple proceeded to buy an attractive new house of their own.

Within the next twelve months, three things happened. Costs of houses over the United States as a whole went up 8 percent. A large government project having to do with space flight was discontinued along the seashore 10 miles from where the Hartes now lived. And Bob and Bea decided to move to Wyoming, where Bob had found an irresistible teaching job he could combine with the outdoor work he loved, as a park ranger each summer.

The Hartes promptly put their house up for sale, only to discover that because of the local drop in employment, three other families within the small subdivision had just done the same thing. And they had done so at prices reflecting eagerness to sell.

Only by persistence plus some time and money for repairs, and at considerable personal inconvenience, did Bob and Bea sell their house at what was seen locally as a fair market price. The general rise in building costs almost canceled out the effects of the weakness in the local market.

The price their house fetched equaled the full cost of building it a year earlier—what they had paid for it, less the 6 percent the seller had tacked on to cover the real-estate commission. When that selling cost is skimmed off the top of a sale price, it is elementary economics that buyer or seller, or the two jointly, must absorb a loss of that amount. It is a little like the difference between retail and wholesale prices.

The Hartes, having bought in a seller's market and sold in one favoring buyers, were in effect springing for two commissions, for a

12 percent loss. Add normal costs of moving in and closing (twice) and you see an out-of-pocket of about $6000.

"You could call it the Case of the Vanished Down Payment," said Bea, when everything was added up.

"I'd rather think of it as a year's rent," Bob replied. "Is $500 a month so bad?"

Well, yes it is, at the income and expenditure level we're talking about here. Add insurance and the usual amount of first-year landscaping and it all comes to twice what rent would have.

There we have Bea-and-Bob and Uncle Jim, two ways to go wrong, damned if you do and damned if you don't.

Yet we all know that one's own home is a fine investment, at the right time and place. It's necessary only to apply figures like these, as Bob and Jim did not, to boost the odds in your favor.

Living on a Savings Account

I**F YOU HAVE** a wad of money, one way of living on it is by buying an annuity. You get a guaranteed income, usually for life.

Another way calls for putting the money in a savings account and withdrawing some fixed sum each month. If the interest you can get is 5 percent a year, compounded quarterly, then each $10,000 in your account will permit you to withdraw:

$41.32 per month forever; or

$75 per month for 16 years and 1 month (during which time you'll have taken out not only your initial $10,000 but also $4501.50 interest that it earned); or...

$125 per month for 8 years (including $2115.25 interest).

Although these results are based on a single and apparently arbitrary rate of interest, you may find them quite generally applicable as bases for multiplying or dividing them to fit your own numbers.

What can make a 5 percent figure widely useful is that over a period of time that figure tends to be pretty close to the true rate of interest. True rate, in this connection, means the rate you're told you are getting minus the going level of inflation. Which difference means, in turn, how much the dollars you're getting are really worth when you go to buy something with them.

Appraise a Bank Bargain

IN THE NAME of patriotism, a savings-and-loan company in Nassau County, New York, offered a bicentennial special. Just deposit $1,000, wait only eight and one-half years, and you'd receive a historically appropriate $1776.

But some spoilsport with a pocket calculator wondered how much better this was than the offer made by competitors, and even by this same institution, before the bicentennial promotion came along. Even though the actual numbers fluctuate from year to year, anyone inclined to accept such an offer unquestioningly would do well to perform a self-inoculation by learning how this astute critic put the offer under a mathematical microscope.

The standard offer in those days was 7.75 percent interest on any $1000 deposit left for six to ten years; and, as the ad always explained, this was compounded daily—so on an annual basis, it really came to 8.06 percent.

The spoilsport quickly noted that such a deposit must grow to 1 plus 0.0806 during the first year, becoming $1.0806. Multiplied by itself eight times (that is, raised to the eighth power) 1.0806 becomes 1.8592. In eight years, one dollar becomes 1.8592 dollars. And $1000 grows to $1,859 and a few cents.

So with the patriotic plan, you'd wait half a year longer and collect $83 less.

At this point, the savings institution confessed that the patriotic special offer actually paid only 6.65 percent interest. It was their advertising department that thought it all up, they said hastily.

Strategy for Checking

WITH BANKS simultaneously fighting for your business and seeking ways to make the maximum profit from it, a little protective arithmetic may be in order.

Does it, for example, pay you to maintain a $500 minimum balance in order to avoid a $5 monthly service charge? If to do so you

have to tie up an extra $500 that could be earning 6 percent somewhere instead of your bank's 1 or 2 or nothing, you're losing the difference on that $500. Divided by 12 that says you're giving up $2.81 to save $5, and that's good.

It may be even better to seek out another bank or other savings institution, possibly a credit union, that omits service charges. Then you can keep the $2.81 too.

Here's another, somewhat more sophisticated, calculation that may apply where you live. For the example, let's assume an interest level that evaporated recently but may be back tomorrow: a savings-and-loan that pays 5-1/4 percent compounded daily and a credit union that offers 7 percent compounded quarterly, both reckoning by average balance and neither requiring a minimum balance nor levying a service charge.

Let's say that experience and a look through your monthly statements suggest that although your balance sometimes approaches zero it averages about $1000. What will that earn you in a year?

Express the S&L's rate as a decimal (0.0525), divide by the number of times compounded each year (360), add 1, then raise to the 360th power.

Multiply by the number of dollars—$1000 here—and you'll find it grows to $1053.90 in a year.

If the S&L is kind enough to use a 360-day year for dividing and then shift to 365 days for compounding, you'll pick up another 64 cents.

You'll find that an equivalent calculation for the credit union, using 4 for dividing and also for the power, comes up with $1071.86. Even when its higher interest is offset somewhat by the less frequent compounding, it offers you a bonus of $18 a year.

Factoring Inflation into Income

Cʌʟᴄᴜʟᴀᴛɪᴏɴs having to do with savings and income over a substantial period have come to mean very little unless they take account of inflation.

This is especially true of any scheme to give yourself a fixed monthly income by withdrawal from a single deposit. Withdrawals that today promise to be ample might be hopelessly inadequate in 20 years.

For realistic planning, indexing is the answer. Applied to interest income, that means subtracting the rate of inflation from the rate of interest you expect to receive and then using that difference in your calculations instead of using the stated interest rate.

You may have noticed that when the inflation rate remained near zero over a long period, conservative investments were likely to pay about 4 percent and home mortgages were only a little higher than that.

During some periods back in the 1980s, when inflation was averaging 10 to 12 percent, there were often quite conservative investments around that paid 15 percent or so. And this was more or less the prevailing rate for home mortgages.

For realistic planning for a constant level of effective (meaning constant-dollar) income, it is probably not prudent to assume that, over the long haul, the actual return on an investment will be more than 4 percent. A withdrawal plan based on a 4 percent return will let you adjust regularly the dollar amount taken, so that the purchasing power of your income remains unaffected.

After a period of years in which inflation is substantial, your monthly draw may become many times what it was initially. But its buying power is what counts. If you agree to make 4 percent your estimate of true interest rates, your course is simple. Spend one-twelfth of 4 percent of your principal each month, recalculating at least once a year.

A rather more precise approach would be to adjust your draw each year at the time announcements appear in the newspapers of what the inflation rate has been for the year. An easily available and conservative choice would be the cost-of-living adjustment announced each year for Social Security benefits.

In case of doubt, it should be noted that no one has so far accused the federal government of overstating an inflation rate. Most consumers and quite a few economists believe that by such moves as eliminating car purchases and housing from the index a considerable understatement of the rate has been introduced.

How Much Can You Take Out?

You may be planning to live, at some time in your life, at least partly on money withdrawn regularly from accumulated savings. Perhaps in an investment, such as a mutual fund.

Then applying our principles for factoring inflation into calculations about future income take on added importance.

Unless you do some figuring first, you may eventually run into a problem. That's what happened to the worried fellow who wrote to a syndicated investment advisor: "...our $50 monthly withdrawal from a mutual fund has forced the sale of 90 of our 590 shares!"

Value of his shares, he added, was about $7,000.

To pinpoint our man's problem and establish a basis for making a withdrawal plan, let's begin with a little arithmetic. That $50 a month is $600 a year, which divided by $7,000 is 0.0857, or 8.57

percent. As in most times, he couldn't get interest that high from any relatively safe investment.

A more realistic approach would begin with a look at going rates of return. If they're running around 6 percent, you should be able to set up a withdrawal plan at that level with a reasonable hope of seeing your investment hold its dollar value. Let that alarmed investor multiply his $7,000 by 6 percent and scale down his withdrawals to $35 a month, $420 a year.

If the value of the shares he holds, or the dividends the fund's stocks are paying, should rise sharply and stay that way for a while, then he might consider upping his withdrawal.

Any of us in a situation like this should also keep a weather eye peeled for the effects of inflation.

Taking out enough income to keep investment constant may not be prudent. It actually involves eating into capital, since the purchasing power of a fixed number of dollars continues to shrink.

One calculation to counter this is to use a realistic interest rate in specifying withdrawals. Don't use the nominal rate of interest you're told you're getting. Subtract the rate of inflation from the rate of interest, and use that. The dollar figure of your capital will grow, thus keeping its purchasing capacity constant.

Casing Inflation

WANT TO know in a hurry what inflation does to your dollar? A simple calculation can tell you in seconds.

Subtract the annual inflation rate from 100 and express the result as a decimal—so that, for example, 6 percent inflation gives 0.94. And the 14 percent we were beating our heads against not too long ago becomes 0.86.

Raise that 0.94 (or 0.86) to the power of the number of years you are interested in knowing about.

In the absence of a scientific calculator that finds powers—even fractional ones!—in a flash (using the key labeled x to the power of y, or something like that), remember that this is easily done with most hand calculators by pressing the times key once and the equals key one time less than the number of years. Thus, for five years of inflation, you'd press times once and then equals four times.

With this procedure, you find out what the given rate of inflation does to one dollar in the chosen number of years. So multiply the result by the number of dollars you're dealing with.

   If you find that in a mere five years, 14 percent inflation (and we did have a taste of that) will knock a dollar down to 47 cents and reduce the purchasing power of $5000 to a measly $2352.14, you've got it right.

   And note that even though we've all had some chastening experiences with inflation, it's still shocking—and a useful warning—to see that a decade of it at 14 percent would shrink the value of $5000 almost down to one G— $1106.51 to be precise

   Sailor, take warning!

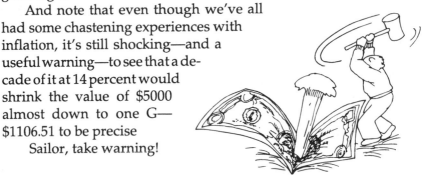

That Old Bugaboo, Average Rate of Increase

To COMPARE how slowly your salary is rising to what has happened to the price of bread, you need to figure rate of increase. This is a tricky factor to deal with because the rate for any period must be based on the new level at the beginning of that period. It's much like what goes on with compound interest, something that most of us find rather a sticky wicket.

   If you were earning $2 an hour thirty years ago and you're making $12 now you would be inclined to call that a 500 percent increase. Which it is, since 500 percent of 2 is 10. And then if you weren't thinking too hard you might go on to divide 500 by 30 and say your earning power in dollars has been going up 17 percent a year on the average. Which it has not.

   It's even worse if this state of affairs has gone on with annual inflation running 8 or 10 percent. You've been taking a beating.

   A widely applicable way of figuring an average rate of increase starts with dividing the new figure by the old. (That's dividing $12 by $2, in our example.) Having done that, you take the root that is equal to the number of years you're talking about. (In our 30-year example, that means the 30th root.) Now subtract 1 from this and multiply the result by 100 (by moving the decimal point two places to the right) to make it a percentage. You have found the true average rate of increase.

That is equivalent to what happens when a salary (or anything else) starts at 2 and is increased by 6.15 percent each year for 30 years.

Since your gain in earnings turned out to be not 17 percent but just over 6.15, it's no wonder you were falling behind in the face of inflation of 8 percent or more.

Finding a root such as the 30th, by the way, can be done by any of several methods covered in this book.

One is logarithms, if you want to acquire the needed tables and renew your grasp of this rather fascinating concept.

Or you can skip the tables and bring up any log you want instantly with a scientific calculator. If you have such a device, however, you may wish to skip the whole logarithm bit and manage roots the quick and easy way—with a stroke of the key labeled x to the y power.

And finally there's the method that doesn't send you running for either a table of logs or a device that costs more than a couple of dollars—iteration with a simple calculator. That is, you guess at the answer, test your guess with the calculator, then narrow it down with further guesses till it's correct to whatever number of decimal place you fancy.

There's more about this in the calculator chapter, especially in the section on cube roots and more.

The Mortgage that Pays YOU

ONE GREAT idea for some people, for which the time seems—at least as I write this—to have come at last is the RAM, or reverse annuity mortgage.

It's true that through the late 1970s and the 1980s, laws to permit, encourage, and control these novel financial instruments were passed in many of the United States. But only in the nineties have many mortgage lenders offered them, except in the form of pallid imitations.

In case an acceptable RAM is to be had in your neighborhood by the time you read this, you might like to calculate how one would work for you in time of need.

That time most often comes after retirement. And, as is true with so many other calculations and decisions taken in this period, life expectancy should be an important part of the scenario.

Suppose that you, like someone I know, live in a house that is worth $100,000 and on which the mortgage has been paid down to $30,000. You could sell your house and move into something cheaper, but you love the place. At the same time, you resent sitting on this $70,000 equity while scraping by.

What could an RAM do for you? It could permit you to use up to perhaps 80 percent of your equity by taking it in the form of monthly income for a stated number of years. I would regard this as prudent only when that number of years is considered in light of the life expectancy of the owner or owners.

How much money? To find out you'll have to decide how many years you want the RAM to run and either the amount you're willing to commit (up to the possibly 80 percent of equity in the house that a lender might permit) or the amount of monthly income to go for.

Assume you opt for the maximum amount and for a ten-year period. This is how to calculate your monthly income from it.

1. Express as a decimal the annual interest rate available (usually this will be close to that prevailing for ordinary home mortgages) and divide by the number of payments—normally 12—you wish to receive per year. Add 1. Raise this total to the power equal to the total number of payments you are to receive. (If monthly, this will be 12 times the number of years you've chosen.) Now subtract 1. Multiply by the number of payments you are to receive each year.

2. Multiply by the amount of equity available to take from your house—probably 80 percent of the difference between what the house is worth and anything owed on it—by the annual interest rate expressed as a decimal. Dividing this by the result of the preceding paragraph gives the amount of the monthly income you could receive.

Let's use as an example the house mentioned above.

With interest rates hovering around 12.75 percent (as they have done and, heaven forfend, may do again) and income to be taken monthly for 10 years, the computation looks like this. Dividing

0.1275 by 12 and adding 1 gives 1.010625. Raised to the 120th power this is 3.5547. Subtracting 1 and multiplying by 12 gives 30.6564.

When the product of 0.1275 and 80 percent of $70,000 is divided by that 30.6564, you find the monthly income: $232.90.

Similarly, you can make the calculation for some previously chosen amount of income.

If you wish to supplement your present income by $200 a month for 10 years, you make the first part of the calculation just as in Paragraph 1 above. Then divide by the interest rate in decimal form. Multiply by the monthly amount—the $200 you've chosen—to find how much equity you must have available to do this—and how much you'll owe after 10 years.

Lazy Person's Withdrawal Chart

IF YOU deposit a chunk of money and draw out so much each year, how long can you keep it up?

The lazy (or shall we say quick?) way to tell is with a chart like the one below. Reading it, you'll find that if you deposit $10,000 in a 5 percent savings account, you can withdraw $800 a year for 20 years.

Annual Withdrawal	Number of years assets will last at Annual Growth Rate of							
	5%	6%	7%	8%	9%	10%	11%	12%
5%								
6	36							
7	25	33						
8	20	23	30					
9	16	18	22	28				
10	14	15	17	20	26			
11	12	13	14	16	19	25		
12	11	11	12	13	15	18	23	
15	8	8	9	9	10	11	12	14

The $800 a year you are asking to take out is 8 percent of your investment, so you look down the left-hand column till you find 8. Then you read the figure opposite 8 under the heading 5 percent, since this is the return your deposit is collecting. It's 20, so that is how many years you can keep it up. You'll note that if the account paid 6 percent you could continue for 23 years; if 7 percent, 30 years.

You can use this table in reverse. Knowing the rate of return that is being offered, choose a period of years from the column under it. Then read in the left-hand column the annual withdrawal percentage this combination will permit. Multiply this by the amount of your investment to find the number of dollars you can take out each year.

(Say you can get 9 percent interest and want to spread the income over 15 years. Finding 15 in the 9 percent column, you read 12 in the column to the far left. This tells you that you can withdraw 12 percent of your original investment each year.)

What about the blank areas in the table? Any periods of years greater than the ones shown would, if combined with the withdrawal rates opposite them, let you withdraw money forever. Your capital would be growing faster than you were taking it out.

Inflation could knock your picture in the head. What if the amount of income that looks just right today proves hopelessly inadequate in an economy where everything costs five or ten times what it does today?

You can protect yourself against this self-deception by how you choose your interest-rate column in the table.

Instead of using the one that represents your expected rate of return, choose a more realistic one that is equal to the stated rate of return less the predicted annual rate of inflation (and less your income-tax rate as well).

If you invest to receive 12 percent, and can expect income tax to take one-fourth of that, you'll have only a 9 percent net return to work with. If you then assume 5 percent inflation over the years ahead and subtract this 5 from the 9, you'll see that the true return may be nearer 4. The 4 percent column may be your wisest choice for long-range planning.

Money Now or Money Later?

It CAN BE useful to know what a sum of money to be received in the future is worth today. This is a rather quick process for figuring that out. Since it has a variety of applications in almost everyone's financial life, you'll meet it several times in this book.

One way you can use it is to discover how much you would have to deposit in savings today to have some specified amount at a given time in the future.

Another way is to apply the procedure repeatedly to different numbers of years and rates of interest, thus generating a table for, say, $100 that simple multiplication will convert for other amounts.

GOOD GOLLY—
I THINK
I'M RICH!

Start the process by stating the interest rate as a decimal (15 percent is 0.15) and dividing by the number of compounding periods per year. Add 1. Then raise to the power of the total number of compounding periods. (That's the number per year multiplied by the number of years.) Now divide the number of dollars you're concerned with by this result.

As an example, suppose the interest rate offered is 15 percent, to be compounded quarterly, and you want your account to reach $2000 in five years.

Divide 0.15 by 4 (since quarterly compounding means four periods per year), add 1, raise to the 20th power (four quarters times five years) and divide $2000 by this result of 2.088. You must deposit about $958 (more precisely, $957.78).

By the way, with some calculators you'll find it easier to use the negative of the power (-20th power in this case). Then you can multiply by the wanted dollar amount, instead of going through a division as the final step.

You'll find your handy-dandy little calculator is more friendly when approached in this way.

Prepay That Mortgage?

WHEN INTEREST rates on mortgages soar into the 10 percent range, bankers develop a feeling of nausea when contemplating loans they hold that were written in the 6 percent days. Often they offer homeowners discounts of 5 percent to 15 percent to pay them off at once.

When this happens the bank can then use the money to earn 10 percent on new mortgages or even more than that for consumer-credit deals that bring in 14 to 18 percent.

If you are paying on one of these older mortgages, how can you figure out whether you'd gain by accepting the discount offer? This assumes, of course, that you have the cash or other resources to make the payoff. Your question is, will you gain more by paying the bank or by using the money for some other safe investment. Suppose your mortgage is now paid down to $17,000 and has six years to run at 5-3/4 percent. Your monthly payments are $279.75 for principal and interest.

At this point, your friendly banker offers you a 10 percent discount if you'll take this low-interest investment off his hands by paying it now in full.

It happens that you have the cash available to accept the offer. Should you?

To figure that out, you'll have to know first at what rate of interest you can invest, or already have invested, the $15,300 ($17,000 less 10 percent, most rapidly calculated by multiplying $17,000 by 0.9) if you don't use it to pay off the mortgage.

Using that rate, you treat the payoff amount as if it were a mortgage debt, calculating the monthly payments that it would require over the given period of years.

You can do this by either of the two methods given in this book for payment calculations. Or you can look it up in a table. Let's assume that your cash is now drawing 8 percent a year.

By any method, you'll find that the monthly payoff for $15,300 at that interest rate is $268.28.

So that's how much you could draw each month for the next six years, to use to make the mortgage payments. Since it is not as much as the payments, you'll be richer by $11.47 each month if you accept the offer.

It's only if you can depend on getting more than 9-1/2 percent for your money that you gain by keeping the mortgage alive. At 10 percent, you'll be ahead by $3.69 a month.

Chapter B
SOME PERSONAL THINGS

Figuring Your Life Expectancy
Decisions: His Way or Hers?
Does College Pay?
Getting Pregnant
Does a Second Job Pay?
Junior Doses
Chart Your Biorhythms?

Figuring Your Life Expectancy

PEOPLE at a life insurance company have rated an assortment of factors and arrived at this formula for calculating your life expectancy. They thought you might like to give it a whirl.

But maybe not. Digging into a story on a similar subject in my days as a very young editor (anybody here remember *LOOK* magazine?) I discovered a fact that I tactlessly imparted to my office-mate, a gloomy Britisher. At his age, the years he had coming exactly equaled the number already used up. He didn't speak to me the rest of the day.

Expectancies are so much rosier today that you may find reality less upsetting than my Brit did. So try this.

Start with 72 and subtract 3 if you're male, add 4 if you're female.

(Now you know why a man should marry a woman 7 years older than he is if they aim to finish in a—forgive the expression—dead heat.)

If you live in a city with 2 million or more population, subtract 2. But if you live in the country or in a town of ten thousand or less add 2.

If your work is at a desk, subtract 3; but add 3 if it includes regular hard physical labor.

If you get in at least half an hour of vigorous exercise five times a week or oftener, add 4; for only half that much exercise, add only 2.

Add 5 for living with another person; take off 1 for each 10 years you've lived alone since you were 25.

(It is only sensible to refine some of these items by interpolating. You might reasonably add 1 if you live in a city of 25,000, say, and subtract 1 for residence in a city of half a million. You might also consider how much of your lifetime you expect to have spent in these environments. Use similar logic on many of the items.)

Subtract 4 if you sleep more than 10 hours a night.

Would you describe yourself as an intense, aggressive, angry person? Take off 3. If you're easygoing and relaxed, add 3.

If you consider yourself happy, add 1; unhappy, take away 2.

Subtract 1 for having had a speeding ticket within the last year— because this suggests you might be somewhat prone to accidents, recently the fourth-ranking cause of death.

Subtract 2 if your income exceeds $50,000. (I'll bet that's a surprise and gives you something to think about.)

Add 1 for having finished college, 2 more for holding a professional or graduate degree.

If you are over 65 and still working, add 3.

Add 2 if any grandparent lived past 84. Add 6 if all four of your grandparents reached 80. Subtract 4 if either of your parents died of heart attack or stroke before 50. Subtract 3 if any parent or brother or sister has had cancer or heart trouble before the age of 50 or has had diabetes since childhood.

For smoking more than two packs a day, subtract 8; one to two packs, 6; one half to one, subtract 3.

Take off 1 if you drink the equivalent of a quarter-bottle of liquor a day.

Take off 1 for each 10 pounds you're overweight.

Add 2 if you have annual medical checkups (if male), annual visits to gynecologist (if you're female).

If you're now 30 to 40, add 2; 40 to 50, add 3; 50 to 70, add 4; and add 5 if you're more than 70.

If you come out with a total of 70 to 71, you figure to outlive the average man. To beat out the average woman, you'll have to come up with around 78 or 79.

Decisions: His Way or Hers?

RATING diverse things "on a scale of zero to ten" has become something between a cliché and a parlor game. But psychologists and counselors advocating no-fault marriage have turned this trick to useful ends.

Given a husband who is looking forward to watching football on TV and a wife who thinks a Saturday afternoon picnic and excursion would be a better way to use some sunny hours, how's the matter to be resolved? Why should it go his way rather than hers—or hers rather than his?

Of course, if he doesn't really care what she wants or she doesn't care whether he enjoys himself, then there is no solution. There's no marriage either, when it comes right down to it.

So let's assume both these people want the resolution that will bring the most mutual satisfaction.

How does he rate staying home on this particular afternoon? Undoubtedly there are some things he'd rather do. Play golf, go hunting, visit the French Riviera, possibly; but, for one reason or another, these events that he might rate 9 or 10 are not feasible on this particular afternoon. Though he's not an avid football fan, this promises to be a spectacular game; and he's tired enough after the week's work to welcome sitting. He rates it 7.

As for the picnic, well it certainly beats going to the dentist (zero), visiting his in-laws (1), or even painting the back porch (2). But it doesn't beat them by much, since it involves week-end traffic. He rates it 3.

Looking more closely at the picnic-excursion, the wife finds she can rate it no higher than 7. Although getting away from the house is high on her list at this point, and so is a bit of togetherness after the working week, she doesn't like the highway part either.

And staying home isn't the worst thing she can think of. There are a few chores it would be good to get off her mind. There's a book she's been trying to find time to read. And the football game won't last all afternoon. A semi-idle day around the house falls into a neutral 5.

That 5 added to his stay-home 7 adds up to a combined-satisfaction score of 12 for staying home. Her picnic rating of 7 added to his of 3 comes to only 10. The verdict is stay home today, by a margin of 2...and with a minimum of hard feelings.

This simple technique has some fringe benefits.

In the process of rating the intensity of their affection for their choices, both parties may find that only some of the aspects are significant. Perhaps our man has no objection to a picnic; he just doesn't want to drive on the highway or miss the football. His wife doesn't object to the football-watching or insist on a long excursion. The couple may find that a brief picnic at a nearby park with return home by game time will please them both at levels of 6 or 7 for a score of 13 or so, a clear mutual gain.

Even if that doesn't happen, another self-leveling ingredient can be expected. Having won this decision without rancor or martyrdom to mar the day, our man may be expected to approach the next such decision in the best possible spirit. He is likely to up his own pleasure rating for his wife's preference because he is indeed pleased at the prospect of her pleasure in it. And so the act of losing today enhances the loser's probability of winning tomorrow.

My editor expresses doubt that an arguing couple would actually pause to quantify like this. At your next opportunity, prove him wrong.

Does College Pay?

A NEWS magazine once cited the compiler of this compendium in a favorable sort of way. The following appeared in the "Essay" feature in *Time* :

"In an irreverent study of the numbers game called How to Lie With Statistics, Author Darrell Huff coined the word 'statisticulation'—the art of lying with statistics while seeming objective."

So no doubt it is ungracious to mention that it was in *Time*'s sister publication, *Money*, that I found a fine double-barreled example of statisticulation. Quoted therein was a student who had used World Almanac figures to "prove" that college doesn't pay, even though those figures show a graduate's median earnings as $2,104 a year higher. The estimated $22,000 the ungrad saves by skipping college will compound (at 7 percent) to a sum far greater than the difference in lifetime earnings.

(If these numbers seem low, note that they are for tuition and fees alone and in an unspecified kind of college—and of course our example would still have to pay some kind of bed and board, in school or not.)

The figures are impressive. While the nongrad is sacrificing $42,072 in current earnings (he works four years longer but makes $2,104 less each year) he has that $22,000 compounding away.

Result: He has nearly half a million when he retires. Added to the similar sum he's earned over the years, he might claim lifetime income of more than a million. The college graduate, however, has only a little over half a million to point to.

Convincing this may seem to be, but it is mathematical non-sense—adding oranges to apples. Or worse, totaling watermelons and peanuts.

For a more illuminating comparison let's say grad and nongrad live at the same level until retirement. To keep up, the nongrad will have to draw on his tidily invested $22,000 at the rate of $2,104 a year. My pocket calculator tells me he can do this for just under 20 years before the money runs out.

How can you use a simple calculator to estimate the length of time a chunk of cash, invested and compounded, will go on tossing out a given amount each year? Here's what the basic formula tells you to do.

Add 1 to the interest rate expressed as a decimal. (For 7 percent interest, that's 1 plus 0.07 equals 1.07.)

Raise this to the power of the number of years—20th power for 20 years, and so on.

Divide 1 by this result.

Subtract this from 1.

Divide this by the rate of interest as a decimal.

Multiply by number of dollars to be taken out each year.

But there's one catch. To evolve a procedure that will work with an ordinary calculator, we've required you to know in advance how many years to use in the second step, although that's precisely what you're trying to find out.

What to do? Apply a technique that will often serve you well when a direct method is out of your reach: guess and test.

Using our figures, you'd quickly narrow it down to a guess of 20 years when you found that this period requires an initial investment of $22,290, close enough to the specified $22,000.

It seems, then, that it pays to go to college?

Now that we've acquired a handy formula and used a math trick of wide applicability, let's fire the other barrel. This is the one that exposes the statisticulation behind this and many another correlation.

Even if the figures make it seem that college people do make more money, we have turned up no evidence that it's *because* they went to college.

Bright kids and rich kids are the kinds most likely to go to college. Right? Surely it is also likely that—with their brains and their connections—they are the ones likely to make the most money, with higher education or without.

Getting Pregnant

Knowing the probability of a couple's achieving pregnancy in a month and there's a known average for that—one chance in five—is there a way to tell how likely they are to get a baby started within, let's say, a year?

Clearly you can't just multiply the 30-day number by 12. That could produce an absurd result of more than 100 percent.

This kind of problem is best figured backwards. If there's a 20 percent chance of becoming pregnant in one month—the figure found by research in both Belgium and the U.S. for perfectly fertile couples in their twenties— that's an 80 percent probability of failure.

Express this 80 percent as a decimal: 0.80.

Raise this amount to the power of the number of months—12th power for a year.

(Multiplying 0.80 by itself 11 times isn't too tedious with paper and pencil if you round it off as you go. Of course, with a simple calculator it means merely entering 0.8 and pressing the times key once and the equals key 11 times.)

Now that you've found the probability of *not* becoming pregnant in 12 months (0.07), subtract it from 1 to find that the probability of pregnancy in the course of a year is 0.93, or 93 percent.

For a couple in their late thirties, it has been found that there is about an 8 percent chance per month. So it's 0.92 that you'd raise to the 12th power and subtract from 1, to find a 63 percent chance of pregnancy within a year for normally fertile couples pushing forty.

Five-year figures come in the same way, using the 60th power instead of the 12th. You may find the results a shock, but you'll see why the world is heavily populated.

Does a Second Job Pay?

As I RECALL a novel of the 1930s depression called If I Have Four Apples, it was about the vain efforts of a social worker to convince some employed wives that they were going in the hole every hour they worked.

The whole process of persuasion fell apart on the grounds of simple arithmetic of the kind that begins, "If I have four apples and take away…"

Now, half a century later, with sexism quite rightly suspected around every corner, let me begin with a disclaimer. I'm not about to apply arithmetic to wives' jobs only. Having determined, on whatever grounds, which of the couple shall hold a job, the tests below will help answer the question of whether the partner should do likewise or would be better employed in keeping the home fires blazing.

You can apply the routine as well to monthly or annual figures, but for convenience let's assume weekly pay.

Begin with the additional income the job will produce. Actual salary, not take-home.

Add the value of the fringe benefits, as nearly as you can calculate them. If the first job in the family does not provide adequate medical insurance, what comes with the second job may be worth quite a bit. Value it according to what it would cost you to buy the equivalent independently.

Face a grim fact. A man or woman who is already regularly employed will probably be in a better career position to support self and any children if the other dies or becomes disabled or if the relationship ends.

If a second job permits the prime earner to carry less insurance while the degree of security is kept constant, the reduction of premiums is an immediate saving. But consider only premiums needed for protection in the form of term insurance, remembering that straight life insurance includes an accumulation of savings as well as current protection.Those are the main money gains. Now consider the losses.

Very often, the biggest of these will be increased taxes, not only federal income tax but also social security and very often state and local income and unemployment and workman's-compensation taxes. The income-tax rates on this second income are going to start where the percentage levied on the previous total family income left off…and may jump rapidly from there. Social Security tax is a significant factor when the incomes in question are not very large. It can be bigger than income tax.

For many people, the biggest additional expense created by a new job is transportation. Bus or train fares are usually easy to estimate with precision. For added miles to be driven in a car owned anyway, a reasonable estimate might be around 20 cents a mile; but if an extra car must be owned because of the new job, twice this much may be reasonable. Or an item-by-item addition may produce a sharper estimate.

Lunches, unless included free with the job, can be an important new expense.

This applies also to additional clothing and its maintenance. Taking on, or adding to, outside employment often means paying handymen and plumbers and window-washers and household workers to do what formerly were do-it-yourself chores. Some men or women or couples do so many cost-saving chores around the house that they almost amount to a second job and pay accordingly. Value added to a house by substantial improvement or remodeling—landscaping, a new room, a garage—can be equivalent to earning thousands of tax-free dollars a year. If taking on a new outside job is at the expense of such activities, their value must be included in the reckoning.

Child care, in or out of the home, can be a major new or inflated expense. Deduct from it, however, any tax savings generated.

Taking on more employment may, in the absence of new household help, lead to buying more convenience foods—quick-cooking but costly. There will be less time to shop for household bargains. More frequent dining out is likely. Costs of these effects of taking a job should be included in estimating the new budget. Although it is not necessarily a clear fiscal plus or minus, here's another factor to consider. There's more stress when either a husband or a wife begins to work longer hours to bring in more money, and this may lead to a conviction that a reward is deserved. This can take the form of more costly vacations or luxury clothing in no way related to the job.

These rewards may be precisely what you're taking the job for, so it may not seem to make sense to deduct their cost in estimating your gain.

But however pleasurable these rewards, they are going to reduce the amount you can stick into a well-chosen mutual fund and gloat over when the year is done. For that purpose, be firm and subtract their estimated cost.

How does this work out in an actual situation?

Let's try it. A printing company has some $14,000 jobs open. Your cousin Lulu, who lives near the plant, is considering taking one of them. So is the wife of an industrious young office manager in a nearby town. How well will it pay each of them to go back to work?

Lulu's husband Fred, 65 but retired more by reasons of health than age, draws nearly $8,000 a year in tax-free Social Security benefits. Investments bring in another $4,000. Since

their income is too small to be taxed and they own their car and small house clear, Lulu and Fred hope they can live on this, though it doesn't look to be easy. More income would be welcome, at least until Lulu reaches an age where she can collect a spouse's Social Security benefit.

The only useful fringe benefit that comes with the job, Lulu notes, is medical insurance roughly equal to the Blue Cross policy she has been paying $170 a quarter for. Gross value of the job, then, is $1223 a month.

Income tax on their joint return will increase from nothing to about $1900, Fred finds by borrowing and filling out a copy of last year's Form 1040. The Social Security tax will be nearly as much, so there goes some $300 a month off the gross.

Bus fares will run about $20 a month. Since Lulu plans to brown-bag much the same lunch she'd eat at home, there's no added cost there, but it seems prudent to allow $60 a month for additional clothing. A similar amount, she thinks, will take care of any small added food costs produced by more hurried shopping or preparation. There is no question of household help or child care. Fred can easily take up the slack so far as chores around the house are concerned.

A little arithmetic now establishes that the job will pay Lulu about $800 month net, adding two thirds to the family net income. It's well worth considering.

Now what about that young Mrs. Popple, down the street? Since her husband's job already provides comparable medical coverage, there's no fringe benefit there for her. With one good salary already in the family, the $14,000 will be almost fully taxable and at a level averaging 30 percent. Adding the same Social Security levy that Lulu has calculated she would pay, there goes some $5000 off the top.

Fourteen miles each way daily in the new car that Mrs. Popple has decided she will need can be estimated to cost $200 a month—though it will actually come to more than that if the small station wagon she has her eye on doesn't help pay its way with some other uses.

Mrs. Popple next notes what a hole a conservative monthly $40 for lunches (if she holds to her intended starvation diet), $90 for additional clothing, and $100 for hurried shopping for convenience foodstuffs will make in the family budget.

Because of its income-tax deductibility, a monthly $400 outlay for child care will actually put the Popples only $240 more out of pocket if there are no drastic changes in the law.

But a modest amount of a cleaning woman's time (merely four hours twice a week or eight hours once a week at an optimistically estimated $6 an hour and transportation) comes to another $200 a month.

With all these costs added together, Mrs. Popple runs into a bafflement. When she attempts to deduct them from the salary offered, she finds they amount to more than the salary.

Even if she can find a car pool instead of buying a car, and even if she adds to her own working day by eliminating the cleaning woman, Mrs. Popple is going to find herself working for a couple of dollars an hour. That's not much, considering the various stresses and often costly emergencies that pop up when both parents of young children work full time.

There may, of course, be good and even overriding reasons for Mrs. Popple to take a $14,000 job—interesting work, future, she's going nuts taking care of the house and kids all day. But, financially speaking, she'll have to regard it as a hobby.

Junior Doses

IT'S NOT always a safe practice to calculate a medical dosage for a child by using a formula to modify the adult dose. However, with drugs that are in common use and not considered especially dangerous, the methods given here have been used by doctors, in the absence of other information.

The simplest of them is called Clark's Rule. To use it, divide the weight of the child in pounds by 150, then multiply by the adult dose. A 50-pound child will get one-third the dose that would be prescribed for a 150-pound adult.

Young's Rule is a trifle more sophisticated. It instructs you to divide the child's age by his age plus 12 and multiply this by the adult dose. A child of eight would get a dose that is 40 percent of the one recommended for an adult.

Under what's known as Cowling's Rule you would divide the age of the child at its next birthday by 24 and multiply this by the adult dose. With this method, an eight-year-old child (going on nine) would receive 38 percent of the adult allowance. This works out to just about the same as under Young's Rule, a helpful confirmation.

An ultra-cautious approach would seem to be to try each rule for size and then use the one that provides the smallest dose.

Chart Your Biorhythms?

THERE are many remarkable rhythms in nature and human life. Many are quite well established as facts, even though much of their nature and origin remains mysterious. The most widely discussed and applied of these are called biorhythms.

Whole books about them appear from time to time. A pocket

calculator designed to chart them is on the market, at least in Europe. Computer programs to the same end are widely available.

But not at all necessary. You can quickly figure these rhythms with that handy computing tool, the pencil.

In this theory, each human life is influenced by three cycles that begin at birth. Differing in length by several days, cycles peak at their own times, putting each of us at a different phase of a characteristic on any given day.

These are rhythms of extreme precision. Thus they offer a tool of which the least that might reasonably be expected is a striking reduction in human strife, inefficiency, and accident rates of all kinds. The biorhythm theory is wonderful if true, and interesting to play with in any case.

This theory has emerged from the work of a Dr. Wilhelm Fliess, who though not universally admired today by Freudian scholars was a friend of Sigmund Freud's and practiced in Berlin at the beginning of the century. From a twenty-year study of 80,000 case histories, he found what he believed to be confirmation of something suggested by Hippocrates in the Fifth Century B.C., persistent rhythms in human life, beginning at birth.

One of these rhythms is best thought of as PHYSICAL. It is an ebb and flow of strength and endurance. Its length is 23 days.

A second is EMOTIONAL. Connected with moodiness, creative ability, intuition, sensitivity, ease of expression, and ability to influence others, it runs 28 days from peak to peak.

A third cycle, of 33 days, is INTELLECTUAL. It governs intelligence, concentration, mental alertness, and memory.

Waxing and waning of each of these kinds of power produces critical days, especially in connection with the first two of the cycles. On some days, the proponents say, a person may be seven to eight times more accident-prone than on others.

If you're an airline pilot, for example, information by which you could be grounded on your critical days could obviously be of enormous value; and a similar argument might be made in the case of a surgeon—and patient. Or of a judge or an athlete or a business executive or anyone facing a critical decision.

What makes calculating these cycles most important—assuming the theory is valid—is that each cycle is held to begin on the day of birth and continue in precise length for the life of the person. That has to mean that the exact position of anyone in respect to each cycle may be found at any time in life. It is necessary only to add multiples of any cycle length to the day of birth.

The biorhythm theory has gained enough respectable backing to get it serious consideration in some major enterprises. An enviable safety record for the high-speed transportation system of the Swiss city of Zurich is claimed for application of the Fliess theory to the scheduling of personnel.

In a preface to one of the earlier books on biorhythms, a Dr. F. Wehrli reports with considerable excitement, but few very useful facts, on experiences with the Fliess theory at Kurhaus Villa Montana at Locarno, Switzerland. He says that for fifteen years, it has been the practice at this clinic to "take into consideration our patients' biorhythmic situations." Dr. Wehrli concludes that the "facts and figures" of this experience "offer proof positive of the important relationship of man's resistance to the ebb and flow of his biorhythms, and that in the 10,000 and more cases of our experience where biorhythms were employed, they generally proved to be a valuable aid."

Perhaps, on the basis of all this and anything you may have read on the subject before, you're a believer. Perhaps you're merely curious and would like to observe how the idea proves out in your own life.

Anyway....

To calculate your own biorhythms, you must first determine where you stand today in relation to each of the three cycles.

Find precisely how old you are by reckoning the number of days from your birth to the end of your first year, adding 365 times the number of complete years since, then adding the number of intervening leap years, and finally adding on the number of days that have gone by so far this year.

(Remember that leap day comes in every year evenly divisible by 4 but not by 400.)

Divide this total number of days by 23. Discard the whole-number answer, keeping only the remainder.

Suppose you find that today (April 15, let's say) you are 22,815 days old. Divided by 23, that comes to 991 cycles, plus a remainder of 22 days. Forget the 991.

Since 22 days of your 23-day (physical) cycle have passed since the preceding cycle ended, you can find the beginning of your cycle by subtracting 22 from today's date—or by adding 1 to it. Having marked those two dates on a chart, draw a smooth curve that begins to rise at the initial date and then falls, crossing the baseline at the midpoint and rising to it again at the end. There a new 23-day cycle begins. And so on for life.

For your emotional (some call it sensitivity) cycle, go through the same process, this time dividing by 28.

For your intellectual cycle, divide by 33.

PHYSICAL CYCLE

EMOTIONAL CYCLE

INTELLECTUAL CYCLE

1 2 3 4 5 6 7 8 9 10 11 12 13 14 15 16 17 18 19 20 21 22 23 24 25 26 27 28 29 30

DAYS (FULL MONTH)

The remainders are 23 and 12 respectively for the person whose physical cycle was just charted.

Some of the writers on biorhythms suggest that you can combine the three cycles for any day by adding the magnitudes of the curves above the baseline and subtracting the amount below. You could even do this just by eye. However, you'll have to decide for yourself just what three unlike characteristics could be said to add up to. A sort of diffuse performance capability, perhaps.

Just to show how it looks, I've charted it. The chart shows the rhythms for our example, who did his calculations on that annual day of reckoning, January 1.

In figuring your cycles you'll very likely want to use your pocket calculator, if you have one around. The only difficulty is that these devices are too fussy. They insist on operating to as many decimals as they have room for, instead of quitting with a quotient plus remainder.

So you must turn the decimal back into a remainder. For the 22,815-day-old fellow just described, the routine goes like this for the physical cycle:

Enter 22,815. Divide by 23. Subtract the whole-number quotient 991. Multiply by 23. Answer is 22. That's the number you're after.

DAYS (FULL MONTH)

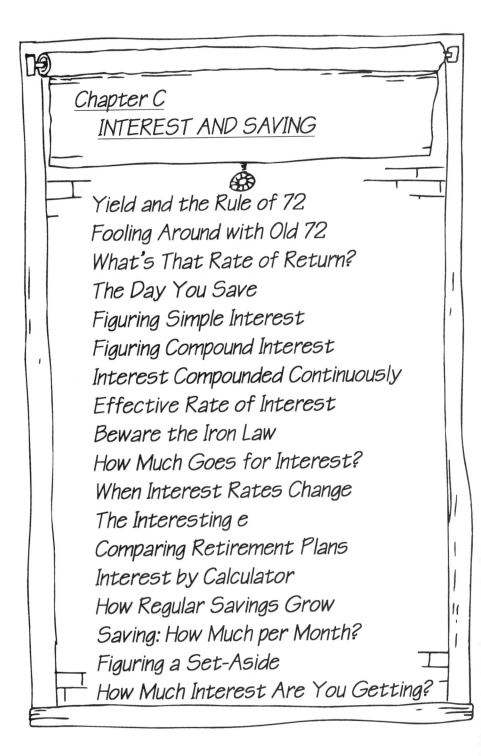

Chapter C
INTEREST AND SAVING

Yield and the Rule of 72

T<small>HE GOOD</small> thing about the old banking Rule of 72 is its simplicity. It lets you calculate instantly how many years your money will require in order to double itself at any specified rate of interest. And it lets you do this notable deed without recourse to a chart or an electronic device or even a pencil—by merely dividing 72 by the promised interest rate.

Not so good is the fact that it loses some of its usefulness when applied to the abnormal interest rates that turn up from time to time. And as daily or hourly or even continuous compounding has become more usual, this quick trick has become less accurate for estimating how many years it takes for money to double.

But it's still useful, especially if you apply fine tuning. What follows are some rather precise approximating tricks you can apply with a minimum of math.

(Given a scientific calculator, or a computer, you can be even more precise about it, as the next part of this chapter will show.)

For starters, use annual yield if you know how much it is, instead of the nominal interest rate. Savings institutions often give both figures in their advertising. If you must use the interest-rate figure, you may find you're getting closer approximations when you use 70 instead of 72 in applying the rule.

Another thing: The rule we've been talking about is helpful only for questions of doubling. What if you want to know the interest rate or time period required to triple your money or to increase it tenfold? Use these new "Rules of..."

For increasing your money by half, Rule of 40. For fivefold increase, Rule of 160. For sevenfold, Rule of 195.

Use these numbers in the same way as the original Rule. That is, to find how long it will take for a $100 investment to grow sevenfold to $700, with daily compounding, use the Rule of 195. That is, divide

195 by the interest rate and discover that for 15 percent (you should be so lucky—or, more likely, you're the borrower), it's about 13 years.

Care to evolve a few Rules of your own? Just factor the number of years you're after and add together the corresponding Rule numbers given above. (For 6 years, note that 6 is the product of 2, 2, and 1-1/2, so add 70, 70, and 40 to derive a brand new Rule of 180 for a sixfold increase.

Fooling Around with Old 72

IT'S QUICK—the rule that says dividing 72 by the rate of interest will tell you in how many years of compounding you can double your money. But for a precise answer, you'll have to drop the rule of thumb and go to a formula.

Same goes if you're after precision on some increase other than doubling—tripling your money, say. The penalty is that this routine calls for a table of logs or a scientific calculator.

Anyway, here's how to find out how long it will take to multiply your money by any factor you care to name.

Step 1. Find the log of the number by which you want to see your investment multiplied. That is, find log 3 if tripling is the aim, log 4 if quadrupling. Either common or natural log will serve.

Step 2. Express the interest rate as a decimal (11 percent becomes 0.11) and divide by the number of times your invested money is to be compounded in a year. (For daily compounding, the number is 365, or 360 if your bank says so.)

Step 3. Add 1. Now find the log of this number (same type of log as you chose previously) and multiply it by the number of compounding periods in a year—the same number you used at the end of the preceding step.

Divide the result of Step 1 by this result.

For finding how long it will take to triple an investment compounded daily at 15 percent annual interest, the calculation goes like this.

Common log of 3 is 0.477. Save it for later.

Dividing 0.15 (for 15 percent interest) by 365 and adding 1 gives 1.00041, of which the log is 0.000178.

This, multiplied by 365, is 0.06513. Dividing this into 0.477, obtained in Step 1, reveals that it will take 7.3 years to triple your money.

What's That Rate of Return?

YOU HAVE money on deposit in an interest-bearing fund. Your periodic statement comes, telling you that your earnings for 28 days were $18.80 on an initial balance of $2812.59. What rate of return is that?

You need only divide the number of dollars earned by the amount of the beginning balance to find the yield is 0.0067.

Multiplying this by 100 (moving the decimal point two places to the right) converts this to percentage form—as 0.67 percent or about two-thirds of 1 percent.

Call this the periodic yield, the rate of return for a 28-day period.

However, you'll have to convert that periodic yield into an annual one to permit meaningful comparisons with other investments, on which you'll ordinarily be quoted the annual return.

It might appear that you could get this figure by multiplying that periodic rate by the number of periods in a year. But doing that would neglect the fact that in each period you are entitled to interest on not only your initial principal but also on every cent of that delightfully accumulating interest.

What you would get by that method is a fairly good approximation but one that wound underestimate the true figure every time. Multiplying 0.67 percent by 365 and dividing by 28 gives an estimated annual yield of 8.73. Round this upward a bit and call it "around 9 percent."

That's not terribly far from the true figure of 9.1178 percent, most handily obtained with a scientific calculator or a computer.

You get this exact figure by adding 1 to the short-time yield of 0.0067 and then raising that total to the power of the number of periods. In this instance, the number of periods is 365 divided by 28.

The result you have at this point is what each dollar of initial investment will have become at the end of a year, just over $1.09. Subtract the 1 that you added earlier and you convert this to the annual yield.

When the calculation starts, as this one did, with a rather odd number of days, it is most easily done with a scientific or financial calculator, or with a computer.

In the more usual instance where the period is a simple fraction of a year, any calculator will serve. Because you're raising to the twelfth power (for a month) or the fourth (for a quarter) instead of the mix of whole number and decimal you produced when forced to divide 365 by 28.

The Day You Save

Because the ultimate figures are truly impressive, financial institutions love to tell you how much you'll have after so many years if you save so much a month or quarter or year.

Although they don't say so, they usually give you figures that assume you've made the deposit with them at the beginning of the period in question.

But what if you put up the money at the end of the quarter or year instead—as many of us will because that is the last possible moment to get certain tax savings?

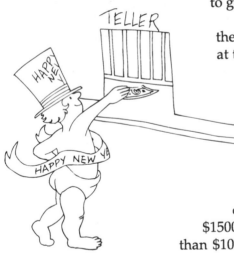

Either way, you'll have put in the same amount. But depositing at the beginning of each year instead of the end will mean quite an increase in the amount you'll have for retirement.

A common arrangement at a good rate might be 7-3/4 percent interest compounded daily. After 25 years, this turns $1500 saved each year into more than $100,000. (You'll find the way to calculate this given elsewhere in this book.)

The amount will be some 8 percent more if you deposit at the start of each year rather than the end.

If you deposit savings quarterly, the increase you get from doing so at the beginning rather than the end is about 2 percent.

This possibly puzzling difference is one of the effects of compounding interest and doing it more than once a year.

Figuring Simple Interest

I GUESS they call it simple interest because it IS simple—if what you're asking is how much you'll owe if you borrow so-many dollars for so-many years at a known rate of interest.

But it can be utterly confusing when your unknown is one of the other three of the four factors:

> amount of *Principal*,
> amount of *Interest*,
> interest *Rate*,
> and *Time*.

If you know any three of these factors, you can find the fourth by using an arrangement of a basic formula, I=PRT.

To find the amount of *interest*, the formula says, multiply *principal* by *rate* by *time*. For example, how much interest will you owe if you let an indebtedness of $2200 go for a month at 18 percent, a common credit-card problem? The product of $2200 (principal), 0.18 (18 percent annual interest rate), and 1/12 (of a year) is $33.

(Note how easily this turns into a quick mental calculation, if you remember that the common 18 percent carrying charge is 1-1/2 percent a month. Just divide $2200 by 100 and increase by 50 percent.)

To find *principal* amount, divide *interest* amount by *rate* multiplied by *time*. How much need you invest at 10 percent simple interest to earn $100 in six months? Divide the interest of $100 by the rate of 0.10 and by the time of 1/2 year to get $2000.

To find *rate*, divide *interest* by *principal* multiplied by *time*. If $3000 is to produce $22.50 interest in 1 month, what must the rate of interest be? Divide the interest amount of $22.50 by $3000 and by 1/12 year. Result: 0.09, or 9 percent. (Note that the easy way to divide by 1/12 is to multiply by 12.)

To find *time*, divide *interest* by *principal* and by *rate*. How long must you leave $300 at 8 percent for the interest to come to $4? Divide $4 by $300 and by 0.08. Result is 0.1667 years, which decimal multiplied by the number of days in a year gives you 61 days (60 if you use the banker's rounding of a year to 360 days).

Figuring Compound Interest

COMPOUND-INTEREST problems ordinarily give you five factors to deal with. You know four of them and want to find the fifth.

1. There's the initial amount—how many dollars you start with.

2. There's the annual interest percentage. If you're asked for the rate as a decimal fraction, divide the percentage by 100. (Just push the decimal point two places to the left, so 15 percent becomes 0.15.) That's how we'll do it in what follows.

3. There's the number of compounding periods in a year—4 for quarterly compounding, 12 for monthly, and so on. This is the number by which you divide the annual interest rate to find the periodic rate.

4. There's the number of years the compounding goes on.

5. And there's the number of dollars that initial amount will reach by the end of the period.

If you start off knowing four of these things, you can discover the fifth.

How much will it grow to? To find how much compounding will make an investment grow, add 1 to the rate, raise this to the power of the number of compounding periods, and multiply by the initial amount.

(Example: For an investment of $2000 at 15 percent compounded monthly for 20 years, raise 1.0125 to the 240th power and multiply by $2000 to find that it would reach a magnificent $39,430.99 at the end of the period.)

What size investment? To find how much you'd need to start with to accumulate a chosen amount, add 1 to the interest rate (in decimal form), raise to the power of the number of compounding periods, then divide the sum you're going for by this result.

(You should find that $39,430.99 divided by the 240th power of 1.0125 is $2000. Not surprising, since we've chosen our figures to make this a mirror image of the previous exercise.)

What interest rate? How do you choose an interest rate—one that will turn a given sum into a pre-chosen amount in a specified time? Divide the end amount by the beginning amount. Find the root of the figure just obtained that is equal to the number of compounding periods—the fourth root, for example, if there have been four periods, as in quarterly for a year. Then subtract 1.

Turn this into an annual rate by multiplying by 12, and into a percentage by multiplying by 100.

(Example: For the rate needed to turn $2000 into $39,430.99 in 20 years of monthly compounding, divide the larger number by the smaller, find the 240th root, and subtract 1. Multiplied by 12 and by 100, this becomes the annual 15 percent that would be required to do the job.)

O Lord, how long? To find how long it will take to turn one sum into another through compounding at a known rate, divide the amount you aim to end up with by the amount of the initial investment. Find the logarithm of this result by table or calculator. Add 1 to the periodic interest rate (expressed in decimal form) and then find the log of this sum. Divide the first log by the second to find the number of compounding periods.

(In our example, you'd divide the log of 19.715,495 (which is 2.981,405) by the log of 1.0125 (which is 0.0124,225) to get the answer of 240 monthly compounding periods, or 20 years.)

A couple of final notes. In this last caper, you can use either natural or common logarithms, as long as you stick to one kind. The quickest way to obtain them is with a scientific calculator (with power and reciprocal keys to speed up everything). But a log table plus a simple calculator, or even a pencil, will do the job.

Even without a log table, you can crunch your way through by trial-and-error with a calculator of any kind. Same goes for finding a root, such as that 240th. Several methods are discussed in some detail elsewhere in this chapter.

Interest Compounded Continuously

THERE'S something pretty fetching about the notion that your hard-earned savings are earning interest for you every second, day and night.

That's why some savings institutions have come to advertising continuous compounding on the money you've deposited with them.

Is this a gimmick or something worth while?

That question is best answered with another question: Compared with what? (As a disillusioned lady is said to have replied when asked how her husband was.)

So, specifically, what will $500 become in 15 years at 8 percent interest?

The standard compounding method tells us to add 1 to the rate (after dividing the percentage form by 100 to turn it into a decimal), raise this to the power of the number of compounding periods, and multiply by the initial amount. Applying it to these figures, we found that if the compounding is quarterly, the $500 will grow to $1640.45.

How can we apply the same calculating procedure to continuous compounding that produces infinitely small compounding periods—and therefore an infinitely large number of them?

What will happen if, for that infinitely large number, we arbitrarily substitute one that is only comparatively large? And then recognize that what we have obtained is an approximation that is somewhat on the low side, and round it up to the next whole dollar?

Since we're taking unauthorized liberties, we might as well choose a number that's easy to work with: 100. The result we get is $1659.26, which we round to $1660. Continuous compounding is significantly better than quarterly, but not dramatically better—or so our approximation method says. A sort of checking run, using 1000 in place of 100, produces $1659.98.

The real way—no approximations—to figure continuous compounding is actually simpler. But I don't know any really expeditious way to do it without a scientific calculator. If you can lay your hands on one of these (and at a sale price that I've seen fall as low as ten dollars, why not?), all you have to do is multiply the interest-rate-as-a-decimal by the number of years, raise the mathematical growth constant e to that power, and then multiply by the number of dollars you started with.

This mathematically impeccable route brings us to $1660.06. Our quick-and-dirty method is proved quite good enough for everyday use.

Effective Rate of Interest

As INTEREST-RATE calculations go, this is a short and simple one. All the same, it's a bit of figuring that has often led to confusion, mostly because it offers a temptingly fertile field for deception by installment-sales operators.

It can hide just about anywhere. If someone tells you of a car loan on which he's paying half the interest rate you know others are paying, you can suspect he's actually been quoted (though not in writing in recent years) the interest charge as a percent of the full initial amount of the indebtedness.

There is a quick route to the truth.

To find the effective rate of interest you are paying, or getting, you need only divide the total amount you are putting out for interest by the number of years and by the average amount that you will owe during the course of the loan.

Typically, you'd work it out this way. Multiply the number of payments to be made by the amount of each payment and subtract from this the original amount of the loan. This difference will be the total amount of interest to be paid.

Dividing this result by the amount borrowed and by the number of years produces the effective rate. Multiplying by 100, a matter of pushing the decimal point two places to the right, turns the rate into a percentage.

Which brings us to a wicked law.

Beware the Iron Law

CLASSICAL economists began talking about the Iron Law of Wages long ago. But it must have been a disgruntled borrower who more recently formulated the tongue-in-cheek Iron Law of Interest for the edification of lenders. This pseudo-law says: Whatever rate you're actually charging, always express it so that it doesn't sound like more than 6 percent.

Soaring interest rates of the seventies and eighties made that a hopeless goal for quite a time but the unprincipled principle survived.

If you consult a car salesman about credit for your purchase, you may be told that an 8 percent loan can be arranged. Or perhaps "8 percent per year." Or, if he's a little more choosy about his language, "8 dollars per hundred per year."

If you accept the offer you may be surprised not to find any figure even faintly resembling 8 percent on the Truth in Lending statement that the law now says you must receive before the deal is concluded. If you read it, you will find that you are paying an APR, or annual percentage rate, of 14.5 percent. Nearly twice what you thought you heard at the oral stage of the deal.

What you have is an add-on loan. The amount you are to pay has been figured by adding the quoted cost (multiplied by the number of years if the loan is to run more than one) to the face amount.

For a one year loan, the $100 becomes $108. This divided by 12 is your monthly payment.

The gimmick lies in the fact that you are repaying by the month. If you were to keep the whole $100 until the end of the year and then repay it, plus $8, that would indeed be 8 percent simple interest.

But you're not borrowing $100 for a year. You are borrowing $100 for one month, then paying $8 of it back with $1 interest. During the second month, you have only $91 of the lender's money. By the last few months, you have only a small amount of it to use.

So, on the average through the year, you had the use of only about half the face amount. Having paid $8 interest for the use of an average of $50 or so for a year, your interest rate is bound to be up towards 16 percent.

It could be worse. There is a close relative of the add-on loan that is called the discounted loan. The difference is that, with this one, when you borrow $100 for a year at $8 per $100, the lender collects his interest off the top—handing you $92. With this kind of loan the true annual interest rate turns out to be even closer to double what the salesman made it sound like. More precisely, it is 15.7 percent.

The reason for the even higher rate is, of course, that this time you're paying for the use of about half of $92 instead of half of $100. And the fee is still $8.

Figuring the exact rate of interest on either an add-on or a discounted loan is not particularly easy.

Books about such things ordinarily offer you one of several methods of approximating the rate. Since the methods are not precise, the answers you get vary quite a bit. It's easier, and just as accurate, to say simply that the true rate of a discounted loan is almost twice what it sounds like when stated in dollars per hundred per year. For an add-on loan, it's only a little less than that.

I haven't been able to construct, or find, a formula for calculating this rate precisely, so I've had to resort to a sort of backward, or checking, method for arriving at the figures in the preceding paragraphs—the 14.5 percent one for the add-on and the 15.7 for the discounted.

The method calls for making an estimate and then checking it. This is handily done by the use of a pocket calculator, but would be so tedious without it that if you don't have one of these things around you will probably have to be satisfied with the estimate. For a discounted loan, a good estimate is simply twice the stated dollars per hundred. For an add-on it's about one-tenth less than double. Then follow these steps if you want to refine your estimate.

1. Express the estimated rate of interest as a decimal and divide it by the number of payments per year.

2. Add 1.

3. Raise this to the power of the total number of payments.

4. Find the reciprocal—that is, divide 1 by the number just obtained.

5. Subtract 1. Drop the minus sign or ignore it from here on.

6. Divide by the interest rate, expressed as a decimal.

7. Multiply by the number of payments per year.

8. Find the reciprocal.

9. Multiply by the amount actually received.

10. Multiply by the number of payments to be made.

For an add-on loan, the result should be very close to the total of your initial indebtedness—that is, the face amount plus the add-on interest dollars. For a discounted loan, the result should be the face amount.

Now run one or more slightly larger or smaller estimates through those same steps until the result is as close as you can make it.

How Much Goes for Interest?

Bʏ ʀᴇᴀᴅɪɴɢ this graph directly, you can discover how much of your money goes to pay interest during the lifetime of a home mortgage—if the mortgage is for precisely $25,000.

For each of seven interest rates you see a trio of bars. The first of each set of three is for a 20-year mortgage. The second is for 25 years. The third is for a 30-year mortgage (or deed of trust, which is used in some areas to serve the same purpose as a mortgage does in others).

The figures are for interest only. Payments on the $25,000 principal are on top of these.

If the rate of interest or term of years on your mortgage is different from any shown, you can use interpolation or extrapolation to find the interest figure with reasonable accuracy.

Try extrapolating to find the figure for 10 percent interest on a $25,000 mortgage for 30 years. Since the interest at 9 percent is $6,375.60 more than it is at 8 percent, it should be about that much more at 10 percent than at 9 percent. Adding that figure to $47,417.60 gives $53,793.20. Since extrapolating in this situation does not produce extremely precise results, it would be sensible to round off the result and call it "about $54,000."

(If you have a calculator handy you can get a more precise answer by using one of the methods given in this book for calculating mortgage payments, multiplying by number of payments, then subtracting the $25,000 principal. Going this route, you'll arrive at an answer in the neighborhood of $53,984, indicating that you were wise to round off the extrapolated answer.)

By interpolation I can find from this graph how much interest I am putting into a 20-year $30,000 mortgage I've been paying on, since the last previous low-interest days, at 6-1/4 percent.

I find the midpoint between $17,986 (forget the pennies) and $19,736 by adding up the two and dividing by 2. But the result is for $25,000, so for my $30,000 debt I increase it by one-fifth. This is most easily done by multiplying by 1.2. Result of $22,633 rounds off to $22,600.

(Of course, with an actual existing mortgage like this, where I already know the payments, the other route is simpler and may be more precise. Twenty years times 12 months times $219.28 less $30,000 gives me the quite exact figure of $22,627. But the amount I actually mail to the bank is $219, rounded off this way for convenience with the difference to be made up at the time of the final payment. Using $219, I find an interest figure of $22,560, not as close to the true figure as the one I found from the chart here.)

When Interest Rates Change

H*ow can* you make a shrewd long-term financial arrangement in the face of changing interest rates?

You can get some help from this routine for contending with variable interest rates. It's almost as simple as the formula for finding the future value of money when dealing with a constant rate of return.

Suppose you'd like to have $1000 in a certain account five years from now—perhaps for a gift to help someone start college. You know or estimate that yield for the first year will be 15 percent, dropping to 14 the second year, then to 12, 10, and 8. How much must you deposit now?

For each year express the interest rate as a decimal and add 1 to it. Multiply all these sums together (1.15 times 1.14 times 1.12 times 1.10 times 1.08 equals 1.744,364).

Divide the future value—the amount your plan calls for at the end of the five years—by that result. You'll find that to have $1000 then you'll have to make a deposit today of $573.27.

If you're doing such a calculation by paper and pencil, you may prefer to round off that first product to three decimal places. You can do that without distorting the result by more than a few nickels, 12 cents in this case.

The Interesting e

Sᴇᴄᴏɴᴅ only to pi in usefulness is the mathematical constant called e. At 2.718 it's a bit smaller than pi. It describes the rate at which many things grow. Money, for example, or a tree.

If your savings are in an account that is compounded continuously, you need only raise e to the power of the interest rate expressed as a decimal to discover how much each dollar will turn into in a year. If you were so lucky as to receive 100 percent interest (rated pretty skimpy in the loan-shark trade), you'd raise e to the 1st power. Since the 1st power of a number is just itself, you find that your account would grow in a year to $2.72.

To calculate something more realistic, say $1000 at 7 percent, raise e to the 0.07th power to discover that $1000 will expand to $1072.51 in a year of continuous compounding. (If your bank uses daily compounding instead, the difference will be negligible.)

But how can you find the value of e, especially if you want it to more decimal places? Many calculators now have keys for it. Or you can look it up if you have the right book.

(Just so this volume will be a right book, I hasten to say that the value is approximately 2.718281828. And, by the way, don't be fooled: this is one of those instances where what at first blush looks like a repeating decimal turns out not to be. Carry it out any further and it quits repeating.)

To calculate it for yourself, recall first that a factorial of a number is just that number multiplied by all the smaller ones. And that a reciprocal of a number is just 1 divided by that number.

Reckon e by adding up the reciprocals of the factorials of 0 (which equals 1), 1, 2, 3, and so on. When the total stops changing you have e to as many places as your calculator can go.

Comparing Retirement Plans

A READER of a newspaper column presented a problem in financial strategy to its author to solve. Suppose we let his reply outline an answer to many other strategy problems.

The company that employs the advice seeker has offered him the opportunity to augment his retirement plan by $1000 a year at 65. He is now 38. Accepting the proposal will cost him $3220 now.

What our man wants to know, of course, is whether this is the best investment he can make with his $3220.

One alternative investment he might make is common stocks or their equivalent in a mutual fund. History says that over a period of years this kind of investment will pay an average of about 9 percent.

Another possibility is purchase of industrial bonds. Very safe ones, rated A to AAA, could be had at the moment in question that would pay 8 percent or more. A 10-year government bond would do about the same. This is probably even safer, although we can't be sure what rate of return will be available when the money must be reinvested after 10 years.

The first question to be answered is how much will $3220 grow to in 27 years at 8 or 9 percent compounded annually?

General ways to deal with compounding are covered elsewhere in some depth under figuring compound interest. There's no need to repeat the general method here, since we can use the special, and especially simple, routine applicable to annual compounding.

It goes like this.

Express the rate of interest as a decimal (8 percent becomes 0.08). Add 1 to this. Raise the result to the power equal to the number of years of compounding. Multiply by the amount being invested.

In our example, we raise 1.08 to the 27th power and multiply this by $3220.

It turns out that by the magic of compounding $3220 grows in 27 years to $25,721 at 8 percent interest. At 9 percent, it becomes $32,989.

Taking the more conservative of these figures and rounding it downward to $25,000, we find our man would have a sum that would give him $2,000 a year income if he invested it at 8 percent.

This is double the additional pension he'd draw if he took the alternative route. And the capital is his forever, in the end becoming part of his estate.

Interest by Calculator

MAYBE you haven't tucked into interest problems since you left school and some admirable people invented pocket calculators. Since these relate to mortgages and other time payments that most of us are plagued with from time to time, it may be useful to squint at them specifically as seen by your calculator.

Simple interest: This tells you what you'll owe a month or a year from now if you borrow money today. To find it, multiply the interest rate by the time, add 1, then multiply by the principal amount. For $250 borrowed for 6 months at 12.5 percent simple interest, first turn that rate percentage into a decimal by moving the point two places to the left. Then divide that 0.125 by 2, add 1, and then multiply by the $250 to find that you'll owe $265.63 when the loan comes due.

When interest is compounded, it becomes more exciting—and more worthy of your electronic helper's talents. The formula for compounding tells you to add 1 to the interest rate (divided by 100 to express it as a decimal), then raise this total to a power equal to the number of periods involved, then multiply this factor by the amount of money. Just remember that "number of periods" means how many times compounding occurs, and the interest rate you want is the one for the period used for compounding...not the annual rate.

How much will $1200 be worth if you leave it in savings to be compounded semi-annually for 5 years at 6-1/2 percent?

Thinking of 6-1/2 as the decimal 6.5, divide it by 100 (because it's a percentage) and then by 2, because compounding is to be twice a year.

Now add 1, and the display will read 1.0325. Multiply that by itself 9 times to raise it to the 10th power, because there are 10 half-yearly interest periods in 5 years. The resulting factor, multiplied by the amount of money, should tell you that your $1200 will swell to $1652.27.

How Regular Savings Grow

Lᴇᴛ'ѕ ѕᴀʏ there's a new baby at your house and you want to start right now to save for his, or her, education. How much will a few dollars a month come to in 18 years?

Or suppose your household has resolved to cut its grocery bill by shrewd shopping—and save the difference for a family trip to Europe in 10 years—or a retirement fund 40 years from now. How can you find out how many dollars the savings will amount to?

Just multiplying the monthly amount by the number of months won't give you even an approximation. Your money will be drawing interest, and interest on interest, during the whole period, swelling to a much larger sum than you might expect.

To take the extreme example, suppose you can save $50 a month for the next 40 years. That's $24,000 actually deposited.

By the way, that's no impossibility. While helping my wife write a book on supermarket strategy, back when a dollar was worth twice what it is now, I learned the surprising fact that a great many families were spending at least $50 a month more than they needed to, just by bad shopping methods.

But when I wanted to tell readers how much this would come to by retirement after 40 years of family spending, I discovered that nobody at my bank or savings and loan had any idea how to figure it. They merely supposed that the computer at headquarters far away could do it. Well, once I got hold of my first mini calculator, I found it wasn't all that difficult to compute for myself.

Let's first take the method, then apply it to our example.

1. Express the annual interest rate as a decimal. Divide it by the number of times compounded per year. Add 1.

2. Raise this result to a power equal to the total number of times compounded.

3. Subtract 1. Multiply by the number of deposits per year. Then divide by the annual interest rate (in decimal form).

4. Multiply by the amount you're saving.

Applied to a monthly saving of $50 for 40 years in a bank that pays 6 percent interest compounded monthly, the procedure with your calculator is:

1. Enter 0.06, divide by 12, add 1 to get 1.005.

2. Raise this to 480th power (40 years times 12 months per year) by method explained below. Result: 10.9575.

3. Subtract 1 and multiply by 12, to get 119.4894.

4. Multiply by $50, then divide by 0.06, to discover that your savings have swelled to nearly $100,000. My mini calculator gives precisely $99,574.54. Because of early rounding off, yours may disagree by a few dollars.

(Now, about that raising to a power. It's an instant operation with a scientific or financial calculator, but quick enough with any model. Just turn that 480 into its factors—any handy set of numbers that multiply to 480. Say you use 8, 6, and 10. That means you hit the times key, then the equals 7 times; then the times, then the equals 5 times; then the times once more and the equals 9 times.)

Saving: How Much per Month?

How much must you put aside each month to accumulate the price of a Mercedes in ten years or a beach condominium when you're 65?

To find out, begin by deciding first the size of your goal and how long you're prepared to take to reach it. Find out, or estimate, the rate of interest you can count on. From this you can figure out precisely how much you must save each month.

If the procedure that follows seems familiar, there's a reason. Up to the last two steps it is the same as for figuring how much a known sum saved each month will come to over the years.

1. Divide the interest rate, expressed as a decimal, by the number of times the institution where you are placing your savings compounds per year.

2. Add 1.

3. Raise this total to the power equal to the number of times compounding will occur. (This is the number of times compounded per year multiplied by the number of years of your program.)

4. Subtract 1.

5. Multiply by the number of deposits you're to make per year.

6. Divide by the annual rate of interest.

7. Divide by the amount of your dollar goal.

8. Find the reciprocal of the result. (This means divide 1 by the number you've found. On an electronic calculator, you can do this by pressing the divide key once and then the equals key twice. Advanced calculators let you do this in a single step with a key marked 1/x.)

Let's run through an example. You want a $100,000 stake. You've got 24 years to get it together. You are prepared to bet that bonds, stocks, or real estate investment will bring you a 9 percent return with quarterly compounding, but since taxes in your income bracket will take one-third of the interest, you treat this as a net of 6 percent.

Expressing that percentage rate as its decimal form, 0.06 and dividing by the 4 times a year that quarterly compounding will occur, you have 0.015. Add 1 to this and raise the sum to the 96th power (compounded 4 times a year for 24 years).

From this 4.1758, subtract 1. Then multiply by 12 for the number of times a year you plan to add to your investment. Dividing by the annual interest rate of 0.06 gives 635.1607; dividing this by your goal of $100,000 or bust, produces 0.00635, of which the reciprocal turns out to be 157.44.

So you'll have to sock away $157.44 a month to reach an accumulation of $100,000 in two dozen years at the realistic rate of return we've postulated.

In serious planning of this nature, involving a long span of years, it's important to take into account what inflation may do. In short, ask yourself what purchasing power the $100,000 in our example may have after 24 years.

A way to get a handle on this is to subtract your estimate of the coming rate of inflation from the nominal interest rate at the same time that you subtract the income tax rate. If you see inflation as likely to be around 2 percent, this reduces your true rate of return in the example from 6 percent to 4.

In practice, you may want to use current figures for nominal return, tax bracket, and inflation—instead of trying to guess for the long period. Then recalculate at intervals as these variables change, and adjust the amount of your monthly deposit.

Figuring a Set-Aside

W HAT PROPORTION of your income must you set aside each year in an interest-bearing account to accumulate a fund equal to a year's income?

That's what this table is about.

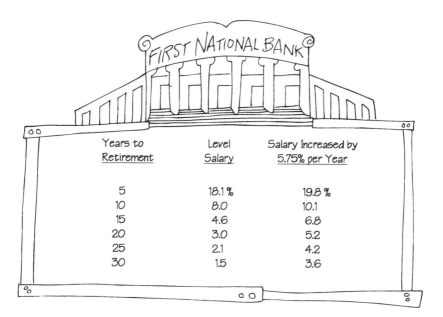

Years to Retirement	Level Salary	Salary Increased by 5.75% per Year
5	18.1 %	19.8 %
10	8.0	10.1
15	4.6	6.8
20	3.0	5.2
25	2.1	4.2
30	1.5	3.6

You might want to use it, as my family did when we gave ourselves a sabbatical year in Europe...traveling, studying, researching, writing stuff that couldn't be expected to produce immediate money.

If you expect your income to remain level through the period, use the "level income" column. (If you take "level" to mean rising at the same rate as inflation, you won't have to allow for inflation in your reckoning.)

The other column is the one to use if you can expect a fairly normal 5.75 percent increase each year on the average.

Here's an example of what the table tells you. If you want to have a fund at retirement 20 years from now that is equal to your final year's salary—your earnings having risen 5.75 percent a year—you'll have to set aside 5.2 percent of each year's income.

This assumes your money is earning 5 percent. You may be able to do much better than that, but remember that the 5 percent means after state and federal income taxes have taken their bite.

How Much Interest Are You Getting?

ALMOST everybody knows how to figure a certain percentage of something, but reversing the calculation is somehow ultimately confusing.

You have $41,106.75 invested in a fund that pays interest monthly. Payoff is at whatever rate the fund is able to earn currently. You receive an interest check amounting to $216.67.

You may want to know what rate of interest that represents so you can decide whether to switch to some other form of investment. Basically, to find an interest rate you just divide the amount of return by the amount invested, and move the decimal point two places to the right to change to percent.

But there is usually one complication. For purposes of comparison, you want to know the annual interest rate, while the return is more likely to come monthly or quarterly.

So, to work out the example above, you must divide the monthly income by the amount of the principal and multiply by 12, the number of months in a year. Move the decimal two places and your result should be just over 6.3 percent.

Sometimes you need to be even more precise than this. If your investment pays, as some do, from the first (or perhaps the second) business day of one month to the next, the number of days for which you are getting interest may vary markedly from one month to the next.

Then this is what you might go through.

Let's say you've just received your check for February. The accompanying slip tells you that you own 4,108.071 shares and their present price is $10.02. Date of the slip is March 3, and date of the previous slip was February 3. What's the interest rate?

You can find it most directly by multiplying the number of days in a year by the number of dollars in the check, then dividing by the product of three things: The number of shares, the price of each, and the number of days in the month in question.

Steps with an electronic calculator are: 365 times 228.24 divided by 4108.07 divided by 10.02 divided by 28 equals 0.0722804. So you find you're getting just under 7-1/4 percent interest.

Chapter D
INVESTING

I REALLY THINK YOU'VE GOT IT!

Finding Earned-Growth Rate

It's a valuable measure of true growth that can be used for any kind of business, from food processing to steel. It's called the EGR, or earned-growth rate.

To determine it, subtract the dividends per share from the earnings per share and divide by the book value (also known as stockholders' equity) at the beginning of the year.

The result will be the percentage by which the value of each share of common stock rose during the year. The price of the stock may not reflect this rise immediately, but it will tend to do so in the long run.

Tailoring an Investment

One man's winning investment can be another man's loser. Or woman's, of course.

An important aspect of tailoring an investment to your own needs is considering the impact of your personal income-tax bracket. This can be pretty tricky when you must consider the effects of a tax that comes at the end of a long period rather than year by year. Or take into account an alternate investment that is not taxed at all—but bears a lower nominal rate of return.

For purposes of illustration assume that your tax bracket (federal, state, and local combined) comes to about 40 percent and, so far as you can predict, will remain about the same.

You might find yourself considering three options.

One option is a single-premium deferred annuity for 20 years at 9 percent.

Another is a 20-year corporate bond at 10 percent.

The third is a tax-exempt bond paying a mere 7 percent.

What you might want to know, for each option, is how many after-tax dollars you'll have at the end of 20 years for each dollar you invest today.

For the single-premium deferred annuity, begin by expressing the percentage as a decimal (so that 9 percent is 0.09) and adding that to 1. Raise this sum to the power of the number of years and then subtract 1. What this will tell you is the amount that original dollar will have earned before you pay the tax on those earnings.

In our example you've raised 1.09 to its 20th power and then taken away 1, to get 4.6044. But if your financial environment is that 40-percent bracket you'll get to keep only 60 percent of the earnings. Multiplying the previous figure by 0.60 and adding 1 (for the dollar you put in to begin with) gives 3.7626. That $3.76 is the after-tax amount to which each of your annuity dollars will have grown in 20 years.

Next question. Can you beat that 9-percent single-premium deferred annuity with a 10-percent corporate bond?

To find out, express the percentage rate the corporate bond pays as a decimal and multiply by the fraction of your income you retain after taxes (0.10 times 0.60 in our example), add 1, and raise this sum to the power of the number of years—20 in this case. Your buck will have grown to 3.2071 dollars that you get to keep. The annuity did better than that.

But if you are dealing with a 12 percent bond instead, you'll discover through a similar calculation that the figure becomes a more salubrious 4.0169.

For a tax-exempt bond the calculation is even quicker. Just express the interest rate as a decimal, add it to 1, and raise that sum to the power of the number of years.

Suppose the bond offers 7 percent. Again express the percentage rate of return as a decimal and add 1. Raise this 1.07 to the 20th power. Since this comes to 3.8697, you've found how many dollars one dollar will grow to in 20 years at 7 percent interest compounded (or reinvested) annually, if there are no taxes to consider. Just about $3.87.

By using these three formulas you can compare the three different kinds of investments, whatever the interest rate and whatever your own tax bracket.

By modifying any of the formulas slightly you can deal with a bond or other investment that pays oftener than once a year. To apply the formulas above, just use the number of compounding periods instead of the number of years; and divide the interest rate by the number of periods in a year.

That probably sounds more complicated than it is. It just means that for the bond that offers semiannual payments at 10 percent a year for 20 years, you'll use 40 as the number of periods and 0.05 as the rate per period. You'll find that with semiannual payments your dollar grows to 3.2620 in 20 years. For a $10,000 investment, that's an improvement of $549.

Local Taxes and Investing

Iₙ ₜₕₑ ꜰɪᴙꜱₜ place, whether to put money into U.S. Government E bonds should depend in part on where you live.

If you have to pay local income tax on bank interest but not on earnings from these bonds, you'll want to start by figuring the net return after local tax (state, county, city).

To do this, subtract the tax bracket this local agency has you in from 1. Then divide this into the rate of return the investment gives. Use decimals to do this.

For instance, suppose your state income tax bracket is a rather high 7-1/2 percent. As a decimal that's .075, which subtracted from 1 equals .925. When you state the E bond rate of, say, 6 percent, as the decimal .06 and divide that by .925 you get .065 approximately, or 6-1/2 percent equivalent.

You can see that if at the time bank and savings & loan accounts yield between 5 and 6 percent, that's pretty good. On the other hand, if you are dealing in sufficient sums and long enough periods of time to buy certificates of deposit from these same institutions at yields of 7 to 8 percent or a trifle more, those bonds are not all that appealing as an investment.

Keogh? Ira?

Wₕₐₜ ᴅᴏ ʏᴏᴜ gain by putting money for retirement into an IRA or Keogh Plan fund instead of just tucking it away in an ordinary bank account?

One way to tell is by looking at this little table an insurance company worked out.

Another way is by calculating it yourself for your own precise situation. Begin with the amount you'd like to put away each month. Multiply that by the tax bracket you're in (federal plus state, if your state allows deductions similar to the federal ones). The result is how much you reduce your tax bill for the year.

But keep in mind, with this and all other such calculations, that the money you take out after retiring will be taxable income. The idea is that by then your earnings will be far smaller and your exemption larger, so your tax rate, if any, will be much lower. This is in addition to what the table tells you about how much you profit by the delay in taking taxes out of your earnings.

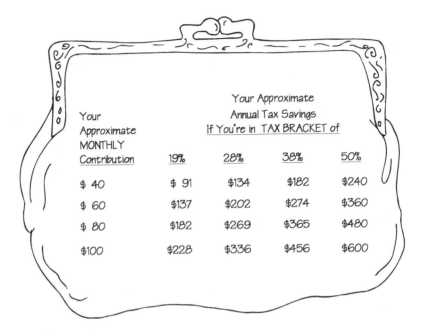

Your Approximate MONTHLY Contribution	Your Approximate Annual Tax Savings If You're in TAX BRACKET of			
	19%	28%	38%	50%
$ 40	$ 91	$134	$182	$240
$ 60	$137	$202	$274	$360
$ 80	$182	$269	$365	$480
$100	$228	$336	$456	$600

A different way of looking at the matter is by figuring how much more you'll have in the end for the same out-of-pocket cost if you do your saving under a Keogh or IRA plan.

Annual Income: $15,000		Tax Bracket: 22%
	QUALIFIED Plan	NON-QUALIFIED Plan
Gross Contribution	$1,500	$1,500
Income Tax 22%	0	-330
Net Contribution After Taxes	$1,500	$1,170
Gross Investment Yield	5%	5%
Net Investment Yield After Taxes	5%	3.9%

This table applies to a person with an income around $15,000, or to a family with several thousand dollars more, which would produce the same amount subject to taxes.

	QUALIFIED Plan	NON-QUALIFIED Plan
5 Years	$ 8,702.87	$ 6,571.07
10 Years	19,810.19	14,527.54
15 Years	33,986.24	24,161.20
20 Years	52,078.88	35,825.87
30 Years	104,641.19	56,520.71

Most convincing of all is the third little table. It shows how much you'll have after various periods if you put aside $1,500 each year and are in a 22 percent (federal plus whatever) tax bracket.

Using a tax-deductible plan instead of an ordinary savings account bearing the same interest rate, means the difference between having $104,641 after 30 years and having little more than half that amount.

Compounding Your Money

THIS TABLE offers a quick way—meaning hardly any figuring—to find out how much money you'll have someday if you put aside a fixed amount each year.

If the sum you plan to save is $100 a year, you can read the result for any of a number of periods and interest rates directly from the table. You can see that $100 invested each year and compounded annually will grow in half a century to $128,030.

For any other amount, divide the figure in the table by 100 and multiply by the number of dollars to be put aside each year.

The table works backwards too.

Suppose your goal is to accumulate $20,000 over 30 years and you can assume an 8 percent net return annually. Where the 30-year and 8 percent lines intersect you'll find $12,234.

Divide the wanted amount of $20,000 by that and you'll get a quotient of 1.63. Multiplying that by the table's assumed $100, you'll find your annual investment must be $163.

A regular investment of $100 per year invested at	Will, compounded annually, grow to the sum shown after						
	5 Years	10 Years	20 Years	25 Years	30 Years	40 Years	50 Years
5%	$580	$1,321	$3,472	$5,011	$6,976	$12,684	$21,981
6	598	1,397	3,899	5,816	8,380	16,404	30,776
7	615	1,478	4,387	6,767	10,107	21,361	43,498
8	634	1,565	4,942	7,895	12,234	27,978	61,967
9	652	1,656	5,576	9,232	14,857	35,820	88,844
10	671	1,753	6,300	10,818	18,094	48,685	128,030

Retirement-Fund Accumulation

THIS IS one for regular savers.

For each dollar you put into a tax-sheltered IRA or Keogh retirement fund each year for the next 25, how many will you have available for use when the time comes?

If it's a fund that pays 12 percent interest—a ridiculously optimistic level in some eras but normal in fairly heavy inflation—compounded twice a year, you'll have $145.16.

At $1000 a year, it'll come to a nice $145,160.

That's the kind of thing this table figures for you.

That's for semiannual compounding. For other compounding periods, interest rates, and periods of time, you can figure it yourself with a hand calculator by methods given elsewhere in this compendium.

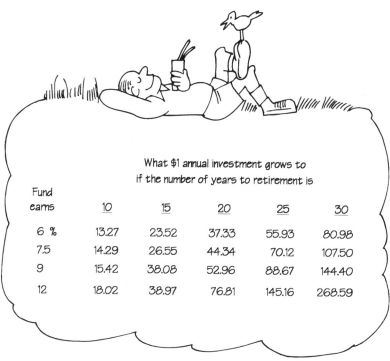

Fund earns	What $1 annual investment grows to if the number of years to retirement is				
	10	15	20	25	30
6 %	13.27	23.52	37.33	55.93	80.98
7.5	14.29	26.55	44.34	70.12	107.50
9	15.42	38.08	52.96	88.67	144.40
12	18.02	38.97	76.81	145.16	268.59

What Does an Annuity Cost?

Have you run through the procedure for calculating how big an annuity a given sum will buy? Reversing the routine will tell you how much an annuity of any given amount will cost you.

Having decided what monthly income you'll need, you can then learn what it will cost you to buy it. You then multiply by a fixed sum that depends upon your age.

For a straight-life annuity bought when you turn 65, the multiplier has recently been 115.

So if at 65 you want an annuity that will bring you $1,000 a month for life—and die when you do—you'll have to plunk down $115,000.

For an annuity bought at the same age but guaranteed to keep paying not only for your lifetime but for a minimum of ten years no matter what happens to you, the multiplier is 123.8.

How Much Annuity Can You Buy?

A STRAIGHT-LIFE annuity is a monthly income you buy with cash in advance. It is guaranteed to go on as long as you live, and it dies when you do, leaving nothing for heirs.

Because of this limitation it produces a larger income than any ordinary safe investment, especially for someone getting along in years. Such an annuity has little appeal for anyone under retirement age.

You can get a pretty close approximation of what a straight-life annuity will produce in the way of monthly income if bought at the age of 65 by dividing its cost by 115. This will tell you that if, for example, you had $45,000 to invest at 65, it would bring you $391 a month as long as you lived.

By comparison, the same $45,000, if put into an 8 percent savings account or bonds, would produce only $300 a month. But, of course, the principal would still be there for your lucky heirs to collect. If you heartily resent the idea that almost every cent you put into an annuity would be lost if you should die shortly thereafter, you might prefer some other type. One variety, called "ten years certain," not only pays as long as you live but goes on for ten years even if you die the next day. To figure the typical return on this type, if you should buy one at the age of 65, divide by 123.8 instead of 115. You'll find the monthly return on your $45,000 would then be $363.49.

What's an Annuity Worth?

LET'S SAY you want to arrange for yourself a supplementary income of $5,000 a year for the next twenty years...total of $100,000.

Tucking a thousand $100 bills into a mattress and pulling out fifty of them every January would be one way to do it.

It would be a wasteful way, of course, since the part of the money you hadn't used yet would be sitting there drawing no interest. Clearly, it should be possible to "buy" the income we're talking about for less than the whole $100,000 if interest can help.

The question, then, is this: What amount must you deposit today in order to have it last just twenty years if withdrawn at $5,000 each year?

Assume you choose to put the money into a bond fund that will pay 7 percent interest.

Then the calculation proceeds this way.

Express the interest rate as a decimal. Add 1. Raise this sum to the power of the number of years you've chosen. Find the reciprocal (that is, divide 1 by this result). Subtract this from 1. Divide by the interest rate expressed as a decimal. Multiply by the number of dollars you wish to withdraw each year. For the example, the computation runs:

Multiplying 1.07 by itself 20 times gives 3.8696833, which divided into 1 is 0.258419. Subtracted from 1 and then divided by 0.07, this becomes 10.594014.

Multiply this by $5,000 and you'll find the cost of this self-created annuity to be $52,970.

So at 7 percent interest, you can buy now for little more than half the amount that will be returned to you.

Make the same calculation for 8 percent net bond return and you'll find the cost to be less than half.

Decoding the Phony "Average Return"

RATES of return on savings are often quoted in the form of a pair of figures. One figure may be described as the *interest* rate and the other, which is a bit higher, as the annual rate of *return*.

Nothing wrong with that. When interest is compounded at short intervals, and especially when compounded daily or hourly or continuously, you gain an extra edge above the mere annual rate.

Thus $100 held for a year at 8.21 percent simple interest will become $108.21. But if compounded continuously at that rate it will turn into $108.56. You are getting what can legitimately be called an annual rate of return of 8.56 percent. Note that the difference between those two numbers (8.21 and 8.56) is only a fraction of a percentage point.

Suppose you pick up the morning paper and see, as I did one day at a time when interest rates were comparatively high, that a savings institution is offering a certificate paying an annual rate of 8.21

percent, for which an "average annual return" of 9.31 percent is claimed for a 3-year certificate.

How can that be? How can the return be more than a whole percentage point above the rate? The trick lies in the deceptive use of that misunderstood word average. A bank or S&L can offer you literally any average annual return it cares to name simply by choosing a sufficiently large number of years over which to average it.

If the S&L in question had wished to offer you the even more appealing (but perhaps suspiciously tasty) bait of 12.73 percent instead, it could easily have done so by mentioning 10 years instead of 3.

(For 100 years, if you're curious, the average annual return for an 8.21 percent account, compounded continuously, is 3,677 percent. Even for a mere 20 years it's a bit more than 20 percent.)

Calculating these returns for yourself is not difficult if you don't mind dipping into the table of logarithms to do it. You won't need that fat table, of course, if you have at hand a scientific or financial calculator or a personal computer.

In that case, this is what you do.

Enter the value of that widely useful constant called e, which your computer program or advanced calculator will have ready. Or simply type in 2.728. Then express the annual interest rate as a decimal and multiply this by the number of years to find the power by which you will then raise e. Subtract 1. Divide by the number of years. Multiply by 100.

For the 10-year example above, this means multiplying 0.0821 by 10 and raising 2.728 to that power (0.821), obtaining the result 2.2728. Subtracting 1, dividing by 10, and multiplying by 100 brings you to the 12.73 percent average cited.

This is an interesting calculation and sometimes a useful one. What's more to the point, though, is remembering that average annual returns are just so many deceptive and essentially meaningless numbers...unless they are used solely to compare two results for precisely the same number of years.

How Does Dollar Averaging Work?

H<small>AVING</small> chosen a stock in which you intend to invest at regular intervals, should you plan to:

   …buy a fixed number of shares each time; or

   …put up the same number of dollars at each purchase?

There's a trick for investing in stocks over a period of years that follows the rules of arithmetic but somehow seems more like magic. It's a device that guarantees that you can, without any exercise of judgment, buy your stocks for less than the average price they are selling at over the period.

If you invest $1,000

At this share price	You get this many shares
$ 8.00	125
11.50	87
12.50	80
11.50	87
14.25	70
12.50	80
14.50	69
16.40	61
17.50	57
15.00	67

All you have to do is invest periodically, keeping the amount you invest the same each time without regard to the price at which the stock is selling.

Called dollar averaging, it's actually easier and simpler to do than buying a preset number of shares at intervals—because it takes the same amount out of each paycheck.

You can apply this routine just as well to buying mutual-fund shares as to purchasing a single stock.

This table sums up what occurs—assuming you have a thousand dollars to invest every month, or every second month, or whatever.

Add up those share prices and divide by the number of purchasing occasions and you'll find that, on the average, the stock was selling at $13.57.

Yet your 783 shares will have cost you $10,000 divided by 783, or $12.77 a share.

Ain't it wonderful?

How can this be? Well, the thing is, the lower the price of the stock on your purchase date, the more shares you get for a given investment.

And that's why putting up a constant sum of money each time works out better than buying some stipulated number of shares at regular intervals.

Super Dollar-Cost Averaging

IF YOU like the idea of dollar-cost-averaging your investing, you may want to carry it even further—into super-averaging. By this rather cute method, you invest fewer dollars when the price is high, more dollars when it's lower.

To do this, you must first calculate the average cost of the shares you own. Divide this by the price that is currently quoted, then multiply by the amount you've determined to invest monthly.

Say you find that your stock has cost you an average of $9 and the market price right now is $8. Dividing 9 by 8 gives 1.125. If your plan is to invest $150 a month on the average, multiply this by that previous result (1.125) to find that your investment this month should be $168.75.

If, however, the market price had been $10 you'd have divided 9 by 10 and multiplied by $150 to discover that your strategy called for a $135 investment this month.

Value-Averaging Your Investment

VALUE-AVERAGING is calculated to make you an even shrewder trader—and automatically at that—than the dollar and super-dollar averagers. It is best suited to purchasers of no-load mutual funds, since brokers' commissions can make small purchases of stocks costly.

Choose a fund that permits a low initial investment.

Decide the number of dollars by which you wish to increase the value of your holdings each month. (Let's say you choose $200.)

At the end of each month, put in enough to bring your fund's total value to this amount ($200) more than it stood at when the month started. When your holding in the fund has grown by $120 in a month, it is time to put in $80. If it has diminished by $10, you'll put in $210. And if it has leaped by $250, you should—to be consistent—take out $50. Better, take this as a signal that it is time to lift your goal by leaving that extra $50 in place—and perhaps to increase the monthly $200.

What's a Tax-Exempt Bond Worth?

WHEN YOU read about people who have great big incomes and pay little bitty income taxes you wonder how they do it. One way is with tax-exempt bonds.

Will it work for you?

The simplest rule of thumb is: probably not unless your combination of federal, state, and local taxes comes to more than 30 percent.

This is because, historically, market value of securities of approximately equal safety has usually worked out so that the return from a tax-exempt is about 30 percent lower than from an ordinary bond. If an A-rated corporate bond is producing an 8 percent return, a comparable tax-exempt will sell at a price to yield about 30 percent less, or somewhere around 5.5 percent.

And don't forget that you find your bracket for this purpose by adding together all applicable federal, state, and local income levies.

At a time when this tax total works out to around 30 percent or so for you, you may find a formula useful. Here is one for calculating how much an ordinary bond must yield to be worth as much to you, after you've paid taxes on the income, as a given tax-exempt one.

First find how much Uncle Sam leaves you by subtracting your top total tax bracket from 100 percent.

Divide the percentage yield of the tax-free bond by that figure. Result is the level of yield from an ordinary security you'll need for an equal spendable return.

For a 36 percent tax bracket and a 4 percent tax-exempt bond, it goes like this: After taking assorted deductions, you get to keep 64 percent of your income (100 percent minus the 36 percent tax liability). That is, you retain your income multiplied by 0.64. The word multiply makes this sound jollier than it is.

Divide the 4 of the bond by that 0.64 to get 6.25. You have determined that an ordinary bond yielding 6-1/4 percent will do as much for you as a tax-exempt 4.

If you have the opposite question, turn the formula around so it says that tax-free yield is equal to taxable yield multiplied by after-tax percentage.

If you're doing the figuring on an electronic calculator you may be surprised to run into a small hitch. Although all you're doing is dividing one figure by the difference between the two others, this is an operation the ordinary calculator can't perform without some help.

Here's how to give it that help.

Instead of trying to divide one figure by the other, multiply that first figure by the reciprocal of the other.

These are the steps:

Enter 1 and subtract from it your tax bracket expressed as a decimal, using the equals key.

Find the reciprocal—by pressing the divide key once and the equals key twice.

Multiply this by the coupon rate of the tax-free bond.

The answer is the rate that an industrial or other bond that is not tax-free must yield to give you as much after-tax income.

Income from Capital

I'T'S NOT hard to figure out how much monthly income a specified amount of money will produce forever. Just multiply the amount by the interest percent and divide by 100 and by 12 to find the monthly income. For a weekly income, make that 100 and 52.

With the table, it's even easier, for the half-dozen percentage rates it covers.

Amount of capital	Monthly income at rate of					
	5%	6%	7%	8%	9%	10%
$ 10,000	$ 42	$ 50	$ 58	$ 67	$ 75	$ 83
20,000	83	100	117	133	150	167
30,000	125	150	175	200	225	250
40,000	167	200	233	267	300	333
50,000	208	250	292	333	375	417
75,000	312	375	438	500	562	625
100,000	417	500	583	667	750	833

But both the table and your own calculations have a weakness to beware of. They don't mean much until you have subtracted the amount of any tax bite for which you'll be liable on the income. And then there's inflation. Your true investment return is only the amount by which the yield exceeds the current inflation rate. So if you have an 8 percent yield and a 25 percent tax, there goes one-fourth of 8, or 2 percent. In a time of 6 percent annual inflation, the remaining interest is wiped out. You can't take out a dime without, as my grandmother always put it, dipping into capital.

How'd That Investment Do?

THERE's a simple and standardized way to compare one investment with another.

Applied to something like a mutual fund, it works like this.

Begin with what's called its NAV (net asset value) at the beginning of the year. Let's say that was $10.

Now add to the NAV at the end of the year ($12, let's say) all that the fund has made in dividends and capital gains during the year— whether paid to you or reinvested. Say the dividends totaled 30 cents and the capital gain 20 cents. Add these to the $12 to get $12.50.

Subtract the NAV at the beginning from this total. Gain comes to $2.50. To find what percentage this is of what you started with, divide the gain by the initial NAV and multiply by 100.

You've had an increase of 25 percent.

To compare this with some other kind of investment, you should also allow for the tax aspects. These, in turn, depend upon what bracket you're in. Remember, too, that—depending upon what has happened in Washington lately—part of dividend income may be excluded from taxation and capital gains may be taxed at a lower rate than ordinary income.

The Moving Average

ONE KEY indicator that many stock-market professionals watch closely, and often act upon with real money, is the moving average. It's also called the current mean.

These pros believe they have found that market moves commonly exhaust themselves after going up or down by about 25 percent at most from the 40-week moving average of a group of stocks such as the Dow-Jones industrials.

To calculate this 40-week moving average, you add up the closing figures on the index for each of the past 40 weeks and then divide by 40.

After that it gets easier. Each week you add the new figure to the old total and subtract the earliest one. Then divide by 40.

The further the new figure is from the original average, the more likely it is that the stock market is shifting direction. Or so say those who call the turns for a living.

Of course, anyone using this indicator will want to consider all other available information as well. Whatever any investment type you talk to might give you to believe, there's no magic in it.

How Much Does a Treasury Bill Yield?

IT IS OFTEN possible to get a better rate of interest by investing in one or more U.S. Treasury Bills than by putting money into a savings account. Of course, that's if you have a chunk of money available for investment at one time.

You can make the purchase at no charge from any Federal Reserve Bank, or you can pay your bank or a stockbroker a small fee to do it for you.

The way it works, you pay the Treasury the face amount of the Bill—say, $10,000—and get back, in a few days, a discount check that constitutes your interest.

To find the yield, divide the amount of the discount check by the amount the Bill has actually cost you. You can convert this to an annual rate by multiplying it by the number of times a year your money could be turned over.

Consider a $10,000 Bill that brought you a discount check of $150.

Your cost is then $9,850. Dividing 150 by 9,850 gives .0152. If this was a three-month Bill (most are for three, six, or twelve months) multiply by 4, for one-fourth of a year. The result is .0608, which means 6.08 percent annual interest.

If you pay a fee to bank or broker, you really should subtract that from the discount. If it were $10 in this example, you'd divide 140 by 9,860 before multiplying by 4, to find the true annual yield rate to be only about 5.8 percent.

Here's something to con-sider, if you want to be truly canny about your invest-ing—especially if you are going to buy many Bills at once or over a period. If you are paying with a certified or bank check on which you are sacrificing inter-est from the date it is drawn to the date the Fed must have it, add this into your cost.

Compare this loss with the cost of having your bank make the purchase for you.

Is Tax-Free for You?

THIS IS a quick-and-easy routine for finding out if a tax-free invest-ment, such as a municipal bond, is a good choice for someone in your income-tax bracket.

From 100 subtract the highest tax percentage bracket that applies to you. (If you're considering a bond that is issued in your state and is exempt from state tax also, add your combined state and federal brackets.)

Multiply the yield of the tax-exempt security by 100 and divide by the figure from the previous paragraph.

This gives you the yield you'd have to get from a taxable investment in order to come out as well.

So now let's suppose your top dollar of income is taxed at 30 percent and you're considering a tax-free bond yielding 8 percent.

Subtracting 30 from 100 and dividing into 800 gives 11.4. So 11.4 percent is the yield you'll need from a taxable bond to equal an 8 percent return that you receive free of taxes.

What Percent Does a Stock Yield?

You know what percentage you can get on your money if you put it into a bank or some such place. The bank tells you: three or four, maybe, or seven or eight percent on a time deposit.

To compare the yield from a common stock, you have to figure it in the same terms.

To find out what a stock is returning to you for what you paid for it, divide the total of dividends for one year by the cost of the stock and multiply by 100. Thus a stock for which you paid $40 a share, including commission, that brings 25 cents a share in dividends four times a year, is yielding 2.5 percent on your investment.

The stock may be going up in price and making you potential profits that way, as well, but this is not what we're talking about at the moment.

A better indicator of whether a stock is worth hanging onto is yield relative to present value.

If that $40 stock has dropped in value to $25 or so, its annual dividend of $1 represents a 4 percent return on the amount of money you could get if you sold it. The fact that this is just 2.5 percent on what you paid for the stock is a matter merely of historical interest now.

What does a p/e tell?

There are a couple of figures by which the value of a stock is often judged.

One of these, yield, is the relationship between the price of the stock and its dividend for the year. A stock that is selling at 50 and producing four $1 dividend checks a year is yielding 8 percent.

If you are investing primarily for income, yield is pretty important to you.

But if you're primarily hoping for growth in the value of the stock, you'll probably learn more from a calculation of the stock's price/earnings ratio. This is determined from the company's earnings per share (the number of dollars the company earned last year divided by the number of shares of stock outstanding). Current price of the stock is then divided by this figure.

For any stock regularly traded, a broker will have the p/e on tap.

You need to know more about a stock than its p/e to make an intelligent decision about it, of course. But in general a stock with a p/e of 10 or less is considered to be underpriced and a good buy.

P/e and yield, for any given stock, could correspond closely, since the more a company earns the more it could distribute to its owners, the stockholders. But what actually happens is that a fast-growing company is likely to hang onto most of its earnings to use in expansion and in research.

So a stock with a low yield is likely to be one with good growth prospects. A high-yielding stock usually indicates a company that is making money but not necessarily going anywhere. (Maybe it's already there.)

It's useful in interpreting a price-earnings ratio, once you've calculated it, to know something about typical levels in the past.

For example, for that representative group of sound, stodgy, reliable industrial stocks used to compute the Dow-Jones averages, the p/e hit a high point of 17.3 back in 1971, then began a three-year slide that brought it to about one-third of that figure before it started up again.

What are sometimes called the glamour stocks are quite another story. Since they have glamour—that is, promise of rising sharply in value and earnings—they normally sell for far more than they could possibly be regarded as worth on a p/e basis. P/e levels of 20 to 50 are quite common for them. Polaroid has had a p/e as high as 114,

IBM as high as 161. Eventually, those ratios slid precipitately, and so did the selling prices of the stocks.

For what it may be worth in making decisions based in part on price-earnings ratios, here's a rule of thumb that is safely conservative and certainly simple to apply. It is one reached by some analysts in the mid-part of the 1970–80 decade. It declares that, no matter how glamorous a company's prospects, its stock should never sell at a p/e in excess of 25.

Figuring Percentage Gain

I‍F YOU'VE made an investment, in stock for example, you might want to know how well it has done for you in comparison to putting your money in the bank. There are several mathematically equivalent ways to find this out.

The easy formula for figuring it begins with what it's worth now. Just the stock, if you want to calculate the capital gain. Or the stock plus dividends you've received.

Divide the present value by what it cost you (or what it was worth) at the beginning of the period.

From this, subtract 1.

To get a percentage answer, multiply by 100 by moving the decimal point two places to the right.

If the period in question is a year, the answer will be the annual percentage return.

It goes like this. If you own some stock that cost you $32.25 a share a year ago, but is now worth $35.75, you divide the latter figure by the former to get 1.1085, subtract 1 to get 0.1085, and multiply by 100 to find the capital gain of 10.85 percent.

(If the answer is negative, you'll have discovered your percentage of loss. Sorry.)

If you want to know your total gain, and that stock has produced a total of $1 in dividends in a year, you'll add that dollar to the present value and begin with $36.75 instead, finding the total gain to have been 13.95 percent.

(If great precision is important, you should add on not only the dividends but also any interest earned by those dividends during the year.)

Total Return and the Deceptive Average

A LARGE mutual fund has announced that its total return to investors for its first 10 years was 413 percent.

That figure is not deceptive in itself and it does convey some information.

But for comparison with a return produced by a competing fund, this number is useful only if the figure from the second fund is also for 10 years (and preferably the same 10).

For making comparisons generally, you might find it more useful to know the annual rate...on the average. In a careless moment, however, this could lead you to use ordinary averaging methods that don't apply to measuring rates of increase. That is, you might—but you shouldn't—say that if the total return for 10 years was 413 percent, then the average yearly return would be one tenth of that, or 41.3 percent.

A large, glossy, and widely circulated consumer-finance magazine blithely went at it that way once and produced a brew of figures that were, at the same time, highly impressive and totally meaningless.

Being more experienced than the editors with numbers like these, the mutual fund has calculated that theirs was an average annual return of 17.8 percent. The fund's method produced a far less exciting conclusion and it's harder to understand at a glance.

However, it does have the virtue of being correct.

What's the trick for translating the total return for a period of years into an annual average?

First turn the percentage into a decimal by dividing by 100. Add 1.

Derive the root corresponding to the number of years (for example, tenth root if you're dealing with ten years).

Subtract 1.

Return this result to percentage form by multiplying by 100.

Doing this to the 413 percent 10 year return, you'll be taking the 10th root of 5.13, which proves to be 1.178. With 1 subtracted and the decimal point pushed 2 places to the right, you'll find you've confirmed the fund's figure of 17.8 percent average annual return.

Don't be alarmed by having to derive something like a tenth root. You can do it almost effortlessly with (take your choice) a modest scientific calculator, a computer, or a log table.

If you lack the first two and have a not unnatural aversion to the third, there's also a great do-it-yourself approach that can make you feel pretty clever. Call it iteration. Or just call it cut-and-fit.

To find any root, including a tenth, make a guess. Then make another guess or two till you have it bracketed. Then close in with guesses that fall between.

Seeking the tenth root of 5.13, you might guess safely low at 1.1. Testing that guess, by hitting the times button once and the equals button nine times, for a total of 10 hits, will give you 2.59. Not big enough.

Try 1.2. This series of ten punches produces 6.19, too big by a trifle, but when you then try 1.19, you have it nicely bracketed.

So you're not surprised when 1.18 hits it so nearly on the nose that you may not bother to refine it further to 1.178.

Understanding Bond Language

IF YOU are interested in buying U.S. Treasury notes and bonds, and bonds of government agencies, you'll have to learn bondish—the language of the business.

Bonds are described by their maturity dates and rate of interest—based on the par value. So if the bonds you're interested in are described as "4s of June 1999," you know they mature—become payable—on that date and bear interest at 4 percent.

Now about price, which is not so simple. Bonds are quoted in points, which means 1 percent of par value—$10 per point in a $1,000 bond. But what looks like a decimal isn't. It's a 32nd. So when a bond is quoted at an asking price of $99.16, that means it is 99 and 16/32, or 99-1/2. For a $1,000-par-value bond, that means a price of $995.

As you can see, this is a pretty crazy system, but it's what the bond world uses. So if the price is given as 99.8, that means it is cheaper than one quoted at 99.16, since 99.8 is 99 plus 8/32, or 99-1/4. Sorry.

Chapter E
OTHER PEOPLE'S MONEY

The Everlasting Payoff
Credit: How Much Is Too Much?
How's Your Credit Rating?
Real Cost of a Discounted Loan
Calculating the Cost of Credit
Figuring an Early Payoff
Payoff Rule of 78s
When You Make Partial Payment
How Big Is That Carrying Charge?
What, Borrow on My Life Insurance?

The Everlasting Payoff

T HIS ONE comes pretty close to being the prince of personal calculations. It figures out the amount of the monthly payment for any size mortgage at any rate of interest for any period of years.

So it replaces an entire bookful of closely printed tables comprising roughly half a million digits. That's what people in the real estate and mortgage businesses carry around to tell you how much something is going to cost you each time the first of the month rolls around.

Actually, this calculation method is far more versatile than your average hefty book of tables, which will cover only a limited range of interest rates—say 6 to 12 percent if it is designed for the mortgage business.

When you do your own calculating you can handle lower or higher rates, such as may be needed for automobile or other installment loans. You can figure payments for odd rates of interest (7-1/8 percent for example) not given in most tables. And you can get exact results for paying off amounts that are not round numbers—all this with no additional arithmetic.

Although the method is reasonably simple, the actual computation is horrendously tedious if tackled by straight paper-and-pencil arithmetic.

A table of logarithms will quickly cut the job down to modest size, as will a slide rule—if I may offer what amounts to a historic note. But most people these days will use pocket calculators.

Whatever your method, these are the steps:

1. Express the stated rate of interest as a decimal and divide it by the number of payments per year.

2. Add 1.

3. Raise this sum to the power of the total number of payments.

4. Find the reciprocal—that is, divide 1 by the number just obtained.

5. Subtract 1. Drop the minus sign or ignore it from here on.

6. Divide by the interest rate, expressed as a decimal.

7. Multiply by the number of payments per year.

8. Find the reciprocal.

9. Multiply by the amount of the mortgage or debt.

All this may sound pretty fierce, but these steps will become easier to follow after you've seen them applied to an actual debt payment.

I once encountered a loan company second mortgage in the odd amount of $4,791.36. It was to be paid off in three years at what the legally required disclosure form admits is an annual percentage rate of 18.25. Finding the nine steps above reveals the size of the monthly payments.

1. There are 12 payments per year and 0.1825 (18.25 percent divided by 100 to change it into a decimal) divided by12 is 0.0152.

2. Adding 1 gives 1.0152.

3. The previous sum when raised to the 36th power (because there are 36 months in 3 years) becomes 1.7218.

4. When 1 is divided by that number, the answer is 0.5808.

5. Subtracting 1 gives 0.4192 when you ignore or change the minus sign.

6. Dividing by 0.1825 gives 2.2971.

8. Dividing this into 1 produces 0.4353.

9. You now have the amount of each payment per dollar of debt. Multiply it by the size of the mortgage ($4,791.36) to find that the monthly payments will be $173.82.

A couple of these steps require techniques not covered by the little booklets that come with ordinary calculators.

Even with this mere 3-year loan, Step 3 calls for raising something to a fairly high power, the 36th. That's not too hard to do by giving the times key 36 punches (or the times key one and the equals key 35) and, I find, takes under half a minute. But the 360 key pokes for a 30-year loan will take several minutes or longer, since you may have to repeat the process several times before you're confident that you've counted right.

Better you should factor. Small numbers that multiplied together give 360 are, for example, 6, 6, and 10. To find the 360th power the quick way, touch the times key once and the equals key 5 times for that first 6, to get the 6th power. Repeat for that second 6. And then touch times once and equals 9 times for the 10th power.

For a 20-year loan, convenient factors of 240 are 3, 4, 4, and 5. And so on.

A couple of the steps call for finding a reciprocal. Some calculators have a special button for this, often marked 1/x. With most others, you just punch divide and then equals. Test this on your own machine before relying on it. These machines often have ways all their own.

If you'd like another example to use in checking your technique, one of the fat payment books the real estate people use says that a $42,000 mortgage at 7 percent interest will pay off in 20 years at $325.63 a month. (My calculator agrees.)

Credit: How Much Is Too Much?

ACCORDING to one "family resources management specialist," there is a simple formula to warn you when you are getting too far into debt. I've found this kind of warning worth respecting, since piling up unwieldy debt is a pretty sure way to surrender control of your life.

In the absence of the guts to tough it out, this has been known to entrap whole countries, even the richest in the world.

To keep control of your credit, says the formula, first add up all your monthly payments, excluding housing. Multiply by 100 (by tacking on a pair of zeros) and divide this figure by your monthly income (not gross—take-home, or after taxes, pay).

That will tell you the percentage going into monthly payments. The red danger line, says the specialist, is about 15 to 20 percent. If you're over that it's time to regain control by refusing to take on any new debts. Try to get rid rapidly of some you have, preferably ones carrying the highest rates of interest.

Of course, you should keep in mind that our specialist is offering only a very rough rule of thumb for families in normal situations. If yours is out of the ordinary, you'll have to cut the generalization to fit. A family paying rent on a rather expensive residence should be pretty wary about letting its payments reach even 15 percent of income. Another family, owning a home clear or buying one in the late years of a low-payment mortgage, might find it tolerable to put more than 20 percent of income into payments for really worthwhile (and durable) things.

How's Your Credit Rating?

HOW CREDITWORTHY—as they say in that line of business—are you? Each loan or time-payment officer has his own way of figuring that, but many rely heavily on some kind of prefabricated scoring

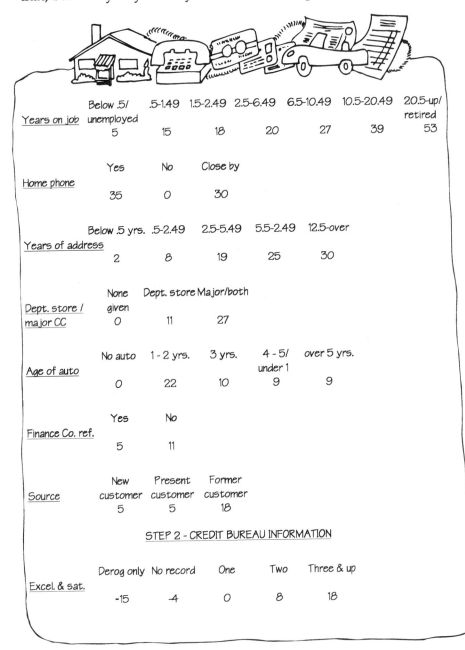

Years on job	Below .5/ unemployed	.5-1.49	1.5-2.49	2.5-6.49	6.5-10.49	10.5-20.49	20.5-up/ retired
	5	15	18	20	27	39	53

Home phone	Yes	No	Close by
	35	0	30

Years of address	Below .5 yrs.	.5-2.49	2.5-5.49	5.5-2.49	12.5-over
	2	8	19	25	30

Dept. store / major CC	None given	Dept. store	Major/both
	0	11	27

Age of auto	No auto	1 - 2 yrs.	3 yrs.	4 - 5/ under 1	over 5 yrs.
	0	22	10	9	9

Finance Co. ref.	Yes	No
	5	11

Source	New customer	Present customer	Former customer
	5	5	18

STEP 2 - CREDIT BUREAU INFORMATION

Excel. & sat.	Derog only	No record	One	Two	Three & up
	-15	-4	0	8	18

sheet. The specimen you see here is a typical example of ones that may have been used many times to rate you.

If you wonder why credit managers use such seemingly crude instruments instead of their own good sense, here's why. Sears Roebuck uses one of these systems and finds it can keep its losses down to 1 percent. But where it has experimented with allowing credit managers to override the rulings of the system, 95 percent of the accounts to which the system had said "no" proved difficult or impossible to collect.

You may wonder at the absence of a salary or income item on the schedule. That's because lenders have discovered that size of income claimed is no help in predicting whether you'll pay your bill.

Suppose you've worked at the same job for 7 years, have a home phone, have lived at the same address 2 years, possess an American Express as well as a department store credit card, don't own a car, have not given a finance company reference, are a new customer—and the local credit bureau has never heard of you.

Your scores will be 20, 35, 8, 27, 0, 11, and 5 for a subtotal of 106, from which you must subtract 4 for lack of a credit bureau record. That's probably passing, since the folks who offer this scoring system say that only by rejecting people who score less can a lender keep his delinquency rate below 4 percent.

Real Cost of a Discounted Loan

IT HAS BEEN a historical, and pernicious, practice (and not just among car salesmen) to quote the cost of a discounted loan in "dollars per hundred per year"—and then to refer to this as the interest rate.

Here's the way to find the true rate, which you might need to know in order to compare it with alternatives.

Multiply the number of installment payments to be made per year by 2 and also by the cost of the loan. (Cost of the loan is the so-called discount, the difference between the amount you actually receive and the sum you agree to repay.)

Now divide by the amount you actually receive. Also divide by 1 more than the total number of payments you are to make.

For a loan of $1000 at $6 per hundred discounted to be paid monthly over 2 years, the calculation works out like this.

At the quoted $6 per hundred per year, the discount amount is $60 per thousand per year, or $120 for 2 years. Subtracted from $1000, this leaves $880 as the amount you receive. At one payment each month, there will be 24 payments in 2 years.

So you must multiply 2 by 12 by $120. And divide by $880 and by 24 plus 1, or 25. Which comes out to 0.1309, or almost 13.1 percent interest. That sounds quite a bit different from "6 percent per hundred." And is.

Calculating the Cost of Credit

THERE ARE three methods in common use for figuring the cost of consumer credit. Even when the stated interest rate is the same, each produces a different effect on your checkbook.

Under the federal truth-in-lending law, you must be told which method is used—as well as the annual percentage rate, which over the years has often been around 18 percent, or 1-1/2 percent per month.

Suppose that by your billing date of August 1, you have paid enough on your bill at an auto-parts house to bring it down to $155. On August 5, you pay $50. On August 10, you charge a pair of tires for $90, and on the 25th a cassette player for $180. It's clear you're going to be billed September 1 for $375 plus interest.

But how much interest?

Read the fine print. If it says adjusted balance method, interest will be simply 1-1/2 percent of the old balance after your payment was subtracted, or 0.015 times $105. This is $1.58.

But suppose you charge your car stuff at a store, or on a credit card, that uses the average daily balance method, excluding new purchases. Your average balance was $155 for 5 days, then $105 for the remaining 26 days of the month. Adding the product of the first two figures to the product of the second two and dividing by 31 days gives your average daily balance as $113.06. Multiplied by 0.015 this becomes $1.70.

If you use a credit card specifying average daily balance plus new purchases, it's quite a different ball game. Now you have 5 days times $155, 5 times $105, 15 days at $195, and 6 days at $180. Average daily balance, when these products have been added together and divided by 31, is $171.13, of which 1-1/2 percent is $2.57.

On large purchases—or even on small ones over the years—it could make quite a difference. Compared to the first method, the second has cost you about 8 percent more interest and the third has elevated the cost of credit by 63 percent.

Pretty slick, that.

On the whole, however, people who run unpaid credit card balances month after month apparently do not concern themselves about where their money goes. So why are we bothering?

Figuring an Early Payoff

L<small>ET'S ASSUME</small> you're making an early payoff of some kind of loan. It's one in which the total amount of interest is stated in dollars and for which the agreement has specified that the Rule of 78s applies.

This widely used rule specifies how much rebate you're entitled to for paying off the loan before it is due. To discover this amount, you multiply the total interest charge by a fraction.

You can find the upper term of this fraction by subtracting from the total number of payments called for the number you've already made—and then multiplying this by itself plus 1.

(That is, if yours is a 24-month loan on which you have made 19 payments, you'll multiply 24 less 19 by 1 more than 24 less 19—or 5 times 6.)

To find the lower term of the fraction, multiply the total number of payments by itself plus 1. For this 24-month loan, that means multiplying 24 by 25.

So your fraction is 30 over 600, or 1/20.

If your loan was $1,500 at 12.75 percent, the 24 payments would each have been $71.14, for a total of $1707.36, of which $207.36 is interest. Using the fraction in the preceding paragraph, take 1/20 of that total interest amount to find that your rebate is $10.37.

Payoff Rule of 78s

THE BUSINESS world has an odd way of figuring how much of the interest on a loan you've paid off at any given time. You can make use of it for any kind of loan or installment payment where the interest has been figured as a dollar amount—and the Rule of 78s is specified.

While the rule most often comes into play when you're making an early payoff, it is also useful for computing (for income-tax or other purposes) how much interest you've paid during a calendar year.

Here's how it goes.

Starting with the last payment that your agreement calls for, and labeling it number one, number all the payments—two for the next to last and so on.

Add up those label numbers for all the payments made during the tax year. Use this total as a numerator (that's the top number of a fraction—remember? I never can.)

Now add up the numbers for all the payments, past and future. Make this the denominator of your fraction.

Multiply the total amount of interest by this fraction. The result is the amount of interest you paid during the year.

Example. If you're being charged $1,000 interest on a 12-month loan and made the first 9 payments during the tax year, you'll multiply that $1,000 by 72/78 to find that you paid $923.08 interest during the year.

When You Make Partial Payment

YOU'VE borrowed $3000 for 6 months. After a month has gone by and you've become acutely aware that the 18 per-cent interest you've agreed on is mounting up at a fierce rate, you find you have some money to spare. You seize an opportunity to pay back $1200 of the debt.

How should this be credited?

What is known as the United States Rule has been in common use since the U.S. Supreme Court embodied it in a ruling in 1839. It says that whatever you pay goes first to interest, with anything remaining being applied against the principal amount.

It works like this.

After a month you owed only $45 interest ($3000 times 1/12 year times 0.18). So $45 of your payment takes care of interest to date, leaving $1155 to be credited to principal.

Subtracting $1155 from $3000 leaves $1845—now the amount of your debt.

If you make no further payments, you'll owe this amount, plus 5 months interest on it, at the end of the 6 months. This is $1845 plus the product of $1845 times 5/12 of a year times 0.18. Which comes to $1983.38.

Adding this to your earlier payment gives a total of $3183.38—or $183.38 interest compared to the $270 the loan would have cost you in interest if you hadn't paid anything on it until you had to.

How Big Is That Carrying Charge?

THE SIGNIFICANCE of 18 percent annual interest, or 1-1/2 percent per month, is that it's a fair approximation of what most of us pay for a good part of the credit we use.

Department stores commonly assess this amount on time accounts, as do most credit-card organizations. Many installment purchases and loans take this rate in most states.

Take a look at your credit cards or their disclosures. You may be bluntly informed that the cost in your state is 18 percent, although it may be half that in some others. You might ask yourself, or your legislators, why some states protect their citizens against shocking rates of interest while yours doesn't. But anyway, to know quickly what such interest is going to cost you in pennies-out-of-pocket on a specific transaction, just keep this handy figure in mind: 18 percent a year (1-1/2 percent a month) is precisely a nickel a day on each $100.

It was a banker that helpfully pointed this out to me, in connection with the fact that 18 percent happens to be what many banks charge for automatic loans to cover overdrafts. Though insidious and potentially costly, this kind of plan seems to me well worth the price—if you repay that automatic loan in a hurry.

Perhaps that's just how you will use it, if you keep those nickels-a-day in mind. Ponying up five cents a day for a week to avoid a rush trip to the bank when you're a hundred in hock is a bargain. It's even more so compared to being hit with a ten- or fifteen-dollar penalty for a single careless overdraft just because you haven't signed up for this kind of plan.

One warning! Don't waste a useful amount, as I did once to the tune of $11, by forgetting that the interest will run on and on until you speak to the bank about it, even if meanwhile you put a million dollars in your checking account.

One of my kids, by the way, tells me that her bank in Maryland charges only 10 percent, which comes to almost exactly 2-3/4 cents per day per $100. That's 83 cents a month.

To make your calculation for one of these Executive Check or Balance Plus overdraft accounts, you need to know one other thing. In what increments are the transfers made? Some banks will move $100 the minute you need a cent, which can make a tiny overdraft disproportionately expensive. I can forgive my own bank for its high rate when I note that it lends me the money in $5 bits so I'm not charged for more money than I'm using.

By the way, the statement permitted under the federal truth-in-lending law, that 1-1/2 percent per month is 18 percent per year is actually true only when you're talking about a single month. If you let your account run, the interest you'll actually pay is 18 percent per year compounded monthly and this comes to 19.56 percent simple interest. If you owe $100 and don't pay it for a year, you'll find that you now owe not $118 but $119.56.

(This is figured by using the same formula that would apply to a savings account at 18 percent compounded monthly. Because 1-1/2 percent is 0.015 and there are 12 months in a year, the calculation is 1 added to .015, that sum then being raised to the 12th power. Run that off on your mini-calculator and you'll get 1.1956. Multiply by $100 and, lo, the result is $119.56.

What, Borrow on My Life Insurance?

I'M RIGHT sick of coming upon the same silly statement in every discussion of whether it's smart to borrow against the cash value of your life insurance.

In one way or another, you're cautioned against doing so on these grounds (to quote a recent magazine article): "If you die while the loan is outstanding, the amount going to your beneficiaries will be reduced by the amount of the loan and interest unpaid."

Sure it will. But how much you leave behind is going to be reduced by the amount of any other type of loan you get instead, so what's the difference? Either way, your heirs aren't going to get it because you already got it.

With this cliché out of the way, you're free to consider a life insurance loan with the knowledge that the only thing against it is that there is no pressure on you to pay it back. If that's a disadvantage.

Going for it is the fact that it is probably easier, and most certainly cheaper, to borrow against your insurance than any other way. Well, it ought to be cheaper—it's your money.

I've been happily paying interest on an insurance loan for upwards of ten years. It's costing me less than 4 percent and I don't see any reason to pay it back with money that has meanwhile been invested to fetch me two or three times that for an easy profit of several dollars a year on each hundred.

Chapter F
GETTING A HOME LOAN

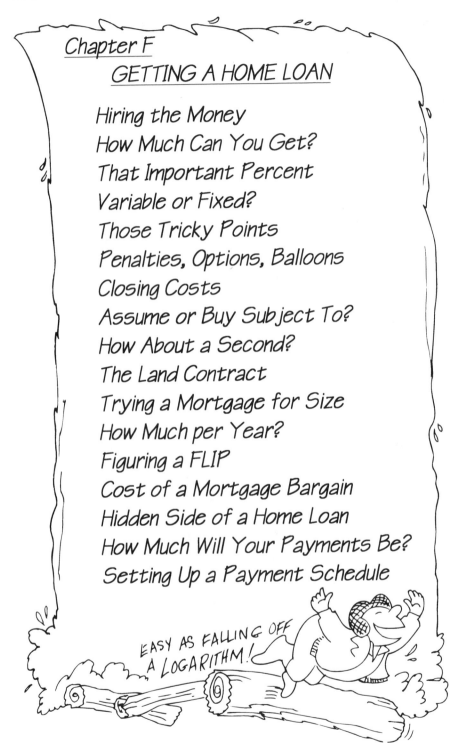

EASY AS FALLING OFF A LOGARITHM!

Hiring the Money

THE MOST costly purchase you'll ever make is probably not your new home—it's the mortgage on it. A $100,000 loan at 8 percent for 30 years will cost you a total of $164,152 in interest alone. Those payments of $733.77 you'll be making for 360 months come to $264,152. Even with a 20-year loan at 8 percent, the total you'll repay will be more than twice the sum you borrowed.

You can figure this out for the mortgage you have, or are contemplating, in a couple of steps—for any interest rate and period of time.

Multiply the monthly payment by 12 times the number of years. From this, which is the total you've promised to put out, subtract the amount of the loan to find out how much is interest.

Here's another way to look at it. On a 30-year loan at 7-1/2 percent interest, you will pay $1.50 interest—as well as $1 in repayment of principal—for each $1 you borrow. And that $1.50 becomes $2.00 as interest rates pass 9-1/4 percent.

Following are the factors to look for and weigh in shopping for the best mortgage deal. The essentials are the same, whether it's called a mortgage or, as in some areas, a deed of trust.

What kind of loan is it? Home loans come in two main types: government sponsored and conventional. In most areas, the latter are more usual today, except where groups of houses are involved.

Best of all are the subsidized kind, if your state has such a program (in California, there's Calvet for veterans). If you're eligible, it can save you a lot of money. VA (sometimes called GI) or Veterans Administration loans are also a bargain for those who qualify through military service. The government partially guarantees them and they're often made with little or nothing down, if the appraiser agrees that the house is worth the full selling price.

Next best may be an FHA loan, insured, but not subsidized, by a government agency. This insurance will cost an additional 1/2 percent, which may be partly refundable. Be sure to include this insurance cost in your comparisons. Banks and lending institutions make the FHA loans, not the government.

The majority of home loans, those without subsidies, are the kind called conventional. However, interest rates, amount of loan offered, and subsidiary clauses on these loans vary a great deal from one bank or other lending institution to another.

There are also private mortgage-insurance agencies. The biggest are Guarantee Insurance Corporation (known as MGIC), American Mortgage Insurance Corporation, and Continental Mortgage Insurance, Inc.

Here is how this type of insurance might help you get your home. You have found the house you want, at $80,000, but can manage only 10 percent down. After failing to get either a conventional or FHA loan for the needed $72,000, you find a savings-and-loan that will lend you the money if an insurance company will cover most of the risk.

Private insurance is obtained. The cost (paid by you, of course) is a small appraisal-review fee, plus a first-year cost of perhaps 1/2 of 1 percent of the amount of the loan. This buys a guarantee to the lender that either you or the guarantor will pay the top 20 percent of your loan. Since the lender still has the house as security for the rest of his money, he is quite well protected—and happy to make the loan that he otherwise might rule out as too risky.

Your loan-insurance premium will decrease each year. You may even be able to persuade the lender to let you drop the insurance after a few years as the loan becomes smaller and you build up a record as a reliable payment-maker.

How Much Can You Get?

BANKS sometimes lend a severely limited percentage of the appraised value of a property, perhaps as little as 65 to 75 percent of the value unless the loan is insured. Savings-and-loans and mortgage companies may go higher.

Appraisal policies of banks tend to be conservative. Rules for estimating your financial capacity for a given loan may also vary among lending institutions. A private lender may be most flexible of all, though likely to require a high rate of interest, unless he is the seller of the house.

FHA has rigid rules that you must meet before you can get an insured loan. Since many lenders have similar qualifying rules, it is helpful to know what some of them are.

First, you must be able to show a dependable income. A second income will count only if it comes from what appears to be a permanent job. Casual moonlighting, overtime pay, plans to rent out a room in the new house—all these may not count at all, even when they should.

Your present indebtedness, credit rating, and standard of living should be within your income. Your total financial obligations must leave room for the new payments. And beware: Your mere ownership of several high-limit credit cards, even if unused, may be construed as such an obligation.

If your children are approaching college age, you may be viewed as in a more stressful financial period than if they were very young or through college.

Your total housing expenses will be considered, including payments, taxes, insurance, repairs, utilities—and even transportation if the house you are buying is in a rural area or far from your work.

Considered are all such factors and these as well: your age (30-40 is excellent, 25-29 good, 41-55 fair, over 55 or under 25 poor); number of dependents (one or two children best); substantial life insurance (about twice the amount of the proposed mortgage).

That Important Percent

Interest rates vary, just as prices of other commodities do. Shop around. The buyer of a former home of ours saved substantially on a loan by shopping carefully so that he was ready to make a decisive move. His own bank was asking 7-1/4 percent, but another bank he dealt with was still lending at the previous week's 7 percent. He closed his loan deal just hours before the second bank boosted its own rates.

What you do for a living can also be a factor when you come to borrow a large sum. Some years ago I found four rates of interest offered within a radius of a few miles. Savings-and-loans wanted 6-1/4 percent; a branch bank serving wealthy depositors (and overloaded with cash) was asking only 5-3/4 percent; an insurance company, 5-1/2 percent. I qualified for the 5-3/4 percent and saved $20 to $40 a month for a long time. I was denied the even more favorable 5-1/2 percent loan because the insurance company wouldn't risk its money on a free-lance writer. They didn't seem to take my point that a writer's income may be somewhat unpredictable but he's never unemployed.

How important to you is a jump from 8 percent, say, to 9 percent? For a typical mortgage each such jump will add $6 to $7 to every monthly payment on each $10,000 of the debt.

Here's another way to look at it. If you find yourself paying 9-1/2 percent when, with more shopping around, you might have been paying 8-1/2 percent, it will take you 25 years to pay off a loan that otherwise would have taken only 20 years. The higher rate will eventually add $18,400 to the payoff cost of a $40,000 loan—the price of a trip for two around the world (if you want to look at it that way—and I do!) or a new car or two.

THE HOUSE THE INTEREST COULD HAVE BOUGHT

THE HOUSE WE BOUGHT

Variable or Fixed?

Should you go for a variable-rate mortgage?

The term can apply to any home loan on which the interest rate (and your payments) will go up or down with changes in the level of interest rates generally. The measuring standard may be any of several widely published rates, such as banks' cost of funds, the prime lending rate for business customers, and so on.

This choice between fixed and variable can be a nasty one to make. It can involve a lot of money—yours. It's a forced gamble with a family's economic security that no one should have to make. For most of us it's a once- or twice-in-a-lifetime undertaking, so we face it with little or no previous experience. It's justified only by the happy circumstance that the values of home ownership generally make it worthwhile.

And, happily, some rules-of-thumb come down on your side.

Rule 1. Begin with the basic, and obvious, principle that if you believe interest rates are going to fall, you should go variable. If it appears that you are borrowing at a period of relatively low interest, you will prefer a fixed rate.

Rule 2. Temper that truism if you expect to keep the mortgage for no more than five or six years, which is not far from the average life of home loans. Since most variables start out, for tactical reasons, with a temporary low rate, the cost for the first few years usually averages out quite attractively. This may be your most important consideration.

Rule 3. If the combined force of the first two rules is not decisive, consider this. It is something lenders may neglect to emphasize; they have a natural bias in the direction of a variable, since it guarantees that the interest they collect will never drop below a profitable rate.

If a time comes when mortgages are offered at a rate well below your fixed one, you can probably refinance yours at that lower figure. You can go to a new source of money, unless your lender meets the competition.

With a variable, however little you may have come to love it, there is usually no way out.

Rule 4. Look for special-incentive variables that might fit your own situation. Ask about plans that combine features of both fixed-rate and variable. Example: A variable providing that if at any time within the first three years the going rate should drop, you could ask to have your rate adjusted to that level—but only once.

Rule 5. Start doing the arithmetic as far as possible before you must make your decision final. Here is how a variable 8 percent 30-year mortgage for $25,000 might go. Get a preview of your position by multiplying each of the following figures by the number of times $25,000 goes into the sum you are about to borrow.

Monthly payments on $25,000 would be about $183. If after five years the going rate for similar mortgages had risen to 8-1/2 percent, your payments would rise to $191 a month. Another 5 years and the rate goes to, let's say, 9 percent. Your payments would then be $198.

Those Tricky Points

A PROSPECTIVE mortgage lender may charge you one or more points, in additional to other fees. This happens when he's not willing to make the loan at a stated rate but finds it undiplomatic or illegal to raise the interest directly.

"Point" is short for percentage charge. You pay it only once. A 2-point charge, for example, on a $122,000 loan means you'll pay (in advance) a one-time fee of 2 percent of $122,000, or $2440, in addition to all other costs.

FHA specifically forbids charging points to the buyer. The seller, knowing he is stuck with the points, must manage to work the charge into the selling price of the house some way or other. Of course, it's usually the buyer who really pays after all.

Points are essentially just another way of raising the interest rate on a mortgage and collecting the whole bonus right now. If you happen to sell the house in a year or two, 2 points will have cost you an additional 2 percent. In other words your 9 percent mortgage will have been an 11 percent mortgage. But if you spread the cost of the points over 30 years of payments, those points will have cost you only a fraction of 1 percent a year.

A useful quick-calc rule to go by: assuming you keep the mortgage 20 years, each point equals 1/8 of 1 percent added to the stated rate of interest.

So when you hold a 20-year mortgage at 8-1/2 percent plus 4 points all the way to maturity, its cost to you is approximately the same as for a 9 percent loan with no points.

Penalties, Options, Balloons

AFTER a lender has gone to the trouble of making a loan, he doesn't want it paid off before he has had time to collect quite a bit of interest. He doesn't want you to pay him off by borrowing elsewhere for less if interest rates decline. So he may stipulate a penalty if you pay off early.

Keep any prepayment-penalty clause out of your mortgage if possible. Then if you want to make bigger payments, refinance, or sell the house, you won't be stuck with a big bill. FHA cut off prepayment penalties way back in 1972.

An open-end clause is a very desirable provision in a mortgage. It lets you re-borrow the money you've paid in and do so at the same interest rate the original mortgage carries.

Let's say you've been paying on a $100,000, 30-year mortgage for 12 years. Balance owed is $79,068. Now you need some extra cash to finance home improvements or tuition fees for your college-bound children. An open-end mortgage will permit you to borrow the difference between $100,000 and $79,068—$20,932. You'll repay this new loan—with interest—by increased monthly payments or by an extension in the length of time the mortgage is to run.

An open-end offers an unfortunate temptation to use up a considerable chunk of money you've safely invested in your home. It is, however, such a convenience that it has spawned a monster with an even greater potential for evil. This is that all-too-familiar home-equity loan, which can offer useful savings but which lacks the advantage of a previously set interest rate that might be a good bit lower.

Caution! Balloon!

A mortgage that is scheduled to come to a sudden end with a great big final payment, appropriately called a balloon, may be the only way to keep the monthly costs low. But unless you know where the cash is coming from, you must be prepared to refinance at what may prove to be a time that is both awkward and expensive. Remember the distress of the widow in the old melodramas as the villainous banker approaches with foreclosure papers.

Closing Costs

ALL THOSE petty annoyances and minor disasters totted up as closing costs can come to a substantial figure. Here are a few that you may have to pay before you move in: credit report, survey, title fee, title insurance, attorney fees, document preparation, recording fee, transfer taxes, escrow fees.

Depending mostly upon where you live, the total amount on even a modest house can come to several thousand dollars that you'll have to put up in addition to the down payment.

You can learn what these costs are through your bank, real-estate agent, or loan source. However, the hardest part may be to pry this information out of them as far ahead of the fatal day as you may feel you need it.

Often, the biggest single cost of closing is for title insurance, a protection for the lender against loss arising from a defect in your ownership title. Similar protection for you will usually come from a title policy purchased by the seller—though in some areas the buyer also pays for this, or divides cost with the seller. You naturally should try to get this arranged to your advantage, unless local custom is too fixed to buck.

At closing time, you can expect yet another financial shock. Any taxes or insurance costs the seller has paid in advance for months yet to come must be refunded to him.

By you, of course.

Assume or Buy Subject To?

Is THE PROPERTY under an existing mortgage? If so, find out if you can assume it. The interest rate may be much lower than you could get on a new loan, say 6 or 7 percent when the going rate on new loans is 8 to 10.

Such an assumption at a favorable rate will put you well ahead in the long run, if you can tolerate the short-run drawbacks. An assumable mortgage very often is not as big as you'd like and has to be paid off more quickly than you might wish.

Consider this situation. A man and wife came upon a 10-year old house they liked very much. Price arrived at after some haggling was $98,400. The existing mortgage, initially $72,000 for 20 years, had been paid down to $52,800. At a modest 7 percent, compared to the going rate of 9, it sounded good. The hitch showed up when they realized they'd have to put down $45,600 instead of the $19,680 (20 percent) they'd hoped to get by with. Happily, the pair had enough savings to handle the difference, and the bargain interest rate convinced them this was the best investment around.

If they hadn't had this much cash, or wanted to keep it intact, they could have tried to interest the seller in taking back a second mortgage for the amount lacking. Even if this second were written at a rate somewhat higher than the 9 percent going rate for firsts, the combination would still have been a bargain if they could handle the comparatively high total monthly outlay. Their reward would be an early mortgage burning, traditionally one of the more joyous bonfires.

Instead of assuming, or taking over, a mortgage, you might buy a house "subject to" an existing loan. This is a perfectly good alternative from the point of view of lender and buyer alike.

However, since the seller remains responsible for any loss if you default (instead of fully getting out from under) he may not be altogether enthusiastic unless you can make it a condition of purchase.

How About a Second?

THE VERY words second mortgage have taken on an unpleasant ring from association with high interest rates and buyers who are in over their heads. Caution is called for, as with any enterprise requiring either new clothes (Thoreau) or borrowing (Franklin, probably).

When is taking on a second justified? Generally, a second mortgage is an acceptable tool when it permits going ahead with a transaction that would be impossible without it and that in all other respects is prudent. Of course, there must be assurance that the double monthly obligation can be met.

The more usual second is the kind called purchase money. The lender is the seller of the house. In order to make the sale he is accepting the down-payment partly in cash and partly in a second mortgage. Because of the same incentive, he may go along with an interest rate on the second that is no higher than the bank or other prime lender has offered on the first.

The Land Contract

Mortgage? Deed of trust? There is another kind of financing most buyers or sellers of building sites or houses never think of. It is called a contract of sale or, in some parts of the country, a land contract. It doesn't involve an actual mortgage or trust deed, but it takes the place of one.

There's really nothing weird about it. It just means buying a home the way you've normally bought a car or a refrigerator: on time payments. The difference is that with real estate there are more safeguards. In some states, partly because of past abuses, legislation and court decisions have made land contracts more protective of the buyer than the alternatives have proved to be. In short, safer than an ordinary mortgage.

Before buying or selling on a contract of sale, consult a lawyer who works with real estate. He can see that the agreement is properly drawn up and can tell you whether there are any hidden dangers in this arrangement in your state. As a buyer, one thing that you are interested in is this: if you should prove unable to make one or more payments at some time, you will not be treated more harshly than you could be under an ordinary mortgage. Make sure that the terms of the contract meet this requirement.

One advantage of a land contract is that, being an installment sale, it can permit the seller to postpone, or spread out, payment of much of the income tax on any profit he makes in selling the house. This concerns you because it may offer an incentive to him to accept a lower price. I've used it as a regular thing, to avoid formalities and the nuisance and cost of a lot of paper work in transactions that didn't call for those money-borrowing instruments called mortgages and deeds of trust.

Not everybody—especially lawyers—may see this my way. I went into contracts of sale in writing a book on housing some years ago and darned if the publishers didn't come back to me with alarm after their lawyers had given the manuscript a routine going-over. Said their legal talent had never heard of such a thing. I never heard of anyone complaining, however, after the book came out.

Trying a Mortgage for Size

THIS IS a special kind of mortgage calculation. It works in reverse. Instead of starting with a mortgage of known size, it uses interest rate, payment amount, and number of years to discover the face amount of the mortgage.

You might want to use it to estimate how big a mortgage you can handle, and from that the price range of houses to look at.

It will serve equally to reveal the original size of an existing loan for which you have the other information.

Suppose you are looking at a 9 percent 20-year mortgage with payments of $965.55 a month. What is the face amount?

1. Divide the interest percentage by 12 to get a monthly figure and by 100 to make it decimal. Store this for later.

2. Add 1 to the previous result, then raise this total to the power equal to the number of months (not years!) the loan is to run. Store this result, too.

3. Subtract 1.

4. Divide by the result saved in Step 1.

5. Divide by the result saved in Step 2.

6. Multiply by the amount of the monthly payment.

Here's how it runs for that 9 percent, 20-year mortgage with its payments of $965.55 a month.

Dividing 9 by 12 and by 100 gives 0.0075. Adding 1 to that result and raising to the 360th power gives 14.73. Subtracting 1 and dividing by 0.0075 and 14.73 produces 124.29. Multiplying by $965.55 shows the face amount of the loan, what it came to in the beginning, to be approximately $120,000.

How Much per Year?

HERE IS a different kind of table from the usual one dealing with home mortgages.

Using this table, you can instantly find what percentage of the face amount of the mortgage you will have to dig up in payments each year.

You can then quickly calculate how much that will be in dollars each year for any size mortgage. Then you multiply the figure in the table by the size of the mortgage.

START WITH THE PERCENT

	6%	6.5%	7%	7.5%	8%	8.5%	9%	9.5%	10%	11%	12%	15%
5	23.20	23.48	23.76	24.05	24.33	24.62	24.91	25.20	25.50	26.09	26.69	28.55
6	19.89	20.17	20.46	20.75	21.04	21.33	21.63	21.93	22.23	22.84	23.46	25.37
7	17.53	17.82	18.11	18.41	18.70	19.00	19.31	19.61	19.92	20.55	21.18	23.16
8	15.77	16.06	16.36	16.66	16.96	17.27	17.58	17.89	18.21	18.85	19.50	21.53
9	14.41	14.71	15.01	15.31	15.62	15.94	16.25	16.57	16.89	17.55	18.22	20.31
10	13.32	13.63	13.93	14.24	14.56	14.88	15.20	15.53	15.86	16.53	17.22	19.36
11	12.44	12.75	13.06	13.38	13.70	14.02	14.35	14.69	15.02	15.71	16.41	18.61
12	11.71	12.02	12.34	12.66	12.99	13.32	13.66	14.00	14.34	15.04	15.76	18.01
13	11.10	11.41	11.74	12.06	12.40	12.73	13.08	13.42	13.77	14.49	15.22	17.52
14	10.57	10.90	11.22	11.56	11.90	12.24	12.59	12.94	13.30	14.03	14.78	17.12
15	10.13	10.45	10.79	11.12	11.47	11.82	12.17	12.53	12.90	13.64	14.40	16.80
16	9.74	10.07	10.41	10.75	11.10	11.45	11.81	12.18	12.55	13.31	14.08	16.52
17	9.40	9.73	10.08	10.42	10.78	11.14	11.51	11.88	12.25	13.02	13.81	16.29
18	9.10	9.44	9.79	10.14	10.50	10.87	11.24	11.61	12.00	12.78	13.58	16.10
19	8.83	9.18	9.53	9.89	10.25	10.63	11.00	11.39	11.78	12.57	13.38	15.94
20	8.60	8.95	9.30	9.67	10.04	10.41	10.80	11.19	11.58	12.39	13.21	15.80
25	7.73	8.10	8.48	8.87	9.26	9.66	10.07	10.48	10.90	11.76	12.64	15.37
30	7.20	7.58	7.98	8.39	8.81	9.23	9.66	10.09	10.53	11.43	12.34	15.17
35	6.84	7.25	7.67	8.09	8.52	8.96	9.41	9.86	10.32	11.24	12.19	15.08
40	6.60	7.03	7.46	7.90	8.34	8.80	9.26	9.72	10.19	11.14	12.10	15.04

Going further, you can closely approximate what the monthly payments will be—just by dividing that last figure by 12.

Say yours is a $20,000 mortgage at 9 percent for 40 years. Opposite 40 and under 9 percent you will find the figure 9.26. That's the percentage of $20,000 which you are obligated to pay each year— $1,852.

Divided by 12, that becomes your monthly payment of $154.33.

Figuring a FLIP

A FLIP IS A special form of graduated-payment home mortgage, offering Flexible Interest Payments. It takes less out of your pocket in the early years of a loan and more in the later years.

Since paying later is always more costly than paying now, using a FLIP increases the total paid. This may be worth while if it permits owning a house that, although affordable in the long run, would otherwise be temporarily out of reach. That's why this kind of loan may be especially useful for a young couple who can count on a rising income.

The next table compares three possible sets of figures with those for a traditional loan, for a $45,000, 30-year mortgage at 9 percent. Even in a period when mortgages are running far from 9 percent, these figures can be useful for making comparisons. Likewise, the use of $45,000, as for a $50,000 house after a 10-percent down payment, makes it easy to calculate the result for a house selling for two, three, or more times $50,000.

For the most extreme version shown, you can see that the initial monthly payments would be nearly $100 less than for a conventional loan, increasing until the sixth year, and remaining level after that. Even for a house where a family might be required to show a $22,656 income for a conventional loan, the same family might be reckoned able to handle a FLIP on an income of $18,132. The same relationship would hold for purchase of a more typical house by a family with a larger income.

The feasibility of a FLIP is based on two factors that tend to raise a young family's dollar income as the years go by—increased real earnings and inflation. Add these expectations together. If they come to as much as the percentage by which the payments will increase each year, meeting the increase should not be a special problem.

(How can you calculate the percentage of increase? Subtract this

Year	Traditional	FLIP 4%	FLIP 5%	FLIP 6%
1	$362.09	$292.64	$280.01	$267.75
2	362.09	310.14	300.33	290.76
3	362.09	328.24	321.65	315.12
4	362.09	347.27	344.05	340.91
5	362.09	366.95	367.56	368.20
6-30	362.09	387.43	392.25	397.08
First-year income needed	$22,656	$19,296	$18,720	$18,132

year's monthly payment from next year's and divide by this year's. Multiply by 100 to make this decimal a percentage.)

Cost of a Mortgage Bargain

W HEN A house is offered with a conventional mortgage carrying an interest rate below the going rate, you naturally wonder how the lender can be so generous.

The most likely answer is that the seller is paying the bank or other lender a fee in return for the bargain rate that is being dangled before you. And since he has to get this additional money somewhere the seller adds it to the price he charges you for the house.

If you were to pay cash or get your own financing you should expect to buy this or a comparable house at a lower price.

The question is—how much lower?

You can get a pretty close idea by using a rule of thumb. It says that for every 1-point loan fee (1 percent of the amount of the mortgage) he receives, the lender can afford to drop his interest rate by one-eighth percent.

So if you are offered a $100,000 mortgage at 10 percent when 11 is normal, that's nearly $1000 in your pocket the first year, shrinking only slowly thereafter. But it means that the seller is paying the lender a fee of about eight points which he had to get by inflating the house price by something like $8000.

It's another case of there isn't any free lunch.

Those being the facts, what should your strategy be?

Unless you have reason to assume otherwise, it's reasonable to regard it as a tossup whether you'll stay with the mortgage to the end, thus taking full advantage of the lower rate you've paid for.

If, as is the more usual case, you should refinance or sell in a few years you'll probably be better off accepting higher interest in return for a lower price.

Hidden Side of a Home Loan

IN DECIDING whether you can afford the home you covet, you'll naturally begin by ascertaining the monthly payment. That's easily done by using the tables or the calculating methods given in this volume. Either will give you the exact monthly figure for each of the usual interest rates and time-payment periods, for loans of various amount. They apply equally to conventional and guaranteed mortgages and to homes purchased on contract of sale.

To determine whether you can afford a payment of a given size you must, of course, add other shelter costs: property taxes, insurance, maintenance. You may pay these directly, or they may be added to your basic payments.

Maintenance is the great variable. For people short of manual skills and interests who have the misfortune to acquire a house in poor condition, upkeep can be a financial disaster. A house designed, built, and landscaped for easy maintenance, on the other hand, may require only the attention that a willing owner can easily give, at a cost that may average but a few dollars a month.

A typical maintenance chore is replacement of the thermocouple in a gas water heater or furnace, which tends to wear out within 10 years. A new part commonly costs less than $5, and a householder of average savvy can install it even if he has never done this particular task before. But put in by a repairman the little thermocouple inevitably is accompanied by a charge of $30 to $50 or so—because the trip alone may have taken up an hour of the repairman's time.

To estimate maintenance costs on your prospective home, begin with a look at history. How much have you (and your landlord, if you've been renting) been spending? How much, if any, more should have been spent for fully adequate maintenance? Judging by the amount of mechanical equipment and its condition, exterior and interior finishes, and status of landscaping, how much more or less will the new house require? By this route, you can arrive at a fairly precise and realistic expectation—and thus avoid a major money trap.

Consider also what is an even bigger figure in many cases: automobile or other daily travel costs to the extent that they will be greater in your new location than in your old.

Many a budget has foundered because that idyllic new home site added 40 miles a day (at 20 to 40 cents a mile) to the daily home-to-job round trip. Other budgets have suffered because an isolated location necessitated an extra car for someone. Car cost, for many families, belongs under "shelter" in the budget.

How Much Will Your Payments Be?

I HAD THOUGHT of including in this volume a set of tables of mortgage payments. They would have told you how much you would pay each month for each size of mortgage at each rate of interest for each number of years.

Let's see now. I could have included at least 25 mortgage sizes, 25 interest rates, and upwards of 30 terms of years. Fifteen or 20 thousand figures! Fifty to a hundred pages of this book, produced the easy way!

Tempting.

But even if I'd done that, you'd still have had to do some arithmetic to fit the table to the exact dollar size of the mortgage you're interested in.

So it seems to me to make a lot more sense to give one or two ways to calculate an installment payment yourself, precisely for your own situation, whether payments on a car or furniture purchase or on a personal loan or a home mortgage.

Here's a real do-it-yourself alternative to digging figures out of a great big table and then interpolating. At a glance, you might deem it hopelessly tedious to tackle without some kind of shortcut device, such as a scientific or financial electronic calculator or a slide rule or logarithms.

Not so. The only threatening element is in Step 2, and it's a devil only when payments will extend for many years.

To find a 360th power (for a 30-year mortgage) you can strike the times key once and then the equals key 359 times for the required 360. But if you note that 6 times 6 times 10 equals 360, you can leap to the same result in a few seconds by...

...punching the times key once, then equals five times;

...doing the same sequence again;

...then hitting times once and equals nine times.

Here are the steps, for payments to be made monthly or at any other interval.

1. Express the interest rate as a decimal (9 percent becomes 0.09) and divide by the number of payments per year.

2. Add 1, then raise this sum to the power equal to the total number of payments you are to make. (For a 6-year loan paid monthly, this is the 72nd power.)

3. Divide the result of Step 2 by itself less 1.

4. Multiply by the amount of the original debt.

5. Multiply by the figure found in Step 1 (or multiply by interest rate as a decimal and divide by number of payments a year).

Here's how it goes for monthly payments over six years on a $5,000 loan.

1. Dividing .09 by 12 gives .0075. Add 1 to this.

2. Raised to the 72nd power, 1.0075 becomes 1.71255.

3. Dividing 1.71255 by 0.71255 is easy on any calculator, since you have the digits before you until you begin to enter the divisor. Result: 2.4034102.

4. Multiplied by $5,000, this becomes 12,017.05.

5. Final multiplication by 0.0075 gives a result that rounds off to $90.13 monthly payments.

(For a final step, multiply that last figure by the number of payments (72) to learn that repaying this $5000 loan in this way costs you $6,489.19. Subtracting the loan amount of $5,000 reveals that almost $1,500 is for interest.)

Setting Up a Payment Schedule

If you're buying or selling something on installments—house, car, whatever—it can be very convenient to have a payment schedule all laid out on a piece of paper or in a little book.

To create one, you'll need to know the amount of the debt, the payment period (usually one month), and the rate of interest. With these you can calculate the payment by using any of the methods given in this chapter or the one on computers.

You can then fill in the figures for each month of each year under column headings of INTEREST, PRINCIPAL, and BALANCE.

To find the first of these, multiply the balance owed by the interest rate, divide by the number of payments per year, and move the decimal point two places to the left. This is the part of your first payment that will go for interest.

Subtract this interest from the amount of the payment and you'll have the sum to be applied to the principal.

Subtract this principal from the amount owed. This is the balance.

Here it is, as I set up the schedule when my wife and I sold our house to our oldest daughter and her husband after building a new one a mile away. I commend, to anyone who can use it, this shrewd form of family planning. It assures that as we grow too old and cranky for anyone else to come near, there will be someone delightful who is stuck with us. Just a little twist on the cynical definition of home as where, when you have to go there, they have to take you in.

Here's the routine for a 30-year, 9 percent, $120,000 mortgage, calling for monthly payments of $965.55. Multiplying $120,000 by 9, dividing by 12, and moving the decimal point two hops gives $900 paid to interest the first month and $65.55 to principal, with a balance owing of $119,934.45. This operation is repeated for the second month's figures, using that $119,934.45 instead of $120,000. And so on.

As you can see, this is quite feasible as a paper-and-pencil operation, for problems covering a short span of years. But if you start that way you'll surely spring for a pocket calculator very soon.

With a programmable calculator, it's a cinch. Putting in pauses for the needed figures lets you make a payment table as fast as you can note the figures down. You're saved even that chore if yours is a calculator that prints its results on paper, or if you are using a computer hooked to a printer.

Everything so far assumes equal payments made every week or month. If the buyer has the option of paying more when he likes, and exercises that option, you can't use a schedule made up in advance. You'll have to do each monthly line as payment is made. The other effect this has is that you'll want to provide an additional column, marked PAYMENT, to the left of the others, to show the sum actually paid. Naturally, it is from this amount that you will subtract the sum attributable to interest as you find how much to credit to principal.

Chapter G
SPENDING

The (Relatively) Painless Budget

THERE ARE whole books full of forms and formulas for setting up a household budget. They can be quite useful, especially if you want to make a real production out of it.

If you're given to impatience, however, you may be glad to know it's feasible to set up a fine, workable system by just following the steps outlined in this limited space. And you'll have that budget in full flow in less time than you might spend reading how to do it.

1. List all income you can reasonably count on for the next 12 months—wages, interest, dividends, gifts, bonuses, rent from property you own. Subtract any expenses connected with getting the income and also subtract federal, state, and local income taxes, whether paid directly or withheld. This will give you your net annual after-tax income.

2. List your fixed expenses for the next year—all those things that are quite predictable and in amounts that aren't up to you. These may include rent, property taxes, insurance premiums, tuition, installment payments for which you are already committed, alimony or support, even such other things as utilities if you can estimate them closely enough. And one final item you probably hadn't expected to list under fixed expenses—savings.

Maybe this calls for some explanation. In the first place, there must be savings in your program—if only in preparation for the inevitable emergencies. In the second place, it's essential to get the habit going even if the amount is pretty small. Equally important, once you get the savings fund up beyond what you must keep for emergencies, you can use it in wondrous ways.

For one thing, it can ward off the never-never, as the English are wont to label their hire-purchase (our installment) plan.

Say it's time to buy a $400 washing machine, which on the never-never, with its accompanying 18 percent piggyback, could actually run you $500 by the time it's paid for. Pull the $400 in cash out of your savings when you buy, then take the amount of the installments out of the household-operation budget till the $500 is repaid, giving your savings a fine boost with it. Why let someone else grab those high-rate interest dollars when your own account can just as well earn them?

Eventually, you should be able to buy even such things as new cars and other durables free of interest costs. As you use up your car, put the amount by which you estimate it has depreciated that month into the savings fund to earn interest till it's time to buy your new car—for cash.

So how much shall savings be? If you have no established habits, try to begin with at least 5 percent of your net income. I don't think you need me to tell you that you'll be happier in the long run if you can start with 10 percent or more.

3. Now list your other expenses by major categories such as: food; household operation; furniture; clothing; transportation; medical care; education and recreation; gifts and contributions; personal (including allowances); miscellaneous and investment. Put down the amount you intend to spend.

4. From your net income, first subtract the total of your fixed expenses you've just listed.

5. If there's money left, check the list again to make sure you've been realistic in each category. Use any remaining money to expand the last category with an investment program in mind. Either now or after your budget has stood the test of a year, you may find yourself in the happy position of being able to expand the savings item substantially. Perhaps recreation, too.

If however, as budgeters more often discover, the total of variable spending is larger than the money available, something's got to give. Cut as much as you can from each item. But don't kid yourself. There's no use taking a new suit out of the paper budget if, in fact, you're still going to buy that suit.

If the budget won't balance now, and it isn't feasible to increase your income, you will have to recognize that your fixed expenses are out of line. Look for long-range changes that will reduce them.

6. Unfortunately, there is a second paper chore connected with successful budgeting. You must keep track of your expenditures. Do this with special care for the first couple of months. This doesn't mean writing down every penny. Whatever may have been the case in B. Franklin's time, records to the nearest dollar are close enough in our day.

And you can take Robert Frost's word for it, endorsed by suffering members of my own family when I become over-zealous, that "Nobody was ever meant/To remember or invent/What he did with every cent."

The ideal time to start keeping records is a month or several before starting your budget. The sad truth, however, is that most people who postpone budgeting while they compile a history never get around to budgeting at all. Better to strike while the iron of impulse is hot.

If you tough it out in the beginning, your rough working budget will refine itself into a flexible and precise tool for helping you get the most out of your money. It may even teach you and your family so much about the arcane art of handling money that you won't need to maintain a formal budget unless you feel like it.

Buy Now? Buy Later?

T HAT PANICKY impulse to buy now to beat inflation in a period of rising prices needs a sober second look.

But so does the urge toward thrift that says don't buy it as long as you can get along without it.

In short, some calculation is called for.

Will the purchase have some immediate and continuing use? That's one vote in favor of acting now. If it will not wear out rapidly or deteriorate from the passage of time, it deserves another pro vote.

With those aspects considered you're in better shape to make a decision on the sound basis of alternate use of your money. Do you have a safe alternate lodging place for your money, one that pays a true return? How much?

This brings you to questions of inflation and taxes.

Begin by supposing you're in a time when inflation is running at an annual rate of 10 percent. Something—a painting, an antique, a building lot—that you can buy today for $50,000 will stand to cost you $55,000 a year from now.

You know that if you do not make the purchase, you can invest your money safely for a year at a yield of 10 percent. But keep in mind that this is the gross return. In your combined state and federal tax bracket of 40 percent, you'll retain only 60 percent of that return, or 6 percent net. That comes to $3000.

So at the end of the year your $50,000 will have grown to $53,000, but it will take $55,000 to make the buy. If you can trust those suppositions, buy now.

It could also cost you something, in property taxes or storage and insurance outlay, to hold the purchase for a year. Consider both the amounts of these and when you'll incur the added costs.

And don't forget to factor in the value of use you can make of earlier ownership. Which brings us back to where we came in.

How Much Inflation?

WHEN INFLATION leads a manufacturer to boost a list price he sometimes takes the opportunity to slip a hidden boost behind the apparent one. What follows may seem a trivial instance, but it has its point.

On the occasion when Luden's Menthol Cough Drops went up from 15 to 20 cents a box, the maker no doubt hoped you'd see this—if you noticed it—as just another nickel in a 20-cent purchase, a price jump of one-fourth or 25 percent. You, however, might reasonably complain that this was a jump of 33-1/3 percent—a nickel being one-third of the 15 cents you'd been used to paying.

But there's more to this tiny inflation story. At the same time that the price went up, the contents of the box went down from 16 cough drops to 15. So how much has the true price risen?

One way to get at this without finding yourself mired down in confusion is through a unit-pricing calculation. A cough drop used to cost 15/16 of a cent. After the increase, it cost 20/15. So we ask, as we did in the simpler calculation above, what percentage of the old price is the increase?

Again, we find out by first dividing the increase by the old price, to find the ratio of increase, and multiplying that by 100 to make it a percentage of increase.

One way to perform the operation comes under the head of Fun With Fractions, complete with that grade-school enemy the Lowest (or did we call it Least?) Common Denominator. I guess it is allowable to remind you how to do it since I've just gone to considerable trouble to recall it myself. Starting with the new price of 20/15ths and subtracting 15/16ths from it (after turning them into 225/240ths and 320/240ths) produces 95/240ths. Dividing that by 320/240ths and multiplying by 100 gets you to 9500/240ths. Dividing by 15⁄16ths is now guaranteed to bring you to 42.2.

Doing it like this, with fractions, is the old-fashioned way, for sure, but it tends to make the principle reasonably clear.

Most calculators abhor fractions, love decimals. So with one of these little fellows, you just divide 20 by 15, from this subtract (15 divided by 16) and then divide what you now have by that same quantity (15 divided by 16). Multiplying by 100 turns this into a percentage of increase—42.2.

So whichever way you look at it, the price of not coughing has gone up a rather formidable 42.2 percent.

When the magazine *Consumer Reports* noted this occurrence, the editors calculated the increase as 41.5 percent. The discrepancy is minor, but it suggests that we might well check our result by using an alternate method to see if we're wrong.

Consider that 240 drops would now be 16 boxes at 60 cents each, for an expenditure of $3.20. In the good old days, you could get 240 of the things by buying only 15 boxes at 15 cents each, for $2.25. The difference is 95 cents which, times 100 and divided by the old cost of $2.25, again comes to 42.2 percent. How the magazine got its slightly different figure is not readily apparent.

But the virtue of rounding is.

If the editors had eschewed unnecessary precision by reporting the rise as "about 42 percent," they'd have been right, and quite precise enough as well.

Puzzle of the Price Markup

Business price markups would be a pretty simple matter if it weren't that they come in two dissimilar forms.

Wholesalers are in the habit of figuring markups as a percentage of what the item cost them. Retailers base their markup figures on what they charge the customer.

This creates no misunderstandings as long as markups are expressed as so many dollars and cents. When an item goes from $3 to $4 it has been marked up $1.

But suppose a wholesaler who would tell you he operates on a markup of 50 percent has passed along a $10 (his cost) item to the retailer for $15. What will the retailer put on the price tag if he also adheres to a 50 percent markup? Will he charge $22.50?

Not if he follows the more usual retail practice of expressing the markup as a percentage not of his cost but of the selling price.

But how can the retailer go about adding 50 percent of the retail price to his wholesale cost when he hasn't yet figured out what that retail price is? He goes about it somewhat backhandedly, by dividing his cost price by the markup percentage—expressed as a decimal.

As a decimal, his 50 percent markup becomes 0.50, since percent means one hundredth and moving the decimal point two places to the left is dividing by 100.

Dividing his cost of $15 by 0.50 doubles it, giving a retail price of $30. The amount of the markup is $15, or 50 percent of the selling price.

Which also is, of course, 100 percent of the amount he paid for it.

It's confusing, but that's the way it's done. For one thing, it sounds better this way.

What's That Price Increase?

One day some time ago, along with other news good and bad, the day's mail brought word that my modest Blue Cross plan had gone to billing every two months instead of quarterly. Reason: "Because so many members have asked for more convenient billing."

I'll admit that my immediate response was to recall reading that on one social occasion the Duke of Wellington, imposing and impressively dress-uniformed, was accosted by a lady. "Mr. Smith, I believe," she said. "Madam," replied the Iron Duke sternly, "if you believe that you'll believe anything."

And if writing six checks a year instead of four is to my convenience, why not make it twelve?

Convenient or not, the new billing was accompanied by a price increase, as the covering literature mentioned incidentally.

How much of a jump is neatly concealed by the billing change? All that is evident at first glance is that the next check I mail will be gratifyingly smaller than previous ones.

Instead of $107.46 quarterly I'll be paying $85.76 six times a year. A little multiplication tells me that my annual cost has just risen from $429.84 to $514.56.

What's that increase as a percentage? This jump of $84.72, divided by the old cost and multiplied by 100, proves to be almost 20 percent at one clip, which could be called whopping. No wonder Blue Cross deemed it worthy of some obfuscation of presentation.

Too bad, though, that frankness wasn't the order of the day. Considering where medical costs went from then on, we all could have used a sharp warning.

Since the same figures applied to my wife's policy, we were being hit by a 40 percent combined increase, weren't we? If there were three more of us, it would come to 100 percent, wouldn't it? You can best put your finger on this bit of confusion by asking 100 percent of what. In a typical month when you buy a thousand things and read that costs went up 1 percent while you were doing it, your total expenditure was inflated by 1 percent—not 1000 percent.

(Don't laugh. Silly as this unwarranted adding up of percentages may be, no less than our nation's most respected Sunday book review once came up with something equally silly. It found that "plant and book manufacturing expenses alone have risen as much as 10 to 12 percent over the last decade, materials are up 6 to 9 percent, selling and advertising expenses have climbed upwards of 10 percent. Combined boosts add up to a minimum of 33 percent....")

A Look Today at Tomorrow's Costs

THIS IS a quick-reference table showing how much you can expect to pay in the future for each dollar the same purchase would cost today.

It tells you that an apartment that rents for $540 today can be expected to cost 6.85 times that much ($3699 a month!) 25 years from now if inflation meanwhile has amounted to 8 percent a year.

Compounding factor at inflation rate of

Years to go	5%	6%	7%	8%	9%	10%
1	1.05	1.06	1.07	1.08	1.09	1.10
2	1.10	1.12	1.14	1.17	1.19	1.21
3	1.16	1.19	1.23	1.26	1.30	1.33
4	1.22	1.26	1.31	1.36	1.41	1.46
5	1.28	1.33	1.40	1.46	1.54	1.61
6	1.34	1.42	1.50	1.59	1.68	1.77
7	1.41	1.50	1.61	1.71	1.83	1.95
8	1.48	1.59	1.72	1.85	1.99	2.14
9	1.55	1.69	1.84	1.99	2.17	2.36
10	1.63	1.79	1.97	2.16	2.37	2.59
15	2.08	2.40	2.76	3.17	3.64	4.18
20	2.65	3.21	3.87	4.66	5.60	6.73
25	3.39	4.29	5.43	6.85	8.62	10.83
30	4.32	5.74	7.61	10.06	13.27	17.45

Regarded in another way, it tells the future value of a present asset.

It's no crystal ball, of course. To make it work you must correctly guess the future course of inflation. In the case of future purchases, the table assumes that the object will not be an exception to the general level of inflation. With non-money assets, there is an assumption that the thing will not deteriorate or otherwise depreciate with the passing of time. The assumption with money is that it will bear interest at about the rate of inflation.

You might like to know the multiplier for some rate or period not shown in the table, such as 30 years at 18 percent (a rate not unknown in the U.S. economy and vastly exceeded at times in many others). It's a neat 143.37. Today's $8000 car would then cost a good bit over a million dollars.

There's a familiar formula by which you can figure this and other rates and periods not in the table.

Just express the predicted annual rate of inflation or rate of interest rate as a decimal and add 1 (so that 18 percent, for example, becomes 1.18), then raise this to the power of the number of years. The 30th power of 1.18 is the 143.37 of the preceding paragraph.

Realism at the Supermarket

In today's supermarket, you rarely have to ask the price of an item. It's marked on the package, tacked to the edge of the shelf, or posted nearby.

What takes figuring is far more important—the unit price. That's the amount you pay for each ounce, pound, pint, quart, or whatever—and somewhere in the receding future, gram, kilogram, or liter, when we join the rest of the world by going on the metric system. In recent years, supermarkets have come to the aid of the shopper by posting unit prices, usually by attaching a sticker to the shelf where the item is displayed. Unit prices are printed right on the label in the case of some items such as meat and cheese. You are given the total price of the package of hamburger, as well as the weight and price per pound. It's a cinch then to compare the price per pound of this ground meat with that of any other type of meat or grade of hamburger.

Likewise, you can quickly compare prices of two brands of canned applesauce, or two sizes of corn flakes. If the ounce price on a small can or package is 2-1/2 cents and 2 cents on the giant family size, it's easy to figure that the smaller one is one-fourth, or 25 percent, more per ounce than the large one.

You're then in a position to decide whether the saving in price is more important to you than any inconvenience in having a large can or box that takes up more shelf space at home and may not get eaten as quickly. There's no saving if the super-size deteriorates because it's used slowly. A part of the item might even have to be thrown out.

It is unfortunate that so few shoppers seem to make use of these posted unit prices. Yet they offer the simplest, most convenient way of cutting food costs. If your grocer doesn't post unit prices (or if they're printed too dimly to read or are too close to the floor to please your knees) you can figure the arithmetic mentally—or easier yet, take along one of those inexpensive pocket calculators to do the job for you.

All you need do to calculate a unit price is divide the total cost of the item by the number of units—pounds, ounces, pints—it contains. Be careful to use the same unit in both cases. The unit of one item may have to be changed—usually by multiplying the number of ounces by 16 to find the number of pounds or pints. Simplify the task by rounding off. If 12-1/2 ounces of an item cost 62 cents, forget precision, and call it 13 ounces for 65 cents, or 5 cents an ounce.

Unit-cost calculations may lead you to discover that a small loaf of compact bread costs little or no more per ounce than an air-filled and apparently more economical sandwich loaf.

Consider an incident my wife reported after a supermarket trip.

Two shoppers arrive at the checkout counter. One woman has a small loaf of a rather dense, widely distributed bread called, I think, Wheatberry . The price is $1.38.

"Golly," remarks the checker, "that's a lot of money for such a small loaf of bread."

Next in line is a shopper carrying a foot-long loaf of family sandwich bread, price 39 cents.

"With the size of my family, I have to buy the biggest and the cheapest," she remarks.

It's a slack time at the checkout counter, so all three take a minute to look over the labels of the two loaves. The small wheat loaf weighing a pound and a half, contains high-protein cracked-wheat, freshly milled wheat flour, high-score creamery butter, honey, and not a single preservative.

That mighty long sandwich loaf weighs but a pound and its contents are primarily white flour, softeners, emulsifiers, mold-retardant and a chemical freshener.

A quick calculation shows the difference in price per ounce to be a fraction of a cent; the woman with the big family was buying neither a nutritional bargain nor an economic one.

Let's say you decide to buy canned sausage and find three comparable brands on the shelf. They vary in price and weight. You can easily run a unit-cost comparison test of the 4-ounce size. A little more than one-fourth is saved by taking a 50-cent can instead of a 68-cent can of the same size—so forget the 68-center.

But what of another brand that contains 5 ounces for 70 cents?

In this case, a direct comparison is easy. Since 5 ounces is one-fourth more than 4 ounces, it should cost one-fourth more than 50 cents—or 62-1/2 cents. At 68 cents, forget it too. It's overpriced by 5-1/2 cents.

Often, informal price comparison is easy. If frozen orange-juice concentrate in the small size is less than half the large-can price you can choose small almost without thinking. You can, that is, if you look at the labels and find that the big can has exactly twice the content of the small one.

Buy Meat Wholesale?

Buying meat in quantity for freezer storage can save you 15 percent or more on your meat bill, home economists find, IF....

IF you're a good judge of the quality of meat as well as of the reputation of the farmer or dealer who offers you a good buy. Watch out for fast-pitch operators who may take a whole page in your local paper and promise an attractively low price per pound without telling you that this price is before cutting and before removal of waste. Extra charges for wrapping may also be added.

We early discovered the importance of taking into account the eating habits of the family. All cuts from steak to stew are best if used in a year or less. Hamburger should be consumed within a few months after it is frozen. If you happen to buy a tough side of beef—and this occurs all too often—you may have to grind the entire side into hamburger to make it edible.

Do you own a freezer? Will you need to rent a locker? Will you have a long drive to pick up meat from a locker? Facing such questions will reveal costs that add to the price per pound.

What do you get from a side of beef? A typical half-ton steer will dress out to about 600 pounds, but only 430 of these pounds are edible meat.

Half of the 430 will become stew meat, hamburger, liver, pot roasts, kidneys. Tender steaks and oven roasts, preferred by most families, add up to about 180 pounds. If kidneys, liver, and soup bones are thrown out or become dog food in your family, Rover is getting some expensive meals.

High Cost of Living Like a Dog

Y OU CAN FIGURE—roughly—that a fully grown dog will require an ounce of food for each two pounds of his weight. That means a terrier weighing 30 pounds or a little more is going to eat a pound a day. You can find yourself going broke buying dog food at the rate of as much as five pounds a day if you acquire a St. Bernard, or any other large breed.

How Much Rebate?

I F YOU HAVE an installment loan of the add-on type and some cash you can spare, there may be a bonus lurking for you in the banker's Rule of 78s (which also figures a couple of other places in this volume).

Your loan is an add-on if the lender has simply added the amount of the finance charge to the face amount of your loan when making it, then divided by the number of months to calculate your monthly payment.

Here's one way to find how much rebate you have coming for an early payoff.

Divide the number of payments already made by the number you were supposed to make in the course of the loan. Subtract that figure from 1.

Now divide the number of payments you have made by 1 more than the number you were supposed to make. Subtract this from 1.

Multiply together those two results and the amount of the finance charge. The result is how much rebate you can get by paying up now.

Here's a sample.

If you've made seven out of 24 payments on a loan that includes $70 interest, your rebate for paying off the whole thing now is figured like this. Divide 7 by 24 and subtract from 1. Then divide 7 by 25 and subtract from 1. Multiply these two results together and by $70. Pay now and save $35.70.

Inflation and the Metric Trick

A READER of the Los Angeles Times wrote to alert us all to what he is pleased to call metricized inflation:

I went to get a gallon of my favorite cream sherry the other day at my favorite wine store. The smiling clerk informed me that, due to government directives, they had to go to metric size bottles. They now would furnish three-liter bottles instead of the gallon size.

"For this they would drop the price 20 cents from the gallon price. I demurred, finally capitulated and bought the three-liter size for $5.75, plus tax.

"On my way home, I mentally juggled the relationship between the liter and the gallon. It slowly dawned on me that the price of my favorite wine had jumped over 30 percent."

Although the gentleman's hasty mental arithmetic is—and not surprisingly—imperfect, his point is valid. Changeovers to metric have offered a lot of temptations. It behooves us all to watch out for concealed price increases.

Here's the calculation for this one.

Divide the new price by the number of liters (3) to get the new cost per liter: $1.92.

Divide the old price by the number of liters in a gallon (3.7853) to find the old cost per liter: $1.57.

Subtract one from the other to find the increase: $.35.

Divide the increase by the old liter price to discover the percentage of increase to be 22 percent. Not 30 percent, but bad enough.

Same routine works, of course, for any packaging change from ounces or pounds or pints or quarts to grams or kilograms or liters, with or without an alteration in the dollar price.

Chapter H
MEASURING THINGS

Human Yardstick
How's the Wind Up There?
Power of a Wave
Weighing with Coins
Comparing TV Picture Sizes
Measuring the Unmeasurable
How Tall Is It?
Height by Trig
How Big Is a Tree?
Speed of Sound in Water
How Much Does It Slope?
Light Through a Prism
Estimating Water Flow
How Far Away Is It?
How Fast the Fall?
How Deep the Well?
The Metric Nickel
How High Is Up?
Measuring Height by Angle
How Big a Pinhole?
How Fast Will It Cool?
How Cold Will It Get?
Recipes for Odd Numbers

Human Yardstick

IF YOU memorize a few of your body dimensions, you'll always have with you a means of making approximate measurements.

For example, your span, or wingspread—fingertip to fingertip with your arms outstretched to the sides—will be 6 feet, give or take a few inches. In most situations, this is easier to use than your height and rather more dignified.

Elbow to tip of index fin-ger will be about 18 inches. Half a yard should be easy to remember. (But check to see if using a finger other than the index won't give you even more precisely a foot and a half.)

If you're a man of aver-age stature your shod foot is probably about a foot long—naturally enough. But if you're not a man or not of average dimension, that may not hold. Measure a shoe now for a more exact figure.

Your stride is most useful of all, as you've probably noted many times. It will be more useful in the future if you measure it now or practice stretching it to a consistent three feet or restricting it to two.

Another good move is measuring the length and breadth of the tip joint of your thumb so you'll know these dimensions when you need them.

How's the Wind Up There?

FOR SUCH purposes as planning a windmill (and—who knows?—you might; there's a lot of cheap energy there to be harvested) it's useful to know wind velocity at various altitudes.

A rule of thumb says wind speeds increase as the one-seventh power of the distance from the surface of the earth.

Suppose you have measured wind speed at the highest point you can reach, say 14 feet above the ground, and want to know what it will be at 60 feet.

Dividing 60 by 14 tells you the greater height is 4.2857 times the lesser.

Then a scientific calculator (or resort to the log table) will find the one-seventh power (also called seventh root) of this to be 1.231. Multiplying this by the measured speed for the 14-foot altitude will give you a close approximation of what you would find at 60 feet if you had some way to get your anemometer up there.

This is a rather arcane bit of calculating—a fractional root, for goodness' sake. So I guess it won't hurt to mention here, and not for the only time in this book, that you can do the trick almost as quickly with an ordinary hand calculator as with the flossiest of scientific models.

Or in a matter of minutes with pencil and paper.

By guessing and testing, in either case.

You know your root must be greater than 1. So you might start by guessing 1.1. When you find that to be far too small, you might try again with 1.2 and quickly discover you're well on your way. Punching in 1.2 followed by times once and equals six times brings you so close that in a couple more tries you'll probably hit on 1.23, which is as precise an answer as you'll ever need.

Power of a Wave

Expressed in horsepower, how much potential energy is there in an ocean wave?

In general, a wave twice as high as another packs four times the energy. But one twice as long as another offers only twice the energy.

To find the potential horsepower in a wave, express its height and breadth in feet and also express in feet the length of the wave between successive crests. Square the height of the wave. (Save this square to use again later.) Divide this square by the square of the length of the wave between crests.

Multiply by 4.935.

Subtract that product from 1.

Multiply by length of the wave between successive crests.

Find the square root of that result.

Multiply by the square of the wave height. (That's the figure you saved during the first step.)

Now multiply by 0.0329.

And, finally, multiply by the breadth of the wave to discover its energy potential in horsepower.

Run through this for a wave that is 1 foot high and 25 feet long and you'll find that the next to last step brings you to a horsepower figure of 0.1638. So for, say, 10 feet of wave breadth you'd have 1.638 horsepower.

Weighing with Coins

As LONG as you have a good mix of coins in pocket or purse you can do an emergency weighing job with surprising precision.

Just rig up any kind of simple balance—such as a pair of paper or plastic bags hung from the ends of a stick balanced over a sharp edge.

With the object to be weighed in one bag, an easily recalled clue to the heft of the coins you place in the other is the 4-to-1 relationship between U.S. coins and weight in grains.

Since a silver dollar weighs 400 grains, a half-dollar is 200, a quarter-dollar 100, and a dime 40 grains.

Nine U.S. cents weigh 1 ounce. An ounce is 437.5 grains.

Comparing TV Picture Sizes

A MAGAZINE given to testing consumer products expressed surprise at the discovery that two inches can make quite a difference in the apparent size of a TV picture.

The editors might have noted that this applies equally to other cathode-ray tubes, including computer monitors.

The magazine was comparing 19- and 21-inch (diagonal measurement) sets.

Well, the difference in these diagonal measurements is, of course, only about 10 percent. But what you're buying is area, or square measurement. So you really should compare the squares of those dimensions. Do it by subtracting the square of the smaller diagonal from the square of the greater, then dividing that quantity by the square of the smaller.

That will tell you you're getting an increase of slightly more than 18 percent when you jump to the 21-incher.

Measuring the Unmeasurable

It's easy enough to discover the weight of an irregular or recalcitrant object. If it's something you can't put on scales or can't keep there long enough—a cat, for instance, or perhaps a baby—the classic trick is to hold it while standing on the scales yourself, then weigh yourself separately and subtract. This is handy also with packed soft-side luggage too amorphous to sit on scales unaided.

But how can you measure the volume of an irregular object? If water won't hurt it, do what Archimedes did. Dunk it under water and measure the amount by which the water rises. Cooks use this trick with butter since it also avoids having a greasy measuring cup to wash. After using this method, you are entitled to flash down the street naked, shouting "Eureka!"

How Tall Is It?

An amateur explorer returned from sunny Baja California to report finding what he was convinced was the tallest cactus of its type ever discovered—the variety sometimes called the boojum tree, I think it was. He was most regretful that he hadn't had some kind of instrument along to measure its height with.

But of course he did have all the instruments he needed, right on his feet and in his head.

He might have started by measuring his shadow, preferably mid-morning or mid-afternoon when it would be fairly long. This he could have done with the length of his boot sole as a module.

Then he could have measured the length of the shadow of the cactus in the same way.

Having found his own shadow to be, say 8-1/2 boot soles long, and having lain down in the sand long enough to prove his shod height to be an even 6 "soles," he'd have had a fairly simple calculation to make after measuring the cactus shadow length and finding it to be 72 "soles."

Since the length of each object would bear the same relationship to the length of its shadow, he'd have had a little proportion like this to scratch in the sand with a stick: 6 over 8-1/2 = ? over 72.

He could look at an equation like this in several ways, but the simplest thing to do with it is cross-multiply. Starting with 6 times 72 is 432, he needed merely to divide that result by 8-1/2 to find that he had a 51-sole boojum tree. (Quickest way to do division like this is by doubling both numbers, since 864 divided by 17 is easy enough.)

Another way to look at the same math is that the fellow's height (6 boot soles) bears the same relationship to the length of his shadow (8-1/2 soles) as the boojum's unknown height bears to the known 72-sole length of its shadow. Since 8-1/2 divided by 6 can be treated as 17 divided by 12, or 1-5/12, this is the number by which 72 must be divided to produce that same result, about 51.

Our explorer could assume for the moment that his boot sole was about a foot long, making his cactus roughly a 50-footer. On return to civilization, he could check the precise length of his boot and correct for any discrepancy from 12 inches.

Height by Trig

Y<small>OU'RE</small> placing a guy wire or a prop to steady a pole. It is to go at an angle of 65 degrees to the ground, and be secured 14 feet from the pole. Without climbing or otherwise measuring how high on the pole the prop is to be fastened, you'd like a means of calculating the length of that prop.

Drawing a little picture of the situation, you'll see that you have a right triangle, of which you know a leg and an angle. From that information, you want to find out the length of the longest of the three sides, the hypotenuse.

The rule to apply is that the hypotenuse of a triangle can be found by dividing the length of a side by the cosine of the angle between it and the hypotenuse. In this case, you're dividing 14 by 0.4226, to find an answer of 33.13 feet. No climbing.

Also no use, of course, unless you have either a scientific calculator or a few pages of trig tables at hand.

How Big Is a Tree?

T<small>REES HAVE</small> quite specific size designations. The unit of measurement is dbh, which means diameter breast high.

A sapling has a dbh of 4 inches or less.

From 4 inches to 10 inches, the tree is called a pole.

A standard tree has a dbh of 1 to 2 feet.

When the diameter exceeds 2 feet, the tree is a veteran.

Speed of Sound in Water

T<small>HANKS</small> to Leroy, you can readily calculate the velocity in feet per second with which sound will move through water.

Combine the six quantities below. But omit the two parts of the fifth step unless you are dealing with salt water—or do not know if it is fresh or salt or what the degree of salinity is.

This procedure uses something called Leroy's equation, which is easy enough to apply, wonderfully mysterious though it is (like the man himself, about whom I cannot tell you more).

1. Start with 4755.2.

2. Add the temperature of the water in degrees Celsius multiplied by 15.067.

3. Subtract 0.1765 times the square of the temperature.

4. Add 0.00085 times the cube of the temperature, still using Celsius, of course.

5. With fresh water, that finishes the calculation. With salt water add 3.9 times the result of subtracting 35 from the salinity in parts per thousand; and also add the result of dividing the depth in feet by 61.

You might want to check yourself on the use of this equation. Try it for water at 20 degrees C., salinity 50 parts per thousand, depth 250 feet. The result you get should be 5055 feet per second. Note, however, that if the water is fresh, the indicated speed would be 4993 feet per second.

How Much Does It Slope?

MEASURING the slope of land can be important for many projects.

Building steps, for example. Or excavating for a swimming pool. Or in applying for a building permit if your town, like the sometimes over-fussy one I live in, demands that you go through an additional process for an application for a use permit—even before you can apply for a building permit—if you're on a site having a slope in excess of 30 percent.

You can measure a slope most precisely with a surveyor's level costing many hundreds of dollars. Or a simpler dumpy or transit level at a few hundred. You'll need a tripod with any of these.

For similar results, with less precision but more speed, you can use a sort of pocket-size monocular called a hand level. Metal types cost several times as much, but you should be able to find a wooden version for just a few dollars. You could make one yourself, for that matter. This kind is often used by farmers doing contour plowing to help them plow furrows that may wander all over yet remain near dead level so rains won't wash soil away.

With any of these, you can sight on a yardstick or a measuring tape stretched along a stick held vertically—or on the convenient but more costly equivalent thereof called a sliding-section leveling builder's rod.

Or you can dispense with the stick, and with a helper as well, by sighting directly into the slope. If the point at which the ground sights out to eye level is 20 feet away and your eye level is 5 feet, you know the land is sloping 5 feet in 20 feet. Divide 5 by 20 and you get 0.25. Your slope is 25 percent.

(Slope measuring is a little confusing. A slope of 20 feet in 20 feet is a 100 percent slope. And a straight-down cliff is an infinite slope, not 100 percent as you might think.)

But you don't need even a sighting level if you have time to rig up a more primitive and bulky device that has served me well.

What you can do is fasten an ordinary carpenter's level, preferably one at least a couple of feet long, onto a photographer's tripod or other type of stand or table.

Since the standard tripod screw is 1/4", 20 thread, you can fix the level tightly in place by drilling a hole through the rim of the level and putting on a 1/4" nut.

You don't have cross-hairs to use in sighting with this kind of home-made device, but you'll find you can sight quite well along the top surface of the level. If you want to refine the procedure a little, you may be able to find a hardware store that carries little clamp-on devices for converting an ordinary carpenter's level into a sighting level.

Even without such fancy stuff, however, some friends and I once applied this method to excavate for a 50-foot swimming pool that took us weeks of pick-and-shovel work. With the help of the level-on-a-photographer's-tripod, we were able to pile up what we dug out to form a precisely horizontal bank, even though we were working on a rather steep side hill.

Light Through a Prism

As a ray of light passes through a triangular prism, the path of the ray is bent. How much will the deviation be? Several steps are required to find out.

The procedure calls for using trigonometric functions, a prospect that may seem grim but need not be. The job can be done with a table of sines and either an ordinary calculator or merely paper and pencil. It's far easier, though, if you use a scientific calculator with its built-in trig functions.

The data you will need are the angle of incidence of the light, the prism angle, and the refractive index of the material of which the prism is made.

Step 1. Divide the sine of the angle of incidence by the refractive index, and then find the arc sine of this result.

Step 2. Subtract this from the prism angle, find the sine of the result, and multiply by the refractive index.

Step 3. Find the arc sine of the previous result, then add the angle of incidence and subtract the prism angle. Result is the deviation you were looking for.

To check the procedure with an example, use 30 degrees as the angle of incidence, 45 degrees as the prism angle, and 1.8 as the refractive index. The deviation should come out 45.3598441 degrees.

Estimating Water Flow

Y OU CAN get a pretty good estimate pretty quickly of how much water is flowing in a stream you plan to dam or use for production of power. To do this, you need two figures.

One is the cross-section area of the stream.

The other is how fast the water is moving.

To get the first one, begin by placing a board from bank to bank just above the surface of the water. At, say, ten to twenty regular intervals along the board, beginning at one bank and ending at the other, measure the depth of the water.

Take an average by adding these depth measurements together and dividing by the number of them.

Multiply by the width of the stream to find the cross-section area. To change this figure to square feet, divide by 12 if one of your measurements has been in inches rather than feet, by 144 if both have been.

Next, find out how fast the water is moving. Drop in some fairly heavy object, such as a corked bottle or plastic jug with some water in it, and measure the distance it goes in one minute.

Multiply this by the area figure and then by 0.83, or 5/6. This is to allow for the fact that water flows faster at the surface than below.

Result is the flow of the stream in cubic feet per minute.

To be sure this is all clear, let's run through an instance.

Your stream is 13 feet wide. Measured every foot, the depths of the stream, in inches, are: 0, 4, 9, 11, 8, 8, 12, 13, 15, 12, 10, 11, 10, 2.

Their total of 125 must then be multiplied by the width of 13 and divided by the number of measurements, 14, and by 12 (to convert to feet) to produce a cross-section area of almost 10.5 square feet.

If you've found the float to move 88 feet in one minute, you multiply 10.5 by 88 and by the correction factor of 0.83. The result is about 767.

Since conditions vary with the season and the year, and since none of your measurements will have been quite precise, it would be safer to think of the result in round terms. In a season like the present, your stream flows at seven or eight hundred cubic feet per minute.

How Far Away Is It?

To measure the distance from here to there—when "there" is 6 to 100 feet away—a tape and someone to hold the end of it constitute the normal way to go.

If you don't have that someone along, you'll need a device for fixing the end of the tape into place. A spike you can push into the ground works pretty well. Even a twig can serve.

If you don't have a tape either—or only a short one and you're in a hurry—pacing can be surprisingly accurate. But only if the ground is fairly even and you know the length of your stride. Best thing is to practice occasionally on a marked distance so you know just how to hold your stride to a fairly exact yard. You can check the precision of your one-yard pace right on the job if you happen to be carrying even the shortest pocket tape. Or if you know the length of your shoe (which, not surprisingly, may turn out to be just about 1 foot) you can use that.

Measure reasonably short distances quite accurately by taking heel-to-toe little steps. If you know your sole length and it's, say, 11 inches, you'll find it isn't hard to estimate for the inch gap between toe and heel.

For quicker results over somewhat longer distances, try using the range finder in a good camera.

How Fast the Fall?

A San Franciscan made headlines nationally by surviving a 29-story drop down an elevator shaft in the Transamerica Building, "hitting the cement floor at the bottom at an estimated speed of more than 100 mph."

How did the reporter know how fast the man was moving?
To find out, you could start with a couple of formulas.
One says the speed, in feet per second, of a freely falling object is found by multiplying the elapsed seconds by 32, although in practice

it varies a trifle from that with changes in altitude and geology. The centimeter equivalent is 975.

The other formula tells how far an object will have fallen after a given number of seconds. To get this answer in feet, square the number of seconds and multiply by 16. (Multiply by 490 instead for centimeters.)

Assuming a reasonable 12 feet of height for each story gives a fall of 348 feet.

Reversing the first formula, divide 348 by 16 and then find the square root, 4.66. That's how many seconds the fall must have taken. Multiply that by 32 to find the man's velocity at the end of his fall. It's 149 feet per second, which multiplied twice by 60 and divided by 5280 gives his mph as just about 102.

That checks so neatly with the newspaper story that we may conclude that the news man, or perhaps a police source, also used the assumption of 12 feet to the floor.

If you want to penetrate this subject a bit more deeply, you'll find more stuff about it in the later chapter on the outdoors.

How Deep the Well?

You can pretty accurately measure the height of a cliff you're standing on, or the depth of a well, by dropping a stone. All you need is the information that acceleration from gravity is about 32 feet or 975 centimeters per second per second ("about" because it varies a little from place to place). Those numbers tell you how fast a freely falling object will be going at the end of the first second. At the end of the second second it will be going twice 32 feet, and at the end of the third second three times 32 feet per second.

I realize this "second second" stuff is a mite confusing but you'll have to blame that on the English language. Even my computer program took an instant dislike to what it regarded as the double-word fault. *Zweite Sekunde* or *deuxieme second* does sound more sensible.

Well, anyway, the speed of our falling object during each second will be the average of starting and ending velocities. So it will cover 16 feet (the average of 0 and 32) during the first second, 48 feet (the average of 32 and 64) during the next, then 80, and so on.

From this it follows that to find out how far an object has fallen you square the number of seconds elapsed and multiply by 16.

So a stone that reaches the bottom of the cliff or well after 5 seconds must have traveled some 400 (25 times 16) feet. This ignores the effect of air resistance, which isn't great for a small distance and an object as dense as a stone.

Also left out here is the time it has taken for the sound to reach your ear. Sound traveling at some 1100 feet per second would need about one third of a second to go that 400 feet to reach your ear. We really should subtract that one third of a second from the 5, calculating16 times the square of 4-2/3 instead. Then the height of the cliff or the depth of the well comes out very close to 350 feet.

The Metric Nickel

As LONG AS you have one coin—a U.S. five-cent piece—in your pocket, you have a key to the metric system.

A nickel weighs 5 grams. So two weigh a decagram. And a collection of nickels can give you any multiple of 5 grams, including the useful 100 grams.

The diameter of a nickel is 2 centimeters. String out five and you have a decimeter. Multiply the length of the five nickels by 10 and that's a meter.

How High Is Up?

With the help of a few figures you can use an ordinary barometer to do the work of an altimeter.

The result you get will be quite close, even though the relationship between pressure and altitude actually involves quite a number of factors.

Since pressure at sea level is 760 millimeters of mercury, you start by dividing that figure by the barometer reading—also in millimeters of mercury, of course. Then find the natural log of the result, from a table or by pushing a key (probably marked LN) on a scientific calculator. Multiply by 7620. The result is a close approximation of your altitude in meters.

For example, suppose your barometer reads 238 mm. Dividing 760 by that gives 3.1933, of which the natural log turns out to be 1.1610. Multiplying by 7620 gives just over 8847 meters.

If, by the way, you ever produce just those figures in an actual situation you'll know not only how high you are but also just where in the world you probably are. The summit of Mount Everest is at 8848 meters.

Measuring Height by Angle

You can determine the height of an object at a known distance away even though it is inaccessible. It might, for instance, be a building on the other side of a river.

You can do this by using any device that will measure the angle between two lines of sight.

Place yourself directly opposite one end, so that the length you wish to calculate is at a right angle to your line of sight. Sight on that end and on the other end, finding the angle between the two sightings. If you've read the angle in degrees, minutes, and seconds, convert to decimal form in degrees.

Multiply the tangent of the angle by the known distance from you to the near point of the object. The result is its length.

Say the object is half a mile away and you find the angle to be 3 degrees, 15 minutes, 7 seconds. Dividing the minutes by 60 and the seconds by 3600 and adding both results to the number of degrees will give you the decimal form of the angle: 3.2519.

By a trig table or scientific calculator, you find the tangent of that angle to be 0.0568. Multiplying by the number of feet in half a mile (5280 divided by 2) you find that the distant object has a length of just about 150 feet.

How Big a Pinhole?

Anyone with the faintest interest in the technical side of photography must sooner or later build a version of the camera that started it all.

It's a simple light-tight box that holds a piece of film, or even photographic paper, at one end and has a pinhole at the other. Because a pinhole—although in its own way it does the work of a lens—has no focal length, a pinhole camera can produce kinds of pictures that an ordinary camera can't. Everything's in focus at the same time.

Before you prick, with a needle, that bit of black paper or foil or whatever you choose to produce the pinhole "lens" in, you'll want to know what its diameter should be.

Here's how to figure it. Real scientific-like.

Start with the wave length of the light you want to work with. If you have no preference, you might use the figure for light in the green region, 0.00055 mm.

Multiply that by the length of your camera, from the plane of the film to the pinhole, again in millimeters.

Take the square root.

Multiply by 1.9 and you'll have the optimum diameter for your pinhole photography.

How Fast Will It Cool?

HOW LONG will it take an object to cool to some predetermined temperature?

For example, in a 75-degree room, how long will it take a glass of tea to drop from boiling to 132 degrees Fahrenheit if you've found it drops to 172 in 5 minutes?

It's a nice exercise in the use of logarithms, if nothing else. And they are worth knowing about, even if back in school when the whole subject baffled you, you could hardly have cared less.

Begin with some simple subtractions.

Subtract room temperature from final temperature desired; from original tempera- ture; and from temperature after first interval, calling these three results A, B, and C. Call the measured time interval D.

You now have three ways to go.

You can get the result by using a pocket calculator that does logarithms. That's the easiest way.

Or you can use a log table. That's almost as easy and it makes a fine introduction to the use of logarithms—or a fine brush-up course. If you're interested.

Or you can use a trial-and-error technique, if you have at hand a calculator with a powers key (marked x to the power of y) on it. No logs needed.

For either log approach (the first two above), subtract log B from log A and multiply by D. Divide this result by log C minus log B.

You can check your use of the log table with the example. You should find yourself multiplying 5 by the difference between 1.7559 and 2.1367, then dividing by the difference between 1.9868 and 2.1367. Result: 12.7 minutes to cool that tea.

The guess-and-check method is both quick and interesting. First you divide A by B. Then divide C by B. Then try raising the latter quotient to various powers until you find the one that most closely approximates the first quotient. Multiply that by D to find the number of minutes the tea will have to cool.

In the example, the two quotients (57 and 97 each divided by 137) are 0.416 and 0.708. You can see that if you square something that is close to 0.7 the result will be about 0.49, which is too big. If you cube it, you'll get around 0.35, which is too little. Now, using the calculator, you try raising 0.708 to a power in between these, perhaps 2.5. This proves to be close, so you now try 2.6 and find you have the answer bracketed. Another try or two brings you to 2.54, the power of .708 that turns out to be equal to 0.416. Multiplied, this 2.54 by 5 (as before) gives 12.7 minutes.

By the way, you might want to take note that this and the companion formula are accurate only where the object in question is not in closely confined space. In cooling, it must not appreciably raise the surrounding temperature.

How Cold Will It Get?

THIS CALCULATION will tell you how cold your coffee will be after sitting a certain number of minutes. Of course, the method applies to all sorts of situations where cooling takes place.

First you'll need a test figure. This is obtained by measuring the temperature of the object at the beginning and again after some convenient interval. From this, you can figure what the temperature will be after any other length of time.

Of course you must also know the temperature of the room.

Then proceed like this. Subtract the room temperature from that of object after the interval.

Subtract the room temperature from the original temperature.

Divide the first of these results by the second.

Raise this to the power obtained by dividing total time by time of measured interval.

Multiply this by the difference between original temperature and room temperature.

Add room temperature.

For an example, take that cup of coffee served to you at full boiling, as it always seems to be when you're in a hurry. Your thermometer tells you that after 5 minutes in a 65-degree room the coffee has cooled to 190. What will its temperature be after another 10 minutes?

You subtract 65 from 190, divide that by the difference between 212 and 65, to get 0.85. You raise this to the power of 15 divided by 5; that is, you cube it.

Multiplying by 147 and adding 65 tells you that after 15 minutes, your coffee will have cooled to about 155 degrees. Still pretty hot.

(Note that there is one precaution you should take unless you have a pocket calculator of the type that has a powers key on it, or are prepared to resort to logarithms. Be sure the total time is a simple multiple of the measured time. Since we wanted to know the temperature after 15 minutes, we measured it after 5. We could have used 3 minutes or 7-1/2 minutes without creating difficulties. But if we had begun with 4 minutes the power called for would have been 15/4, which is not especially amenable to paper-and-pencil arithmetic, although with a scientific calculator it is easy enough to handle.)

Recipes for Odd Numbers

Most recipes are published for four or six eaters. If you're feeding one or two or eight, you may have some figuring to do. Often the easiest way to do it is through using equivalents such as those offered here.

A dash or a few drops or a few grains is taken to be less than 1/8 of a teaspoon.

A tablespoon is 3 teaspoons or 1/2 fluid ounce.

One-fourth cup is 4 tablespoons or 2 fluid ounces.

It follows that 1/2 cup is 8 tablespoons or 4 fluid ounces, 3/4 cup is 12 tablespoons or 6 fluid ounces, and 1 cup is 16 tablespoons or 8 fluid ounces.

Four cups is 1 quart or 32 fluid ounces.

A gallon is 4 quarts.

These equivalents should simplify dividing in two a recipe meant for 4 or adding one half to it to serve 6. For the odd problem, such as serving 7, it's usually better to stay with the figure for 6 or double everything as if for 8. It's safer to make a little less than you should, or a little more, than to get into hasty arithmetic that could lead to a serious error.

Incidentally, if your refigured recipe demands half an egg there's a simple answer. Beat yolk and white together with a fork and then put half into the recipe and half aside to enrich your next omelet. Or just figure that 2 tablespoons is half a large egg.

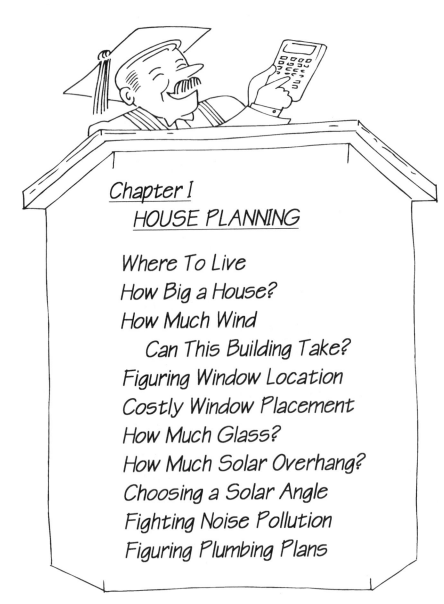

Chapter I
   HOUSE PLANNING

Where To Live
How Big a House?
How Much Wind
   Can This Building Take?
Figuring Window Location
Costly Window Placement
How Much Glass?
How Much Solar Overhang?
Choosing a Solar Angle
Fighting Noise Pollution
Figuring Plumbing Plans

Where To Live

BEFORE you buy a house—or even rent one—you might want to investigate how far it is from various facilities.

The diagram you see here is based on one used by real-estate developers. They have found that most people have pretty clearly established limits of how far they find it acceptable to live from school, church, shopping, and the like. So it should help you figure things about your prospective location while it is still prospective.

You'll note that the diagram is in terms of distances—except for job location, where time is the critical factor. Time and convenience of transportation are involved in the other distances too, of course, so some of them can safely be stretched if those factors are unusually favorable.

How Big a House?

WITHOUT getting into actual bedroom count or square footage, there are some generalizations about housing worth keeping in mind. The idea is to avoid the fundamental mistakes families often commit in their housing decisions.

A study made at the Pratt Institute School of Architecture tells us how families change as years go by.

During the first five years of a marriage, a one-bedroom house or studio apartment is tolerable for most people.

During the second five years, as income grows and children are added, the need changes to a two- or three-bedroom home, with more storage and living space as well.

During the third five years, still more living space is needed, with more bedrooms and an additional bathroom or two.

In the fourth and fifth five-year periods, needs are stabilized, but a change in the family income or job status often occurs.

In the sixth, seventh, and eighth five-year periods, retirement looms, needs dwindle, and a large home may become a liability. Yet it is often just at this period, the study warns, that so many families buy too-large, too-expensive houses.

How Much Wind Can This Building Take?

IF YOU'RE building something in a windy spot, you may need to figure out how much wind the structure will take.

A common requirement in building codes is that a one- or two-story structure must be braced to withstand a pressure of 20 pounds to the square foot.

Beginning with this figure, or one specified for your own area, you can calculate how strong a wind can be (in miles per hour) before it produces excessive stresses. Or you can find how much pressure (in pounds per square foot) will be exerted on the building by a wind of any given force.

For the first calculation you divide the pressure in pounds

per square foot (20, for example) by 0.002558 and then find the square root. To see how this works, try it for 20 pounds. Your answer should be that to produce this pressure on the walls of the building will require a wind of 88.42 miles per hour.

Now try going the other way. How much pressure will a 60-mile wind produce? Multiply 60 by itself (that is, square it) and then multiply it by 0.002558. It seems that a 60-mile wind will produce a pressure of 9.2 pounds per square foot.

How much pressure is generated by a hundred-mile wind? Note that you can do this calculation by merely moving a decimal point four places to the right, thus multiplying the constant, 0.002558, by 100 twice (same as once by100 squared) and getting the answer: 25.58 pounds to the square foot.

Figuring Window Location

NEW ENERGY-SAVING requirements rather drastically limit window areas in houses built today. This makes arrangement and location all the more important.

A window typically influences all four of these important aspects of a room simultaneously: light, view, heat, ventilation. Windows high in a wall are most helpful in lighting a room because they throw light to the far wall. But there's a limiting factor: a window too close under an overhang may not receive much light. Effectiveness of a window in respect to daylight can be judged visually, even in an existing house on a dull day, since the amount of light admitted will be roughly proportionate to the area of sky that can be seen through the window.

To admit light without creating excessive glare, avoid small windows widely spaced. A single window of equivalent area will be easier on the eyes, since glare comes largely from contrast between the windows and the darker wall areas around and between them.

A window as a source of a view is most effective if barriers to sight are avoided at eye level. That's obvious enough; but where designers often go wrong is in overlooking the fact that eye level changes from room to room.

KITCHEN   BEDROOM

LIVING ROOM   DINING ROOM

The four drawings with human figures show what levels are important for various rooms. Between these levels, try to avoid horizontal divisions more than 4 inches thick.

The other four drawings have to do with through ventilation. Each of these structures has the same pair of openings in both its walls, but comfort will vary greatly from one to another as a result of window location.

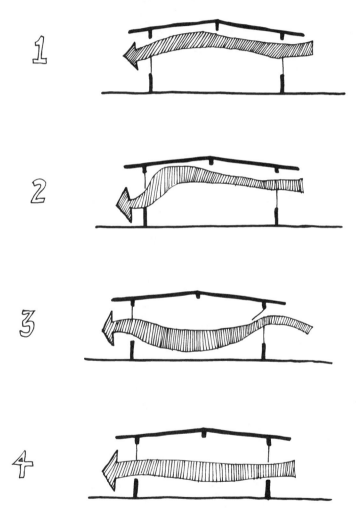

Suppose you supply a pair of high openings, as in the first drawing. Such windows will function fairly well, by carrying off warm air stored at ceiling level.

For a better result you might try placing the openings low in the wall on the side from which breezes come and high in the other. As Sketch 3 shows, deflecting the breezes to body level increases comfort.

Simplest solution of all is to keep openings fairly low in the walls, as in the fourth drawing. When views or light requirements dictate otherwise, a possible solution is to separate the functions—as by placing the windows where you want them and building louvered and screened ventilating openings below.

Costly Window Placement

WHEN ADDING a room or when building a house or cabin, how much more will it cost you to put windows in some arbitrary

WINDOW ON MODULE
6' WALL SECTION

WINDOW OFF MODULE
6' WALL SECTION

PLATES 18 L.F.
STUDS 24 L.F.
JACKS 14 L.F.
CRIPPLES 9 L.F.
——————
65 L.F.

PLATES 18 L.F.
STUDS 32 L.F.
JACKS 14 L.F.
CRIPPLES 12 L.F.
——————
76 L.F.

(17% MORE FRAMING REQUIRED)

location? Compared, that is, with putting them on a module—meaning right where a wall stud must go anyway.

The drawings show how it works out. Putting a 4-foot window in an arbitrary place instead of on module adds 17 percent to the quantity of framing lumber needed for a 6-foot section of wall.

So good planning saves you the cost of that additional 17 percent of framing lumber, and the labor of putting it up. Might even help save a tree too.

How Much Glass?

In BUILDING or remodeling a house in the solar age, window area becomes an increasingly critical factor.

For north walls, and generally for east and west as well, the rule is the less glass you can get by with the better.

South is where to concentrate your attention—and your glass.

A southern window will pay its way in free heat and do far better than that if you can provide means to reduce heat loss when the sun is not shining through it. However, window area beyond what will bring the house to 70 on a partially overcast midwinter day is an embarrassment in excessive heat gains at times, paid for by excessive heat losses at others.

From these evident but regularly neglected principles, some useful calculations emerge. To figure out the optimum window area for a south-facing wall, first find the surface area of the house. This is wall area plus twice the floor area. Divide this figure by 35 and you'll have a good basic number for a very well insulated house in an average climate.

Fine-tune that figure to fit your house.

Double it if the house is poorly caulked, poorly sealed, and poorly insulated. Multiply by 1.5 if it's up to average standards. With unusually tight construction, you can use the figure as is.

Adjust again, for sunshine factor. Multiply by 0.7 if your locality's percentage of sunshine is very high—60 percent or more. For a very low percentage, 30 to 40, multiply by 1.3. For average sunshine, leave the number alone.

You should be able to get that sunshine figure from the local weatherman or from a public utility, a likely source also for the next one you'll want: annual number of degree days.

A degree day is just the number of degrees by which the average (mean) temperature of a day is below the comfort standard of 68 degrees Fahrenheit. Add all those deviations together and you have the figure for the heating season.

If the number of degree days where you live is about 5000, regard it as average—and your calculation is concluded without it. Otherwise, multiply the figure you have by the number of degree days and divide by 5000.

What with all the complications and chances of error— yours or mine or both—we need an example. Let's use a two-story rectangular house 20 by 40 feet with walls 16 feet high.

Adding total wall area (16 times 120) to twice the floor area of 800 gives 3520. Divided by 35, that calls for about 100 square feet of south window.

Now you need to consider the thermal quality of construction—essentially whether this will be an easy or a demanding house to heat. Assuming weather-stripping and insulation are not up to modern standards, you might choose a construction-level factor of 1.9. In doing so, you've decided that your house will demand 90 percent more heat than a well-built and insulated similar one.

For the sunshine factor, if yours is an area getting around 45 percent of the possible sun, you'd assign 1.1.

For degree-days, let us say that the records show 3500.

So that's the rest of the calculation, with 100 to multiply by 1.9 and by 1.1 and by 3500 before dividing by 5000.

It seems that 146 square feet of glass will be about right for that southern exposure. It is well, however, to think of the result as simply 140 to 150, if only as a reminder that this is just a good working approximation based on a series of estimates and is not to be taken as gospel.

How Much Solar Overhang?

WHEN YOU build a new house or modify an old one, proper respect for sunshine is in order.

It is not enough merely to put a lot of glass on the south. You should also protect that glass, and the people living behind it, from too much solar heat in hot months.

What you want is an overhang that will cast a shadow over a window from 8 in the morning till 4 in the afternoon from May through August. You can't do this perfectly, of course. But you can do it for 90 percent of the time merely by making the width of your overhang equal to two thirds of the distance from the bottom of the window to the overhang.

That's if you live in New York City, Pittsburgh, Denver, Salt Lake City, or anywhere else near 40 degrees north latitude. For 45 degrees north (Montreal, Minneapolis, Yellowstone, Portland, Ore.) increase the overhang by about one eighth. Shorten it by one eighth for the southernmost United States, which is at about 32 degrees.

Interpolate for in-between latitudes. Suppose you live in Tulsa, about halfway between latitudes 32 and 40, and you have an 8-foot-high window on the south. Two thirds of 8 is 5.3. Shortening that by one sixteenth gives you just 5 feet for the width of the ideal overhang.

Choosing a Solar Angle

IN PLACING a solar collector to help heat your house—or hot water for it—how should you aim it?

In respect to principal direction, it's south, of course. A variation of as much as 15 degrees from true south won't hurt the efficiency seriously, but if you have a choice in the matter, a little east of south is usually better than west of south.

And keep in mind that you're concerned with true, not magnetic, north.

An easy way to discover true north-south is by observing the shadow of a vertical post. When it reaches its shortest—around noon, standard time—the shadow line is running north and south.

After direction, your next concern is elevation. What angle of elevation will gain you the most heat? You can get it from tables or through intricate calculations, but there is an easy approximation that's good enough for practical purposes. Just add 12 to the latitude of your locality (see any atlas). This works fine for most of the northern hemisphere. For Denver or Philadelphia, at about 40 degrees north latitude, your solar collector should be at about 52 degrees from horizontal.

Here's some consolation, in case the calculation or the actual construction gives you a problem. Variation of 10 degrees either way is not especially important, so the easy angle of 45 degrees—a foot of rise for each foot of run—will be quite acceptable for most of the United States.

Fighting Noise Pollution

Q<small>UIET</small> is a rare boon in the modern world. So it's worth while to know a little about how to protect your ears from sounds you don't want to hear.

From that point of view there are two kinds of sounds—noise originating in the room you're in and noise coming from outside it. Protection methods against the two types are almost totally different.

To muffle sounds within a room you need walls and floor and ceiling that are as soft and rough and porous as you can get. This means heavy carpets and draperies and acoustic ceiling tiles.

However, none of these will do very much to stop sound from going right on through into another room, or to stop it coming into your room. They can most effectively deaden what is being produced right there. In fact, if you soft-surface a room very thoroughly, it will become so dead acoustically that listening to music will become a new problem.

UNITS OF SOUND ENERGY	LOUDNESS LEVEL IN DECIBELS	EXAMPLES OF LOUDNESS LEVELS
10,000,000,000,000	130	Jet engines; threshold of pain
1,000,000,000,000	120	Threshold of hearing
100,000,000,000	110	Airplane motor 18 feet away; thunder
10,000,000,000	100	Loud auto horn 23 feet away
1,000,000,000	90	Subway station; outboard motor
100,000,000	80	Heavy traffic; average factory
10,000,000	70	Conversation; loud music
1,000,000	60	Noisy office or department store
100,000	50	Moderate restaurant noise
10,000	40	Quiet radio in home
1,000	30	Very soft speech or music
100	20	Whisper 4 feet away
10	10	Rustle of leaves
1	0	Threshold of hearing

You can estimate the value of a wall in stopping noise penetration by its weight. Sheer mass is the most important factor in keeping sound out. A poured concrete or stone wall is excellent at doing this, and even hollow masonry blocks are very good.

With ordinary frame walls, where there's not much mass, just putting in ordinary insulation will do disappointingly little. You'll accomplish more by separating the two wall surfaces so there are no mechanical connections between them except the unavoidable ones at floor, ceiling, and wall ends.

Standard way to do this is with staggered studs. Floor and ceiling plates are typically two-by-sixes instead of the more usual two-by-fours, so that every second two-by-four stud can be offset so it doesn't transmit sound across the stud space. An insulating blanket woven between the studs gives some additional muffling.

There is another arrangement that, although a little less effective, permits use of standard two-by-four materials. Using studs turned flatwise, alternate ones flush with each side, permits an insulating blanket, yet does not increase the space occupied by the wall. You might use it for the common situation between rooms where the wall is not required to bear a load.

To figure a floor that won't let much sound get through, you'd again be best off with masonry, such as poured concrete. But fortunately most standard floors are fairly heavy and will do an acceptable job on airborne noises.

The stickler comes with impact sounds—the dread Chinese water torture of having someone pace a bare floor over your head, wearing stiletto heels, while you're trying to fall asleep.

Most effective nonviolent remedy is reducing the impact that produces the sound. Stealing the spike heels will help, but a more feasible solution is installation of a heavy carpet. Or a heavy pad under the carpet. Or both.

A layer of fairly soft insulation board, fastened to the floor before pad and carpet are laid, adds further to the effect. And, of course, each of these does help a little in reducing transmission of airborne sounds from both above and below.

To get further into these important problems and deal with materials and partition systems that have calculated sound-transmission losses, you'll need to know a little about the methods and terms used.

The common unit of sound, the decibel, is most simply described as the least amount of sound your ear can detect.

When the energy producing a sound is multiplied by 10 the sound will seem only slightly louder to you.

To keep the measuring unit, the decibel, proportionate to what your ear tells you, it is on a logarithmic scale. All that means to you is that each perceptible increase in sound level is called 10 decibels.

Here's a typical instance of how decibel measurements work out.

While you're trying to enjoy music in one room, somebody's talking loudly in the next room, at a 75-decibel level. By the time his voice reaches the wall between you and him the loudness will have dropped to perhaps 65 db. Distance helps.

Now on your side of the partition, you will benefit by the masking effect of the music, to the tune of about 15 db. You can tolerate 10 db sound, no louder than the whisper of leaves (as the chart indicates). You'll hear it but you won't really notice it.

SECTION THROUGH WALL SHOWING
HOW EFFECTIVE REDUCTIONS OF
35 db MAY APPEAR TO BE 65 db

Add those two figures to the 5-db drop you can count on as the sound moves from the wall to your ear, and you get a total of 30 db.

The difference between the 65-db sound that will reach the partition and the 30 that you can accept is 35. So you need a wall with a 35-db rating.

For an idea of what kind of wall that means, note that ordinary two-by-four stud construction with half-inch gypsum board on both sides won't do it. But the same wall with a mineral-wool blanket in the stud space will. The blanket makes the difference between a 31- and a 39-db rating.

Figuring Plumbing Plans

Most of what you might need to know about plumbing as an amateur house-builder or remodeler, you can pick up by talking to the people who sell you your supplies.

But the hard part—to me, anyway—is plumbing the mysteries of drains and vents. That's why I was delighted to come upon this inelegant little drawing (now neatened up a little for you) of an old-fashioned house that shows just this part.

As you will see, solid lines indicate drains. Dashed lines are vents. Clean-outs are shown, too. A few minutes with this drawing should give you a good start in designing plumbing for your own job, which I trust will be a simpler collection of pipes than this conglomeration.

Water lines aren't shown. That makes this section drawing easier to follow. And water lines are the obvious part anyway.

I'm throwing in the other drawing because it, too, has to do with figuring plumbing. Those dimensions aren't just suggestions. They're code minimums in many parts of the country, though sometimes I wonder whose business it is how much space I provide around my john.

Chapter J
BUILDING THINGS

Aggregate for Concrete

How big should the biggest stones or hunks of crushed gravel be in concrete? One good rule to figure from is that when you're pouring a slab, the biggest ones should be not more than one-third the thickness of the slab.

Within this limit, however, the bigger the better.

With larger aggregate, you can get concrete of the strength you need while using less cement. Which means fewer dollars.

For the usual 4-inch slab (probably 3-1/2 inches actually, that being the width of the one-by or two-by-fours used for so many forms) you would order aggregate to maximum size of 1 inch. Going to 1-1/2 inches will affect the strength and make the stuff balkier to work with at this thickness.

For driveways, walks, and patios, you'll want concrete giving a compressive strength of 3500 pounds to the square inch when tested after 28 days.

To accomplish this with aggregate of 3/8-inch maximum size, your concrete must contain 610 pounds of cement to the cubic yard.

With 1/2-inch aggregate, this requirement drops to 590 pounds.

For 3/4-inch, it's 540.

For l-inch, it's 520.

Let the aggregate go as high as 1-1/2 inches and the mix may be a little more difficult to place but it will require only 470 pounds of cement to the yard of concrete to meet the strength test.

Figuring Ready-Mix

You won't find a bigger boon than transit-mixed concrete for the man, woman, or child engaged—however reluctantly—in doing a really substantial project.

When it comes to concrete, half of the fun is not getting it there in the form of sand, gravel, water, and cement. It's having somebody bring concrete to you already mixed and dumping it at least close to where it will probably remain into the next century.

What's no fun is to miscalculate your order and run just short— or be stuck with a considerable part of a ton of gooey stuff and no place to put it.

So calculate with care before you place your order. You'll have to figure out how much concrete you need and—with advice from the supplier—what kind.

To find the quantity, first multiply the length of the area by its breadth, both in feet. Then divide this by the number of thicknesses it would take to make a foot. Finally, divide your product by 27, the number of cubic feet in a cubic yard.

Suppose you're about to pour a 4-inch-thick sidewalk that is 4 feet wide and 26-1/2 feet long. Since 4 inches is one third of a foot, you multiply 4 by 26-1/2 to get 106, and then divide that by 3.

Dividing this result by 27 gives just over 1.3 cubic yards.

To include the usual 5 to 10 percent allowance for errors in measuring and for waste in pouring, you'll probably order a yard and a half. Then avoid waste by having a few small wooden forms on hand to use in casting some useful stepping stones with any spare concrete.

If you see any possibility of running short, have a small pile of rocks on hand. You can push them into the slab toward the end of the pour if you need to make bulk.

When You Mix Concrete

CONCRETE is the original handy-dandy plastic, infinitely useful. When you learn to work with it efficiently, you're halfway toward completion of many a major project.

Making concrete that is easy to work, strong, and yet economical calls for figuring rather precisely the amount of each ingredient.

For smallish home jobs, about the handiest measuring device is an ordinary galvanized water pail, although a wooden box is often used.

For each pail of cement, use 2-1/2 pails of sand (preferably wet) and 1/2 pail of water.

The amount of coarse aggregate (stones or crushed gravel) to use will depend upon the maximum size of the stones, which you will probably have specified in ordering the material. This size, by the way, should be not more than 1/3 the thickness of the slab you are pouring—l inch for a 3-inch to 4-inch sidewalk, for example.

For 3/8-inch aggregate, use 1-1/2 pails. For 1/2-inch, use 2 pails. For 3/4-inch, 2-1/2 pails. For 1-inch, 2-3/4 pails. And for aggregate that ranges to 1-1/2-inch, you should use 3 pails.

As you can see, the bigger the aggregate the more gravel you can use for a given amount of cement, the costliest part of the mix. That's why it is good economy to get gravel as large as the rule about thickness of slab allows.

Loading the mixer is best done by putting in the aggregate and half the water, then starting the mixer and then adding the sand, the cement, and finally the rest of the water.

Framing: 16- or 24-Inch?

How much can you save by framing a building with the studs and joists on 24-inch centers instead of the more traditional 16 inches?

This is worth knowing because it is now generally accepted that the wider spacing is perfectly satisfactory for a one-story structure or for the upper story of a taller one.

For the wall-framing part, you'll reduce the labor cost (or hours spent) some 17 percent. Material cost will go down 8 percent. The two combined will drop nearly 15 percent.

BUT if you're covering that framing with plywood and the wider spacing forces you to go to thicker plywood, you'll almost wipe out the saving on materials. Over-all saving will slip to 7 percent.

Works out much the same with floors. Putting the joists on 24-inch centers saves 17 percent but if you must then go to thicker subflooring, figure the saving drops to 4 percent for the whole floor-framing and subfloor job.

The 24-inch framing will save on labor costs for drywall installation and hole drilling for wiring, too.

With these things included, you should wind up with a saving of 6 to 7 percent. The amount in one instance was $219 saved on a $3,175 job of framing walls and floor of a small house.

If there had been no need to shift to heavier wall and floor plywood, the saving would have been half again as much—some $307 or about 10 percent.

Either way, it's enough to make up for quite a bit of inflation while helping to save a forest.

Expansion of Acrylic Sheets

WHEN GLAZING with acrylics—Lucite, Plexiglas—you should calculate the possible expansion and allow for it. This problem comes up most often when replacing a glass pane in a storm door or window because it has been broken—or as a safety precaution.

For each 1 foot that the opening measures, subtract 1/32 of an inch to allow for the thermal expansion.

Figuring Kitchen Cabinets

IF LIKE SO MANY of us you're remodeling your house or itching to, you very likely want to start with the kitchen. It helps a lot if you can draw a picture of what you have in mind. And there's no other detail as complex, and as tricky to draw, as the working walls of a kitchen.

Here's the idea. You begin with the vertical corner line of the room, working to any convenient scale. Mark off location of wall and base cabinets. Those right and left walls take off from the vertical corner line at 60-degree angles (or 30 degrees from the horizontal dotted lines).

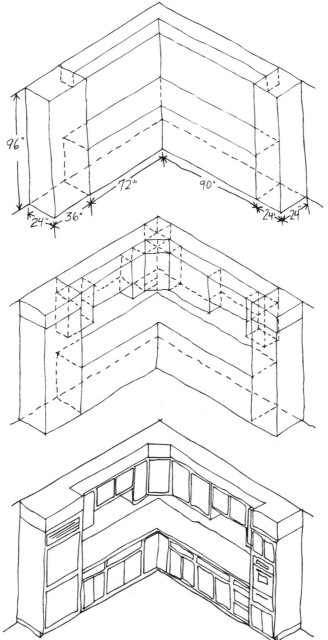

In the second drawing, you see how to block in outlines of base cabinets, oven, fridge, and the soffit. Lines are parallel because your drawing is an isometric, easier for this purpose and more useful, too, than a perspective.

Step 3 adds more details in the upper cabinets. These three steps are most easily done on a single piece of paper, but for the final step you may prefer to tape on a sheet of tracing paper and work on that.

How Stout Is Rough Lumber?

YOU MAY have a choice between using rough and surfaced lumber in some building job. The rough stuff is especially appealing if you happen to have timber of your own or can buy from a small local mill where the rough is cheaper than the smooth.

One really nice thing about rough lumber is that its width and thickness average about what its nominal dimensions say they are. A rough four-by-four is actually 4 inches thick, not the 3-1/2, more or less, of a surfaced piece. This extra wood means more strength. The question is—how much more?

If you're talking about strength in tension and compression, the rough is about one-third stronger. Exact gain depends upon the size of the board or timber you're dealing with, but many decisions can be made on the basis of that approximation.

More likely, you're concerned with bending strength: how a rafter will behave when the load goes on, for example.

With l- and 2-inch lumber the gain will be at least one-third and may be as much as one-half.

Here's how to figure it. First calculate the relative strength of the surfaced lumber by squaring the dimension that is in the direction of stress. Then multiply by the other dimension. Use the actual dimensions, not the bigger nominal ones.

For the rough lumber, do the same, but use the nominal dimensions, or the actual measured ones if they differ and are available.

Thus if you're using two-by-four rafters for a small building and they have been surfaced to 1-1/2 by 3-1/2 inches, you square 3-1/2 to get 12-1/4 and multiply by 1-1/2 to get 18-3/8, or 18.375.

Squaring 4 and multiplying by 2 gives 32 for the rough lumber. Those are your relative strengths. To discover the percentage of gain from using rough instead of surfaced two-by-fours, subtract to get the difference (13.625) and divide this by 18.375 to discover a gain of 74 percent or almost three-fourths.

(If you want to check yourself on the method, take the nominal dimensions of a two-by-four to be 1-5/8 and 3-5/8 inches this time. See if the improvement from going to full 2-inch by 4-inch lumber comes out 50 percent.)

How Big a Beam from a Log?

You HAVE a log. Or a pole. Or any chunk of wood that is circular in cross section.

You want to trim off the four sides to make it into a beam or girder to hold up something, such as a part of a building.

How do you make your saw cuts so as to get the stoutest possible beam from it?

It's no good just taking equal amounts off the four sides. That would give you a square beam, such as a four-by-four, which would not be the strongest you could get.

This is true because what counts most in making a strong beam is its depth—the measurement from top to bottom.

Its width counts less. A beam that's twice as wide as another beam will also be twice as strong—as you might expect.

But one that's twice as deep (high) as another will be four times as stout, since strength varies as the square of the depth. (A lot more use of this relationship and the formula expressing it is made elsewhere in this book, where beams are calculated.)

Unless you have more fondness for a page of calculus than I have you'll be glad just to accept the result of someone else's work.

The strongest beam you can cut is the one whose width is 1.15 times the radius (or 0.575 times the diameter) of the log.

Give it a try for a log or post just 6 inches in diameter. You should find that the strongest beam of rectangular cross section it will yield is one 3.45 inches wide and 4.9 inches deep.

To Build a Winding Stairway

THIS BOOK's no place to try to tell the world all about how to build a spiral staircase. But this drawing impressed me as something that tells its story all by itself and is so useful to anyone with such a problem in building that I wanted to preserve it.

Anyway, the drawing doesn't take up a lot of space and it should get a first-time stair builder off to a good start.

How to Figure Steps

IF YOU'RE going to build a stairway, or even a few steps to a porch, you can find all the dimensions and angles involved in a book of special tables. Or you can figure 'em for yourself very rapidly. Like this, working with four basic dimensions: total height; run; riser; tread.

Total height is the vertical distance you're working with, such as how much above the first floor of a house the second floor is. Since this is generally fixed by circumstances beyond your control, it is likely to be the figure you start your calculations from.

Run means the horizontal length of the whole stairway.

Riser is how much higher the surface of one step is than the one below it. It should be not more than 8-1/4 inches.

Tread, the horizontal measurement from one riser to the next, should be at least 9 inches. It is usually somewhat less than the actual width of the board you use for the step, since the step commonly sticks out an inch or so beyond the riser. This protrusion is called the nosing and you don't count it in your calculation.

Decide first how far down you prefer to go with each step you take. Usual minimum is around 6 inches, which makes for a very easy climb; 8 inches is pretty steep.

Divide the total height by that figure (riser height) to find the number of risers. Adjust it slightly to make the number of risers come out a whole number. (You can't, or anyway shouldn't, try to build a stairway with a half step somewhere.)

Now subtract riser height from 17-1/2 inches to get tread. (That 17-1/2 inches is a figure that long experience has established as making for a stairway most people find a comfortable compromise between ease and speed of use.)

Multiply tread by number of steps to find run. Check to make sure the run is short enough for the space available. From the figures you have at this point, you can work out a couple of others you may want.

One of these is the settings to use on a framing square to cut a carriage. The other is the angle at which this carriage—the slanting piece of framing that holds up the steps—will be placed.

The first is simple. Use the square so the tongue is at the height you've chosen for each riser (between 6 and 8 inches) and the body is at tread width.

The second is not so simple to figure out, but probably you won't need it. It is the angle at which the carriage will go—its incline. You could find it convenient to know this when setting a radial-arm or table saw for cutting the carriage. I've found it a good thing to know when building a special type of stairway having no risers—just treads set into grooves cut at an angle across a 2-by-8 to make a carriage piece without notches.

To discover this angle, first divide the height of the riser by the width of the tread. Then find the angle of which this is the tangent with a pocket calculator of the scientific type that delivers trig functions at the push of a button. Or…

If you have a computer to work with, you can use the program called BASIC (it comes with IBM-compatibles) for trig stuff. Or…

In a math reference book, find the nearest number in the table of tangents (within the section on natural trigonometric functions) to the one obtained by your division. The angle shown for this tangent is the one you want.

I think all this stuff will become clearer when you see it applied to a stair I've built in my house.

Total height, measured from the finished floor surface of my second floor to ditto on the first floor, is precisely 8 feet 9-1/8 inches. Turning that into inches and dividing by various possible number of risers, I find that 13 or less would give me risers greater than 8 inches, making for a pretty steep climb. With 18 or more, I'd get an excessively gentle climb and too many steps.

Wanting a comparatively quick climb, I choose 14 and divide to find that will produce 7-1/2-inch risers. There's one eighth of an inch left over in my calculations, but it has been found by experience that making one riser up to one-fourth inch more or less than the others won't be noticeable to eye or foot. That's about the maximum discrepancy allowable, however.

Subtracting 7-1/2 from the arbitrary, or God-given, 17-1/2 inches, says treads should be 10 inches.

Those two figures tell the marks to use on a framing square. They also tell how to figure the angle by trigonometry. When 7-1/2 is divided by 10, the closest figure to the resulting 0.75 that you'll find in a standard tangent table is at 36 degrees, 52 minutes. So that's my saw setting if I need one.

A stair with 14 risers will have 13 treads, so I find the run by multiplying 10 inches by 13. If this 130 inches, or 10 feet 10 inches, is within the space available, I'm all set to go. If not, I'll have to recalculate, using higher risers, so that the treads will be shorter and fewer.

With rather steep 8-1/2 inch risers, for instance, I'd need only 13 of them, hence only 12 treads, of 9-3/8 inches each. That would reduce the run by almost a foot and a half.

If you're figuring things like this on a newly acquired pocket calculator that doesn't favor fractions, it may not be immediately evident to you how to multiply something like 9-3/8 by 12. The trick is to start with the fraction. Enter 3, divide by 8, add 9; then multiply by 12.

The one question that remains is amount of nosing. This over-hang of each step surface for better footing is usually a little more than an inch, so the 10-inch tread we've figured on is fine if we're using standard lumber. A surfaced 1-by-12 or 2-by-12 will be about 11-1/4 inches wide. Just enough for nosing without any waste.

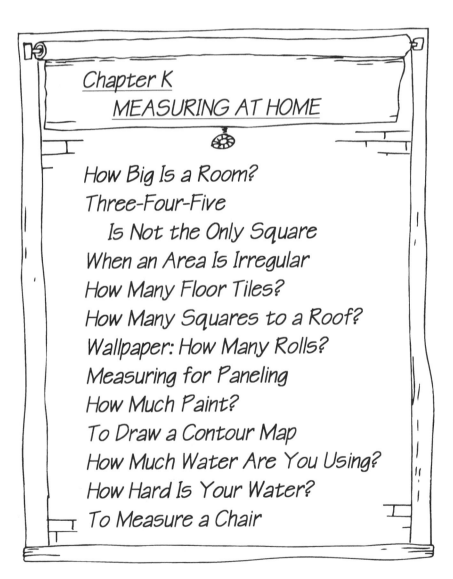

Chapter K
MEASURING AT HOME

How Big Is a Room?
Three-Four-Five
    Is Not the Only Square
When an Area Is Irregular
How Many Floor Tiles?
How Many Squares to a Roof?
Wallpaper: How Many Rolls?
Measuring for Paneling
How Much Paint?
To Draw a Contour Map
How Much Water Are You Using?
How Hard Is Your Water?
To Measure a Chair

How Big Is a Room?

Sɪᴢᴇ ᴏꜰ a room is usually pretty obvious to figure. You multiply the length by the breadth to get the area. If there is a closet or other jog within it that you don't want to include, you figure it the same way and subtract the one from t'other.

If two corners are square but the opposite wall is at an angle, then you multiply the length of the wall between the square corners by the average length of the two adjoining walls.

But if it's a room you're building as part of a house construction or remodeling and the ceiling is too low, you may have another problem. Under most building codes your room is not the size it looks. Some of it may not count at all, legally speaking.

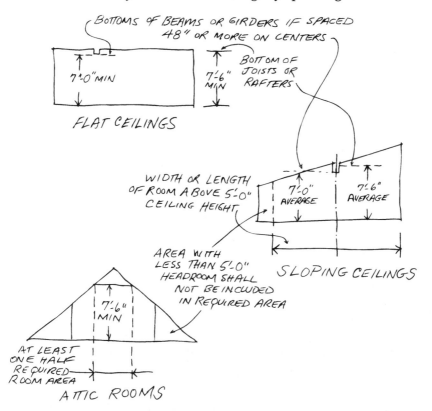

That can be important when the code specifies certain room dimensions, as do most of the various building codes. In general, every dwelling must have at least one room of 120 square feet or more; other habitable rooms except kitchens must have at least 70 square feet; and no habitable room other than a kitchen may be less than 7 feet in any dimension.

Assuming it's any of the code's business what size your rooms are (and I'm not so sure it is) these dimensions seem reasonable. But there's a gimmick. According to code, part of the room doesn't exist—if the ceiling is too low.

The UBC begins by telling you that halls and bathrooms must have ceilings at least 7 feet high measured to the lowest projection from the ceiling. Habitable rooms and those used for storage or laundry must have at least 7-foot-6 ceilings.

But what about a ceiling that slopes, as in so much modern post-and-beam construction? Then, as the drawings help explain, the prescribed ceiling height is required in only half the area. But "no portion of the room measuring less than 5 feet from the finished floor to the finished ceiling shall be included in any computation of the minimum area thereof."

Three-Four-Five Is Not the Only Square

IF YOU'RE laying out a building and would rather like the corners to be square, you may mutter, "Three, four, five."

Because they're handy in size and easy to remember, these three dimensions are in common use by builders and lots of other people. Those dimensions describe the two shorter sides and the hypotenuse, of a right triangle. But they're not the only set of numbers that produces a right triangle. In fact, there's no limit to the number of sets of digits you can use for this purpose...and not just multiples of 3, 4, and 5 either.

You can choose any even number you wish for one side of the triangle.

To find the other side, select any two numbers whose product is half the length of the long side. Square each number and subtract the smaller square from the larger. That's your second side.

Add those same two squares together to find the hypotenuse.

(For example, say you choose 28. Half of that is 14, which is 7 times 2. Square of 7 less square of 2 is 45. Square of 7 plus square of 2 is 53. You have a 28-45-53 right triangle. To check it, add squares of the first two numbers and note that their sum equals the square of the third, the hypotenuse.)

Now, for a working example, let's suppose you've driven four stakes where you'd like the corners of a 20-by-30-foot rectangular building to be. A most convenient triangle for checking squareness would be one having 30 feet for its long side, letting you use the side of the building as the long side of your square.

You can do this with 3, 4, 5 proportions by a little arithmetic. Since the long side is 7-1/2 times 4, multiply 3 by this quantity to find the short side and do the same to 5 for the hypotenuse. But this produces a short side greater than the width of your building layout....

So back to our scheme. Long side is to be 30. Half of that is 15, which can be factored into 5 and 3. Squaring both, then subtracting, you get 16 for the short side. Adding gives 34 for the hypotenuse.

Our numbers are keeping us within the dimensions of the building, so measuring shouldn't be difficult.

You measure 16 feet along the short side and stretch a line to find out whether the distance from there to the far corner is precisely 34 feet. If it is, your building is on its way to being properly rectangular.

The drawings show how these notions are put into practice. In fact, they do a fine job of showing just how to lay out a simple building. This consists of first staking corners, then building batter boards a couple of feet away from which to stretch lines after the stakes have come out during foundation excavating.

When an Area Is Irregular

WHETHER it's to order flooring, buy paint, or pour concrete, you often need to calculate an area. For a rectangle or other regular shape, this is easy enough. But in real life, not all shapes are regular.

They can, however, be broken up into a series of quite regular—and easily calculable—parts with whatever degree of precision the situation demands.

The basic shape for calculation is the rectangle. You'll recall, almost without thinking, that its area is just the length times the width.

The area of a triangle is half the length of the base times the height, the height being the vertical distance from base to opposite vertex, or corner.

(When it's a right triangle, as it usually is, the area is half what you find by multiplying together the two sides next to the square corner.)

What about a trapezoid, a shape you might find in a short driveway? Take an average of the two parallel sides and multiply by the distance between them. If "take an average" seems vague, it just means you add those lengths and divide by 2. Or think of it as the distance halfway between the two .

The area you are measuring, or some portion of that area, may be a circle or part of a circle. For a full circle, remember that you first find the radius, which is the distance from the center to the rim, multiply that number by itself, and finally multiply by the value of pi.

(For most purposes, the value of pi can be taken as 3-1/7 or 22/7, whichever seems simpler. For greater precision, or to eschew fractions in favor of decimals, use 3.1416. With a calculator handy and a wish to indulge yourself with even greater precision, without appreciable extra effort, help yourself to 3.1415927.)

Naturally, a half-circle has half the area of a full circle and a quarter-circle one-fourth.

If these round and three- and four-sided figures don't cover all your needs, here are a couple of others:

A five-sided figure, or pentagon, has an area 1.721 times the square of the length of any side.

Since the area of a regular hexagon, or six-sided figure, is 2.598 (as near to 2.6 as usually makes no difference) times the square of a side, you can proceed by first multiplying the length of a side by itself and then by 2.6. A hexagon with 20-foot sides, like the living area of a house I once built, will have area of 1040 square feet.

Here is another good method of approximating an area that can be combined with this sum-of-parts trick. It also may be used alone, especially when you're stuck with a highly irregular area that may have been laid out long ago b'guess and b'gosh. When the geometric methods prove tedious, map the area with pencil and ruler, and possibly compass, on a sheet of graph paper. Then count squares, half-squares, quarter-squares and so on—and add 'em up.

How Many Floor Tiles?

THIS IS a quick on-the-job way to estimate how many 9-inch-square floor tiles you'll need for a job.

Multiply the length in feet by the width in feet to find the square footage. Double this. Then take off 10 percent.

A floor 12 by 18 feet has 216 square feet. Doubled, it's 432. Taking away 43 leaves 389. That's how many tiles to figure on.

Take the time to make the calculation by a more precise method and you'll find the figure is actually 386. But then you'd probably want to add a few tiles for waste, anyway, or to keep on hand in case you had to replace a damaged tile some day. So the quick estimate turns out to be more useful than an exact figure would be.

In short, it's better than perfect.

How Many Squares to a Roof?

THE SUBJECT isn't square meals but the measurement commonly used by contractors to figure areas for roofing and siding jobs.

A square is merely 100 square feet, or the equivalent of a wall or roof section 10 feet by 10 feet.

To estimate the cost of a siding job, multiply the height of the walls by their combined length and from this subtract the total area of windows, doors, and such things as fireplace chimneys or other masonry features. Divide by 100 to get the number of squares.

For a roof, multiply the actual area of surface and proceed in the same way. You can't use the square footage of the house, because that would overlook the effect of roof slope, which increases the area, and of overhang, which can be substantial.

Wallpaper: How Many Rolls?

IF YOU'RE using American-made wall covering, divide the square footage you've found by measuring the walls (and ceiling if you're papering that) by 30. This will tell you how many single rolls to order. It includes an allowance of 6 square feet per single roll for waste.

For European-made paper you might use 22 as a guess to divide by if you don't know the number of square feet in a roll. But since size varies, better to find out the correct number if you can and then subtract 6 to get the number by which to divide the square footage.

Even after all this it's safest to add another roll to the total to allow for minor miscalculations or excessive waste. Allow more yet if the pattern repeat is unusually large or if your room contains few spaces above doors or windows where scraps can be used.

A reason for being ultraconservative is that if you have to go back to the store for a final roll or two you may wind up with wallpaper from a different batch in which ink colors are not quite the same.

Measuring for Paneling

To FIGURE how many panels of plywood or hardboard or other big-sheet wall materials to buy when doing the walls of a room, first find the perimeter.

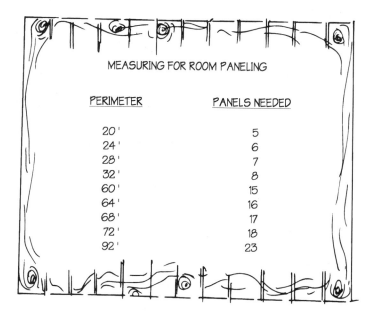

MEASURING FOR ROOM PANELING

PERIMETER	PANELS NEEDED
20'	5
24'	6
28'	7
32'	8
60'	15
64'	16
68'	17
72'	18
92'	23

Find number of panels in the table. Use the next higher number if yours falls between figures on the table.

Deduct for doors, windows, fireplaces.

Typical deductions are half a panel for a window or fireplace, two-thirds for a door.

Subtract the total of deductions from the number of panels originally determined, rounding off to the next higher number if necessary.

How Much Paint?

Figuring out how much paint you'll need for a job is essentially an area calculation, even though several surfaces may be involved. Add up all the areas you're going to cover and divide by the coverage estimate given on the paint can.

Here is a procedure, as worked out by a paint manufacturer, for quite accurately estimating the total area.

1. Measure the rectangular areas and multiply length by width or height.

2. Do the same for triangular areas, such as gables, but then divide by 2.

3. For plain cornices, multiply length by width and double the result. For ornamental cornices, triple the result.

4. For plain eaves, multiply width of overhang by length and add 50 percent. Double that result if the rafters are exposed.

5. For each face of slatted shutters and blinds, multiply length by width and double the result.

6. For columns, measure the circumference and multiply by the height. Add 50 percent if they're fluted.

7. Rails, lattices, and balustrades: multiply the length by 4 times the height.

8. Gutters and down spouts: having found the circumference in inches, divide by 12 to get feet, then multiply by length. For gutters to be painted both inside and out, double this.

9. For picket fences to be painted on both sides, multiply the length by 4 times the height.

10. Add everything together and subtract any areas you've included that are not to be painted—doors, windows, and such.

To Draw a Contour Map

In the beginning, I learned how to draw a contour map with inspiration from one of the most effective incentives in this wicked world—greed.

I had to turn in such a map to get city permission to build my house on a steeply sloping lot. To follow standard practice, I'd have had this prepared by a surveyor who would then have deprived me of three or four hundred dollars.

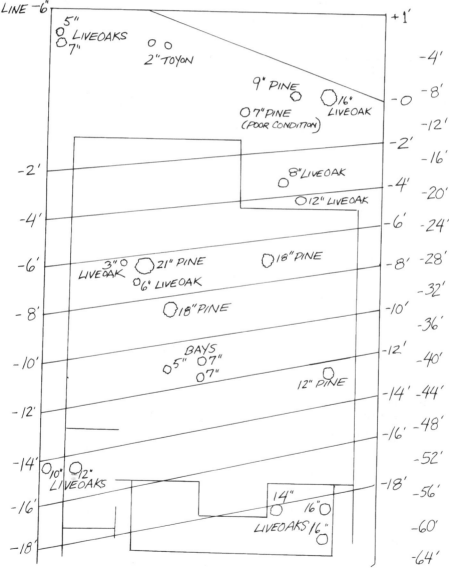

Since I was about to build the house myself—design, plans, concrete work, carpentry, plumbing, all the sweaty stuff—it went against the grain even to contemplate farming out one of the cleanest parts of the task, which I was convinced needn't take more than a few hours. My confidence was enhanced from having gleaned a few tips over the years from a relative by marriage, Gerry Kinsey, who knows a great deal about surveying but unfortunately lived too many miles away to be of personal help in my crisis.

I didn't produce the same kind of contour map a surveyor would have drawn. In the first place, my requirements were special—all I needed was information with which to show a planning commission what my slope was like and how my house would fit onto it. And I didn't have surveyors' instruments. Just a hand-held sighting level (borrowed—but purchasable for as little as $8).

I decided to let my map show 2-foot changes along the critical slope, so I held my sighting level on a 2-foot length of dowel to keep it consistently this distance from the ground.

I started as near the bottom of the slope as a steep gully permitted, along one side of the property, and worked my way up. Each time, with help of a young assistant who placed and held his toe where told, I could find how far it was from my sighting point to a point on a level with my instrument.

That point became the new sighting point on which I placed my dowel.

I marked off the distances along one side of a map of the property that I'd drawn on a piece of ruled (graph) paper.

Then I went back to my original point and sighted across to find the point on the other sideline that was at the same level as my original point. And worked my way up that side in the same fashion.

And then did the same thing up the middle of the lot.

From one side to the other of my 40-foot lot, the slope was uniform. If it had not been, or if I had needed a more detailed map—for landscaping?—I might have drawn several lines within the lot, parallel to sidelines, and sighted them, too. But one line was enough to show the nature of my slopes.

To make my map as useful as possible I then chose a zero point more significant for construction purposes than my starting point. Since easy driveway access was my most important consideration in planning the level of my house, I took as my base point a place in the driveway-to-be that was an even number of feet (18) above my original starting point. I called this 0. Each of my other points then became -2, -3 feet, and so on, down to the original point, which was now -18.

Connecting the points of equal level then gave me the contour map you see. There are better ways to make contour maps (sighting on "break" points—where contours make sudden changes) but this one is the simplest I know, and it serves such purposes as mine well enough. This one not only satisfied a rather stuffy planning commission (after the city engineer told them it would serve its purpose) but also proved to be just what I needed when I drew my house plans and when I did the actual building.

The second drawing you see here shows how I was able to use the contour figures in a drawing of a section through the house, very helpful to me in completing the design.

AVERAGE GRADE
– SECTION THROUGH HOUSE –
SHOWS RELATION TO LOT SLOPE
ALONG CENTERLINE OF LOT

How Much Water Are You Using?

IN MANY places, it's so much a duty these days to conserve water that we've even got the water companies believing in it. It may surprise you, as it did me, to discover that your water meter is a sufficiently delicate instrument to help you measure water wastage.

First, you have to know how to read it. That's easy if yours is digital, whether reading in gallons or cubic feet. You'll notice that with the help of its little dial you can read right down to a fraction of a gallon or one-or two-tenths of a cubic foot.

Same goes for the old-fashioned meters with their clusters of dials, but they're trickier to read—as you've likely discovered. You must tackle them a dial at a time.

If your meter reads in cubic feet and you'd like to have the figure in gallons, multiply by 7.5 to find the result in gallons. Of course, you can change in the other direction by dividing by 7.5. If you want to be fussy, use the more precise 7.48, or even 7.4805194—so says my calculator.

One of these days, you won't have to bother about such calculations. You'll read your meters in liters, which has a nice lilt to it. There are 3.7853 of them in a gallon, by the way.

To use your meter in a way you may not have thought of, watch it during periods of specific water consumption. That is, keep an eye on it for one minute of lawn sprinkling. Multiply what you read by 60 and then by the number of hours you usually leave the sprinkler running. Do you find that lots of gallons go on your lawn and garden (or run down the street) each week?

You could do the same thing for a typical shower and for a tub filling. Also for a cycle of your washing machine.

You can use the meter for a leak test when nobody else is home and you aren't using any water and no automatic water-using device, such as a humidifier, the chickens' drinking fountain, or an ice maker is in operation. Observe or mark the meter reading and then do the same thing an hour later. If a change in reading indicates leakage or drip somewhere and you can't find it, suspect a toilet.

The good old w.c., as flush toilets are known almost worldwide (it's one of the few truly international terms, like O.K. and kaput) most often wastes water at the overflow. If this is set too high, gently bend the float arm so the valve shuts off the water half an inch below the top of the overflow pipe.

If that's not the trouble, you may have a plunger-ball leak, which is not so easy to detect. Try putting a little food coloring in the tank water. If it soon shows up in the bowl, you likely have a ball that needs replacing or putting back into alignment.

How Hard Is Your Water?

IF YOU HAVE a public water supply, the easiest way to find out how hard your water is may be just to phone the water company and ask.

You may discover, however, that this factor varies from time to time, or from one part of town to another.

If the water company can't give you a fairly precise figure, you'll have to get a water-testing kit or have someone test your water for you.

When it comes to judging hardness for purposes of choosing a detergent or determining the softening capacity your household needs, you may have to convert from one measuring standard to the other. The two standards are gpg and ppm. The first is grains of hardness per gallon and the second is parts per million. What is referred to is carbonate hardness—the usual problem with water being calcium carbonate or magnesium carbonate—or equivalent.

To convert grains per gallon to parts per million, multiply by 17.1. To convert the other way, you naturally divide the number of parts per million by 17.1.

So if you find, or are told, that your water is 8.6 grains hard, you can calculate that this means it contains about 147 parts of hardness per million. In each one million grams (1000 kilograms or liters) there are 147 grams of carbonates or equivalent.

Anything under 1 grain of carbonate per gallon is regarded as soft water. Anything over 3.5 grains is hard, though it takes several times that carbonate content to produce the kind of very hard water you sometimes encounter. This is the stuff that so totally refuses to make lather that you have the feeling it is throwing the cake of soap back at you.

To Measure a Chair

"Rule of 3" is what they call it in the furniture trade. It has a lot to do with seating comfort.

It seems the manufacturers start with the standard that a woman is 5 feet 4 inches tall and will be most comfortable on a chair with a seat level 16-1/2 inches from the floor.

However, for every 3 inches of height above 5 feet 4, the sitter—female or male—will welcome another half-inch of height in the chair seat.

Same rule goes for chair depth, the distance between the front edge of the chair and the back upholstery. The seat should be 21 inches deep for that 5-foot-4 woman. And it should increase half an inch for each additional 3 inches of human height.

Since so many chairs are selected, usually unconsciously, by women to conform to these standards, it is no wonder that the man of the house often finds them uncomfortable. He may fit better into a sofa, which probably was made and selected to inflict minimum discomfort on a maximum range of sitters. Thus a common height is 17-1/2 inches.

It probably also follows that if you expect to occupy a chair while barefoot or in slippers, you shouldn't shop for it while wearing heavy walking shoes or high heels.

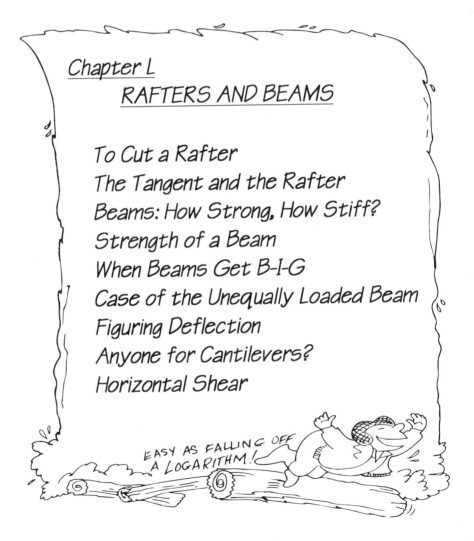

Chapter L
RAFTERS AND BEAMS

EASY AS FALLING OFF
A LOGARITHM!

To Cut a Rafter

W E BEGIN this excursion into trigonometry gently, in the company of those most familiar of old acquaintances named Math. That's right—the Pythagorean theorem and the square root.

You're building something—a cabin, let's say. It's to have a shed roof, meaning a roof that slopes just one way. The structure is 10 feet, 8 inches high on one side, 7 feet high on the other. Those two walls are 12 feet apart.

How long must you cut the rafters, not including any overhang?

Look at each rafter as hypotenuse of a right triangle, of which the sides are 12 feet and 3 feet, 8 inches. Only difficult part of such a problem is that feet-and-inches stuff. Probably the easiest out, until the metric system rescues us all, is turning everything into inches. This produces larger numbers—but with a calculator, who cares?

So the 12-foot horizontal distance becomes 144 inches, which squared is 20,736. That, added to the square of 44 inches (1936), makes 22,672 inches. Since that's the square of the hypotenuse, the next step is to derive its square root—the rafter length we're going for.

You no doubt recall, probably with loathing, that you were required in school to master—momentarily—a routine for finding a square root with paper and pencil. You can reasonably be assumed to having totally forgotten that algorithm, or recipe. So, in the absence of a pocket calculator equipped with a square-root button (or a key for roots and powers generally), it's time for a cut-and-fit procedure described in our calculator chapter.

Since the horizontal distance is 144 inches, you know the rafter will have to be just a few inches longer than that. As it happens, if you make your initial guess the logical 150 (smallest round number greater than 144) and then try 150.6 and 150.57, that third trial will produce an answer so near perfect there'll be no need for further guesses; the square of 150.57 is 22,671. Cut your rafter to 12 feet, 6 inches plus 0.57 of an inch.

That 0.57 looks like about 9/16, verifiable by dividing 9 by 16. So the final figure, outside to outside is 12 feet, 6-9/16 inches. Be sure to add the two overhangs when you make your cuts.

If you now want to figure out how to set your saw to make that cut so that it will be parallel to floor or wall, you'll need a trig table.

As they told us all in high school, the tangent of either of the other angles in a right triangle is what you get when you divide the length of the far side by the length of the near side—in this case 44 divided by 144. Look up the resulting 0.3055555 in the tan column of a trig table. You'll find it comes as near to 17 degrees as makes no difference. So that's where to set your radial saw—there or at 90 minus 17, depending on which rafter cut you're making.

The Tangent and the Rafter

T HERE I WAS, about to build a house for my family to live in. We're a sun-worshipping lot, right down to Ms Blue, our Siamese.

To pick up the sun from the south without getting a view of my all-too-near neighbor's house along with it, mine needed a bank of clerestory windows. These had to be in a wall jutting above the flat roof of a bedroom.

This gives the living room a shed roof, one that slopes in just one direction, rising 4 feet in a horizontal distance of 18 feet.

One little problem I had in building it was cutting off the rafters for this sloping part so that the rafter ends would be plumb. Which is to say vertical.

The answer to this kind of problem, and dozens of others in building, is in some simple trigonometry.

The level distance of this part of my house and the vertical one that holds the windows form a right triangle of the proportions 4 and 18. When you divide the far side of a triangle like this by the near side, you get a proportion, or ratio, called the tangent of the angle formed by the near side and the hypotenuse (or, in this case, the roof).

Everybody learns about tangents in school, and then nearly everybody forgets.

It follows that the tangent of this angle is 4 divided by 18, or 0.222. If you were to look through a table of tangents, you would find that the quantity in the table that came the closest to 0.222 is something like 0.221999, which the table declares to be the tangent of 12 degrees, 31 minutes, also known as 12-31/60 degrees. For the purpose of setting your saw you would think of that as 12-1/2 degrees.

(But sometimes in setting a tool, you're working not from zero but from 90 degrees. In that case, subtract the 12-1/2 from 90 and set the saw at 77-1/2 degrees.)

What do you do if you don't have a table of trig functions lying around? Well, you could wish you owned one of the pocket calculators that have trig buttons on them and sell for only a few dollars more than the simpler machines.

Or you could do a little arithmetic. I don't mean that there is an easy way to calculate trig functions. But there does exist an approximation method that will give you results that are far closer to exact than you'll ever need for cutting a rafter.

In fact there are several such methods, of which one is suited to finding the angle for which you know the number and it is less than 1.

(If the tangent of the angle you're working with is more than 1, shift to the other acute angle of the triangle. It doesn't make any difference to you which one you calculate, since you can then quickly find the other by subtracting from 90.)

For the following procedure I'm assuming the use of a simple electronic calculator, although naturally the steps will work if you follow them by paper-and-pencil. The parenthetical suggestions are to aid someone using a calculator.

Multiply the tangent by itself and then by 0.32383. Subtract from 1 (by changing the sign of that decimal to negative and adding 1). Multiply by the square of the tangent and then by 0.52678. Again, subtract from 1.

Multiply by the square of the tangent and then divide by 3. Once more subtract from 1.

Multiply by the tangent and by 57.29578.

That's it.

Just to be sure, though, let's run through it for that same triangle with its sides of 4 and 18 feet, giving tangent of 0.2222222 on the usual eight-digit calculator.

Squared, it's 0.0493827. Multiplied by 0.32383 and subtracted from 1, it's 0.9840085.

Multiplied by the tangent twice, this becomes 0.0485929. Times 0.52678, it becomes 0.1170622. Subtracting that from 1 gives 0.8829378.

When you multiply that by the tangent twice and then divide by 3, you get 0.0145339, which subtracted from 1 is 0.9854661.

Multiplying once more by tan and then by 57.29578, you get the answer 12.54734, or a trifle more than 12-1/2 degrees. Check.

This trig method is the quick and easy way to find this number—but not if it involves you in a search for a trig table. In some case, a carpenter's alternative may be a better idea. Put one rafter in place and hold a level against it while you pencil in your cutoff line. Set your saw to this angle by eye.

Beams: How Strong, How Stiff?

Strength of a beam refers to how heavy a load it will take before it gives way. For a given species and grade of lumber, this is affected by both width and thickness.

Thickness means the direction of the stress on the beam—the direction in which the load will try to bend it. Bearing capacity of a beam is in proportion to the square of the thickness.

At the same time, it is in direct proportion to the width.

That is, two two-by-fours on edge will have twice the strength of one of them. Naturally enough.

But one two-by-eight on edge will not be merely twice as strong as a two-by-four, even though it contains about twice as much wood. It will be four times as strong. And a two-by-twelve will bear about nine times the load that a two-by-four can carry.

Stiffness is a somewhat different characteristic from strength, even though it sounds about the same. Stiffness refers to how much a beam will bend when loaded, not with how much weight it will handle before giving way. As with strength, stiffness varies proportionately to the width, so two beams of a given size, placed side by side, are twice as strong as one.

But in the direction of the stress, the proportion is not as the square but as the cube. A two-by-eight on edge, merely twice as deep as a two-by-four, will have not twice nor four times, but eight times the stiffness.

Strength of a Beam

WILL TWO-BY-SIX rafters support that carport roof?

For floor joists in a cabin or a room addition—will two-by-eights serve, or must you go to bigger and more expensive lumber?

If there's a building code to be met, you can get the answer from a book of tables. But that's no help when the book is not around, or if there is no book, or if you are departing in any way from conventional construction—as in a post-and-beam design, for example.

In the course of building several houses to live in over the years, I've saved a lot of money in engineers' fees and a lot of nuisance as well, by being able to figure a beam for myself. I can consider many alternatives quickly in trying to arrive at the choice of a girder or arrangement of joists that will cost me the least in dollars and sweat.

Best of all, I can handle unconventional construction methods as well as the more usual ones—post-and-beam, cantilevers, and all that.

To figure the strength of a joist, beam, girder, or rafter, you must first assemble four figures; the actual width and depth of the beam, the distance the beam is to span, and the fiber stress in bending allowable for the species and grade of lumber you have in mind.

This fiber-stress factor is often referred to as "f." The table gives it for many kinds of lumber. Note that there are two columns of figures. The first is for a beam used by itself. "Repetitive" is for a series of three or more members, as would be the case with joists or rafters used in parallel and connected by floor or roof boards nailed to them.

The "f" value is expressed in pounds per square inch. So width and depth must be inches, too, and so must the length of span.

And keep in mind that depth means the direction in which the load will produce the stress—the longer of the two dimensions for lumber placed on edge, as it usually is. This is important, because strength increases as the square of the dimension that is in the direction of the stress.

Multiply the "f" value by the width and by the square of the depth. Divide by the span length in inches. This will give you a basic strength value that is sometimes called Z. It's a value that can be modified to fit all sorts of building situations. This makes it useful in calculating beams that are unequally loaded and also beams that are cantilevered.

For now, though, we'll stick to the condition that applies most of the time: a simple beam, freely supported, with evenly distributed loading. This describes the majority of house girders, beams, joists, and rafters.

SPECIES OR GROUP	GRADE	EXTREME FIBER STRESS IN BENDING	
		SINGLE	REPETITIVE
DOUGLAS FIR - LARCH	Select structural	1800	2050
	No. 1 / Appearance	1500	1750
	No. 2	1250	1450
	No. 3	725	850
DOUGLAS FIR SOUTH	Select structural	1700	1950
	No. 1 / Appearance	1450	1650
	No. 2	1200	1350
	No. 3	700	800
HEM - FIR	Select structural	1400	1650
	No. 1 / Appearance	1200	1400
	No. 2	1000	1150
	No. 3	575	675
MOUNTAIN HEMLOCK	Select structural	1500	1700
	No. 1 / Appearance	1250	1450
	No. 2	1050	1200
	No. 3	625	700
SUBALPINE FIR (White Woods) (Western Woods)	Select structural	1050	1200
	No. 1 / Appearance	900	1050
	No. 2	750	850
	No. 3	425	500
ENGELMANN SPRUCE (Engelmann Spruce - Lodgepole Pine)	Select structural	1150	1350
	No. 1 / Appearance	975	1150
	No. 2	800	925
	No. 3	475	550
LODGEPOLE PINE	Select structural	1300	1500
	No. 1 / Appearance	1100	1300
	No. 2	925	1050
	No. 3	525	625
PONDEROSA PINE - SUGAR PINE (Ponderosa Pine - Lodgepole Pine)	Select structural	1200	1400
	No. 1 / Appearance	1050	1200
	No. 2	850	975
	No. 3	500	575
WESTERN CEDARS	Select structural	1250	1450
	No. 1 / Appearance	1050	1200
	No. 2	875	1000
	No. 3	525	600
WESTERN HEMLOCK	Select structural	1550	1800
	No. 1 / Appearance	1350	1550
	No. 2	1100	1250
	No. 3	650	750

And the modification for such simple beams is: multiply by 4 and divide by 3. (Or multiply by 4/3 or by 1-1/3 if either equivalent method seems quicker.) Call this result W, for the weight the beam can safely carry.

Let's try a rather typical example. You're building a room with a wood floor. The joists must be designed to carry 40 pounds to the square foot. (If this is the "live load" requirement only, you will have to add 5 to 10 pounds to it for "dead load," which means the weight of the beams or joists themselves and the floor boards and coverings they hold up—or the roofing materials in the case of a roof calculation.) That's the usual code requirement and a good safe standard to meet for a solid structure even if you're not governed by a code. You wish to place the joists 24 inches apart, center to center. The distance to be spanned is 10 feet. So each joist will have to carry the weight of an area 2 feet wide and 10 feet long. Twenty square feet times 40 pounds gives a load of 800 pounds for each joist.

Question: Will No. 2 Douglas fir two-by-sixes be strong enough to handle the weight—800 pounds?

The table says that under repetitive conditions, which is what you have with a series of similar joists to which a floor is fastened, the "f" value of No. 2 Doug fir is 1450. Four times this, multiplied by the width (1.5 inches) and the square of the depth (5.5 times 5.5 inches) and then divided by 3 times the span, 120 inches, is 731 pounds.

That's almost strong enough, but not quite. If you were constructing a simple cabin somewhere outside code jurisdiction, you might decide to use the two-by-sixes anyway, especially if you didn't plan heavy furniture or big parties (are any of your friends big parties?) or had noted that 30 pounds has been called an adequate allowance for sleeping rooms in some codes and FHA specs. You would, however, select your lumber with more than usual care to avoid stock with major defects near span centers. More likely, though, you'd stick to your standards or be required to do so. You'd then recalculate either for a higher grade of lumber or a fatter dimension. Or move your joists closer together, so each would have less weight to carry.

Since No. 1 lumber shows "f" of 1750 instead of 1450, it can be considered 300/1450 stronger. The earlier figure of 731 (when multiplied by 300, divided by 1450 and added to 731) becomes 882—same result as you'll get if you calculate it directly.

So upgrading lumber quality will take care of the problem with more than 10 percent to spare.

Going to two-by-eight lumber, while sticking to the original No. 2 grade, should work too. You can recalculate from scratch—or make the change by multiplying 731 by the square of 7.25 (actual dimension of what is called 8-inch lumber) and dividing twice by 5.5. Either way you get 1270.

This is far more strength than you need. But if cost remains comparable you may well prefer to go this route. Recalling that stiffness increases as the cube of the dimension stressed, you calculate that the shift to 8-inch joists—grade remaining the same—considerably more than doubles the stiffness. In a floor, that much added stiffness is a worthwhile bonus.

(How do you make that stiffness calculation? The obvious way is to multiply 7.25 by itself twice to find its cube, then divide by the cube of 5.5. But the easier way, by pencil or calculator, is to divide 7.25 by 5.5 and then cube that result.)

When Beams Get B-I-G

W<small>HEN BEAMS</small> get big, you can call them girders.

Most of what you need to know to figure what size hunk of wood will hold up how much weight is covered in the adjacent item on strength of a beam, but the big ones call for a few special common-sense considerations.

Let's say you're building something that needs a long central support—a beam or girder to run right along the spine of the structure and hold up the joists that hold up the floor...that holds up you and the furniture, and so on. This beam is to sit on rough six-by-six posts spaced 6 feet apart, so you decide you'd prefer that central girder to be 6 inches across, for easy and secure fastening to the posts.

How deep must it be to hold its share of the weight of the floor and the contents thereof?

First of all, you have to find out what that share of weight is. If joists run from girder to foundation wall, then half their weight must be on the girder. If the structure is 24 feet wide, that means 12 feet for the girder to hold up in each direction. Each 6-foot section of girder—the distance between posts—is responsible for an area 6 feet by 24 feet, or 144 square feet.

(Would you use pencil or calculator to do this multiplication rapidly? Not if you're practicing your mental arithmetic. Simplest trick here is to halve one number and double the other, so you need only say 12 times 12 is 144.)

If the floor is for a house, you'll probably want to figure on a loading of 40 pounds to the square foot. That's what building codes usually say. For a roof, it may be much less—perhaps 20 pounds unless there's a heavy snow load to think about. For a garage floor, it may be 100 pounds.

But let's say it's for a room of a house. We multiply the square footage of 144 by the specified 40 pounds to find that each 6-foot span of girder has 5,760 pounds to carry. Not a loading to fool around with.

Will a Western cedar six-by-six, No.3 grade, rough, do it? The strength-of-a-beam table says the fiber strength (the "f" value) of a single beam of the kind of cedar we've specified is 525. That, multiplied by the width of the beam and multiplied twice by its depth (6 inches in both cases since we're dealing with rough lumber where real dimensions approximate the nominal ones), divided by the span, and finally multiplied by 4/3, gives 2,100 pounds.

That's not even half what we need.

What about a six-by-eight on edge? Multiplying the previous result by 8 times 8 and dividing it by 6 times 6 produces 3,711. (If you don't trust that approach, do the new calculation from the beginning.) Nope. Still not enough strength.

If we're determined to stick to No.3 rough cedar, we can only move on to something bigger—say a six-by-ten. That brings us to 5,833 pounds. Just does it.

Why not consider something like a three-by-fourteen instead? Using just over two-thirds as much lumber, it would still give almost as much strength. In general, it's a good idea to make supporting members comparatively narrow and deep, for economy. But for a girder like this, stability is important, too, and the squat shape gives that while still letting it sit nicely upright. And it gives a good surface for attaching the joists.

Case of the Unequally Loaded Beam

Uɴᴅᴇʀ "Strength of a Beam"—which may very well be the most useful single building item in this book—I promised to tell how to modify the calculation to fit an unusual case.

This is the case of the unequally loaded beam. It refers to the timber or girder or joist or rafter that must carry a lot more weight at some known point or points than at others.

There are several different ways the weight may be arranged and still fit some standard methods of calculating. With any of these, you begin by finding what is called Z in the more general strength article. In that article, you were told to multiply Z by 1-1/3 to find how much evenly distributed weight could be safely supported by a simple beam.

For other arrangements, you multiply Z by a different quantity to find W, the weight the beam can carry.

If the ends of the simple beam are firmly held—not merely nailed or bolted in place—and the load is evenly distributed, multiply by 2 instead of 1-1/3. For a freely supported simple beam with one concentrated loading at the center of the span, multiply Z by 2 and divide by 3 to find W.

If the ends of the beam are firmly held in place and the loading is concentrated at the center of the span, use 2/3 as the multiplier; that is, multiply by 2 and divide by 3.

When loading on a freely supported simple beam is at its greatest near the ends of the span and decreases to zero at the middle, just multiply Z by 2 to find W.

With the opposite condition—heaviest load in the middle, dropping to nothing at the ends—don't multiply at all. Z and W are equal.

Figuring Deflection

IT'S NOT always enough to know that beams and joists and rafters are strong enough to do their job safely. Sometimes you want to limit deflection—bending—to some minimum. The purpose may be to keep a plaster ceiling from cracking. Or you may be concerned with meeting the terms of a building code that limits deflection of the wood member in question to 1/360 of the span.

Well, anyway, you figure deflection by first multiplying the width of the rafter, beam, or joist by the cube of its depth. Then you divide by the square of its length. Be sure to use the same unit of measurement (inches, for example) for the length that you use for the width and depth.

(Keep in mind that width means from side to side of the beam as it will be placed, a nominal 2 inches in the case of a two-by-four used on edge, as it most often will be. And depth, then, is the nominal 4-inch dimension that is the direction of the stress in carrying the load.)

Now you need only multiply the result by a certain quantity to find out how many pounds of loading the timber will bear before it will deflect as much as 1/360 of the length of the span.

For the most usual situation—a simple beam, freely supported, with evenly distributed loading—the quantity is 31,290. For a simple two-by-eight joist spanning 10 feet, you would first multiply the width (1-1/2 inches in actuality) by the cube of the 7-1/4-inch depth, to get 572. Dividing by the square of 10 feet (120 inches squared is

14,400) gives 0.0397. Multiply this by the factor of 31,290 to find that the joist will hold 1243 pounds without excess bending.

If the joist is to hold up a floor that can carry 40 pounds to the square foot (a usual code requirement for living rooms in houses) and the joists are to be placed 2 feet center to center, the design load is 2 times 10 times 40, or 800 pounds. Since the joist will handle 1234 pounds stiffly enough, there's no deflection problem. You have a margin of rather more than 50 percent.

If your simple beam with evenly distributed loading is firmly fixed at both ends, the 31,290 multiplier does not apply. You can use the much larger 156,440, but don't without qualified engineering assurance that this is indeed the case; "firmly fixed" means something much stronger, and rarer, than mere bolting or nailing. When in doubt, use the smaller multiplier.

Here are multipliers for other situations:

For a simple beam, freely supported, with one concentrated loading at center of span: 19,555. (But if it is firmly fixed at bearing points, use 78,222.)

Simple beam freely supported, with loading that decreases uniformly to zero at center of span: 43,460. (But if load decreases uniformly from maximum at center to zero at both bearings, use 24,440.)

Simple beam freely supported, with loading that decreases uniformly from a maximum at one end of the span to zero at the other: 31,100.

Cantilevers deflect much more than ordinary beams and rafters that are supported at both ends. So the factors by which you multiply to calculate loads for deflection are smaller.

For a cantilever beam with evenly distributed loading, multiply by 3258.

For a cantilever with concentrated loading near the free end, multiply by 1222. If loading is at a maximum at the bearing point and decreases uniformly to zero at the free end, use 3258.

Anyone for Cantilevers?

A CANTILEVER is elegant and useful. An upstairs deck that extends out from a house, with no posts underneath it, is an example.

A cantilever takes a little engineering, however, as I found out many years ago when I wanted to build a house with my bedroom jutting out over a driveway.

"How," asked the building inspector who had to pass on the plans, "do you know those joists that stick out there are strong enough?"

It was a good question. I didn't really know. So I spent a lot of time looking into the matter, finding out how the calculation could be made, and finally doing it to the satisfaction of the inspector. I am glad to say that the room is still there, jutting out as firmly as ever, and with no objectionable bounce when some heavyweight jumps up and down on the floor.

There are three important procedures for calculating the load-bearing capacity of a cantilevered beam. The one to choose depends upon the way the load on the cantilever will be distributed.

All begin with the basic load-bearing figure that is called Z in the article on strength of a beam. When you've found Z, as described there, you must divide by 3, 2, or 6 to find out how much weight your cantilever can be trusted to support.

If the load on the cantilever will be evenly distributed along the whole of the span, you must divide Z by 3 to find the allowable bending load.

If the load is at its maximum at the point of support of the cantilever, and decreases uniformly until it is zero at the free end, the stress will naturally be somewhat less. You divide by 2.

But when the whole load is concentrated near the free end, the downward push created by this load becomes much greater. You must divide that Z figure by 6.

Horizontal Shear

IN SOME situations, the loading on a wood beam used in construction will cause it to come apart in the middle—parallel to the grain.

This can happen even when the beam is strong enough to carry its load without breaking in two. But it is likely only when the depth of the beam is one-tenth or more of the span for which it is used, and the beam is heavily loaded.

If you have a situation that seems to meet this description, you should use a shear calculation to check on the loading. You could do this after you've used the method described earlier in this section on figuring the strength (W value) of a beam.

What you want to discover now is whether a shear will be produced that is more than 120 pounds to the square inch. This 120 pounds is the amount of sheer stress considered safe when you're using the common grades of most kinds of lumber.

For any simple beam or joist, having found the W value, multiply by 3 and divide by 4 and then by the width and by the depth of the beam.

To put it another way, the intensity of the shearing stress is 3 times W, divided by the product of the width of the lumber, times its depth, times 4.

If the result you find is more than 120, you should change to a beam of different dimension or otherwise modify your design.

The calculation as described applies to all simple beams with symmetrical loads. It goes for beams with evenly distributed loading (the usual thing in house construction) whether the beam is freely supported, merely nailed in place, or firmly fastened at its ends. It works for a beam with its loading concentrated at its center or one with most of the load at the ends.

About the only beams it doesn't apply to are those that have off-center loading or are cantilevered.

Where the load decreases from a maximum at one end to nothing at the other, shear value is found by dividing W by the product of the width and depth of the beam.

Where there is a single load concentrated neither at an end nor at the center of the span, the formula becomes just a little more complicated. Here you first multiply W by 3 times the distance from the point of loading to the more distant point of support. Then divide this by the product of the depth of the beam times its width times twice its length. And remember to keep all dimensions in inches.

For all cantilevers, whether the load is concentrated at the free end or the fixed end or is evenly distributed, the calculation is the same. You multiply W by 3 and then divide by twice the product of the width and the depth of the cantilevered beam.

Chapter M
AROUND THE HOUSE

How Many Degree Days?

D EGREE-DAY figures, so useful when estimating heating bills or insulation needs, are usually to be had from fuel dealers or government or utility-company sources. They were devised in the nineteen-thirties by heating-industry people who required a number as nearly proportional to fuel use as possible for comparing areas and periods.

Cumulatively, at any time during the season, they represent the load for the period to date.

If you need degree-day totals for periods or localities for which they're not available, you can calculate your own. Record maximum and minimum temperature each day and subtract the average of the two numbers from 65.

Like this. In a 24-hour period when day temperature reached 73 degrees and night dropped as low as 45, the average found by adding the temperatures together and dividing by 2 is 59. Subtracted from 65, that's 6 degree days.

All it takes is patience.

Sizing a Wood Stove

A WOOD STOVE that's too large will be difficult to manage. It will produce too much heat or will have to be operated at such a low temperature that a potentially dangerous amount of creosote will be deposited in the chimney.

A stove that is too small will have to be fed too frequently or used at an excessively high temperature if it is to provide an acceptable level of heat.

Those are among the reasons you may find a use for this rough formula for determining ideal stove size.

Begin with an estimate of heat load (or heat loss) for the house.

If you don't have a source from which you can get a more precise figure, use the following to get a good working approximation.

Multiply the number of square feet of floor area by 30. This assumes a house with an average amount of insulation and window area. For a well-insulated building reduce the factor to 25, and for a poorly insulated one increase it to 35.

Your answer will be in Btus (British thermal units) per hour. Multiplying it by 0.172 will tell you the approximate firebox volume, in cubic inches, that your wood stove should offer.

Cord of Wood, Ton of Coal

Though naturally this figure varies considerably with species and condition, burning a cord of wood produces about 26 million Btus of energy.

So does burning about one and one-third tons of coal.

A lot of this energy is wasted, of course. A typical airtight stove might have an efficiency of around 50 percent.

So in calculating the net heat produced, the figure you'd use would be more like 13 million Btus.

Wood or Coal?

Hardwood or soft coal—which is the better buy for burning?

To calculate this you must first know the heating values of the two fuels and the cost of each.

A pound of bituminous coal has a heating value of about 11,000 British thermal units, so a ton has 2000 times this much, or 22 million Btus.

A typical hardwood is good for 27 million British thermal units to the cord.

So dividing the price of a ton of coal by 22 will give you your cost per million Btus.

And dividing the price of a cord of hardwood by 27 will give its cost for each million Btus.

If you pay $170 for a cord of oak you'll get a million Btus for each $6.30.

Coal at $90 a ton will cost you only $4.09 per million Btus.

Of course the true difference in cost will also have to reflect the relative efficiency with which your stove or furnace will burn the competing fuels.

If your modern wood stove can burn wood with 60 percent efficiency, you'll be paying $10.50 for each million Btus you actually benefit by ($6.50 divided by 0.60).

And if your ancient coal furnace has an efficiency of only 45 percent (ask a furnace dealer for his best guess on this), you must divide that figure of $4.09 by 0.45 to discover an effective cost of $9.09 for each million Btus obtained from coal and actually used.

Comparing Furnace Efficiencies

IF YOUR present furnace has an efficiency rating (known or estimated, perhaps with help from a dealer or your utility company) of 55 percent and you contemplate buying one with a rating of 80 percent, divide the old figure by the new one.

Since 55 divided by 80 is 0.69 , multiply your present heating bill by this decimal. The result is how much your bill should be with the more efficient new furnace. If you've been spending $500 a season, you should be as comfortable from here on out with a cozier $344.

What Does Heat Cost?

THESE rough formulas can give you an estimate of the relative costs of different fuels for home heating. The answers are in British thermal units per dollar. (Your best sources for the various base figures you'll need usually are public utilities and dealers in heating devices.)

For a *gas furnace*, multiply the efficiency in percent by 100,000 and divide by the average cost in cents per therm.

An example: If gas costs you 60 cents per therm (equal to 100,000 of those ubiquitous Btus, remember?) and your furnace offers the fairly typical efficiency of 65 percent, you'll multiply 65 by 100,000 and divide by 60 to find you can expect 108,333 Btus of heat for each dollar spent.

For *electric heat,* multiply 3413 by the seasonal efficiency of the heater and divide by the cost in cents per kilowatt hour.

Efficiency of most electrical heating devices is almost a thrifty 100 percent. However, a radiant-heating device that warms you directly, rather than indirectly by warming the room, may give you more for your money. A heat pump may even offer 150 percent. If you are working with a heat pump of this efficiency where electricity averages 10 cents, you'll multiply 3413 by 150 and divide by 10 to get 204,780 Btus to the buck.

If a *heat pump* with more than 100 percent efficiency impresses you as being about as likely as perpetual motion, take note that this device does not merely turn electrical energy into heat. It extracts heat from well water or the surrounding air to make you comfy.

For *wood,* multiply the Btus per cord for the type of wood you're using by the percentage of efficiency of the stove or other heating device, and divide by the price you pay for a cord. Unless your wood is practically bone dry, deduct an additional 15 percent or more for the heat lost in evaporating the water content.

The number of million Btus you can expect from a cord of fully dried wood varies from 13 for pine, 17 for elm, 20 for ash or maple, 22 for oak or birch to 25 for hickory.

A stove with a thermostat should give around 40 percent efficiency, a reasonable approximation to use if you don't have a specific figure for the stove in question. A wood furnace should do even better—say 50 to 70 percent. Therefore, if you burn air-dried (not completely dry) hickory at $125 a cord in a good stove, multiply 25 million by 40 and by 0.85 (as a way of deducting 15 percent moisture content.) And then divide by the cord price in cents.

If those figures—including what in areas like yours may be a wildly optimistic guess for the price of wood if you don't cut it yourself—fit your situation, you're getting about 68,000 Btus for your wood dollar.

How Much Heat Loss?

YOU CAN calculate the total heat loss through a wall, floor, or ceiling of a house for given temperature conditions and heating season. Follow up by feeding in the cost of the energy (from a fuel dealer or utility) and you can find what the heat loss is costing you in dollars per year.

This kind of calculation is particularly useful in estimating how much you can save each year by adding insulation. The formula works equally well for comparing the money saved by one kind or amount of insulation with another.

Here is a list of some common building materials, with the thermal conductivity of each in millions of British thermal units per second per foot per degree Fahrenheit. (You don't really need to absorb that definition, however. You can simply use the number where called for in the calculation.)

A second number you'll need is the difference between the temperature to be maintained inside the wall or ceiling and the temperature outside it. If, for example, you assume 75 degrees inside and 5 below outside, the figure to use will be 80.

glass fiber	6.11
balsa wood	7.41
polyurethane foam	3.94
cellulose fiberboard	6.67
expanded vermiculite	10.40
asbestos	12.90
expanded perlite	8.56
plaster	129.00
window glass	142.00
plaster on wood lath	57.90
cork board	6.94
cored concrete block	30.60
wood	63.90
common brick	16.00

Having obtained the cost figure for fuel from dealer or utility, divide it by the estimated efficiency percentage of your furnace or stove. If you've been given the cost in therms, multiply it by 10 to turn it into dollars per million British thermal units.

To apply this, assume your cost for natural gas is 30 cents per therm (or $3 for a million Btus) and the efficiency of your furnace is 75 percent. Dividing $3 by 0.75 gives the figure of $4 for each million Btus actually benefiting you.

With this information, you can make a sample calculation that will demonstrate how to apply the procedure to any situation you may have. Assume a 5-month heating season with an 80-degree difference between the inside temperature you plan to maintain at ceiling level and the outside one. The ceiling is 40 feet by 50 feet and consists of 3/4-inch plaster on wood lath. It is not insulated.

1. Express thickness of wall, ceiling, or floor in feet. (Divide 3/4 inch by 12 to find a thickness of 0.0625 feet.)

2. Use the table to find the conductivity of the material, and divide this by its thickness. (Dividing 57.90 by 0.0625 gives 926.)

3. Multiply this by area in thousands of square feet and by temperature difference (926 times 2 times 80 is 148,160).

4. Find the number of seconds in 3 months and divide by 1,000,000 (60 times 60 times 24 times 90 divided by one million is 7.776). Multiply the previous result by this and divide by 1000, getting as a rounded-off result 1150. That's the annual energy consumption in millions of Btus.

5. Multiply by your cost per million ($4) to find that about $4600 goes down the drain—or out the wall or ceiling in question.

To compare this heat loss and expense with the result if 4 inches of glass-fiber insulation were added, do the calculation again with a different Step 2. This time divide the *thickness* of each material by its *conductivity*, add the two results together, and find the reciprocal. (If you've forgotten, you find the reciprocal of a figure by dividing it into 1, or by pressing the 1/x key found on many calculators.)

As before, the thickness and conductivity figures for the lath-and-plaster are 0.0625 and 57.9. For 4 inches, or 0.33 feet, of glass fiber, they are 0.33 and 6.11. Dividing 1 by the sum of these quotients gives approximately 18.

From here you could go on with the calculation as before. But if you've already done it for the first material, you need only divide the previous Step 2 result by the new one to find how many times more the plain ceiling is costing than the insulated one. (Dividing 926 by 18 gives about 50. That $4600 figure found earlier, and divided by 50 is $92, which should be the cost of the winter's heat loss through the new ceiling. The difference of more than $4500 is the remarkable annual saving in gas bills produced by the insulation.)

Costing a Heat Pump

Here's a calculation that will mean something to you only if you should get into comparing the cost of heating with a heat pump with that of a gas furnace.

A useful rule of thumb says that if the price for one therm of gas is less than that for 8 kilowatts of electricity, then gas used in a heating device offering 70 percent efficiency is cheaper than power for a heat pump having a COP of 2.5. That COP is an acronym used in the field of thermodynamics for coefficient of performance.

Heat Storage, Heat Loss

This calculation has to do with how much heat can be stored and released by bulk materials.

This is an important element in comfort and energy saving in any house. It is especially significant for a house that has been properly designed to take in a lot of free heat on sunny days. In fact, the first tax credit offered to encourage solar design made it mandatory.

How much heat can a brick or concrete floor tuck away on a sunny winter day to help keep the whole house warm overnight?

To find this you need to know first of all the heat capacity per cubic foot for the material.

In British thermal units per degree of temperature change, it is 28 for concrete or brick, the materials for which you may be most interested in having a figure.

It is 62.5 for water, a most efficient heat-storage medium though not always an easy one to work with since, unlike masonry, it doesn't store itself.

Let's suppose you have a room sitting on an 8-inch concrete slab 20 by 40 feet. You find that it reaches an average temperature of 75 degrees from sun shining on it through a couple of hundred square feet of windows.

Find the cubic footage of the slab (20 times 40 times 8 divided by 12 equals 533) and multiply by the figure of 28 for concrete. The result of about 15,000 is the number of British thermal units of heat that the slab stores with each rise of 1 degree in its temperature.

And of course that's also how many it will give back later in the process of dropping 1 degree Fahrenheit.

If the temperature of the slab goes down to 67 overnight, a drop of 8 degrees, it will supply 8 times the previous per-degree figure, or 120,000 Btus during the night. Is this enough to make up for what is lost to the outdoors during those sunless hours?

To figure the heat loss, multiply the number of hours by the loss per hour by the difference between the indoor and outdoor temperatures during the night.

You might take 15 as the number of hours between one solar heating day and the next.

If it's 65 degrees indoors, 25 outdoors, difference is 40.

Loss per hour? You can work from an approximation. For a small, tightly constructed, and very well insulated house you might use 200 Btus per hour per degree of temperature difference.

(For comparison, I recently estimated heat loss for a generously windowed 2,200-square-foot house of average construction and—by standards of these energy-conscious times—minimum insulation as just under 800. My justification for going ahead and building to this standard is that I'm in a mild California climate where Minnesota-level insulation isn't warranted. Also, much of what otherwise could be deemed excessive heat loss will be balanced by the large solar intake from the very windows producing much of that loss.)

So there are three numbers to multiply together—15 hours, 40-degree temperature difference, and loss per hour of 200—and their product is 120,000.

This being the same figure produced by the previous bit of arithmetic, we have described circumstances in which heat gain and heat loss are neatly balanced.

Trying a Cooler for Size

AN AIR CONDITIONER that is too small won't do its job adequately when it's needed most. One that's too big will cost more than necessary and will complete its cooling job too rapidly to dehumidify the air properly.

The proper thing to do when choosing such a device, you're told, is to have it sized scientifically by means of detailed calculations. But what with various imponderables and personal choices not allowed for by the formulas, you may come out as well with a quick estimate.

Make it this way.

First figure the volume of the room in the standard way—length times width times height, all these dimensions in feet.

If the windows are few and the attic overhead is insulated (or if there's a room above) multiply that cubic-foot figure by 10. With many windows and not much insulation, use 20 instead.

For an in-between situation, choose an in-between number.

Now multiply the result you have by 16 if the longest wall faces north, by 17 if it faces east, by 18 if it faces south, or by 19 if it faces west.

Divide by 60 to find the number of Btus (British thermal units) that will be needed to cool the space.

Meeting Insulation Requirements

On adequate insulation, people's ideas and the building codes have both been changing lately—in the direction of more.

Code minimums shown in the drawing are adequate for mild climates, but thickness should be increased if possible where climate is extreme or heating or cooling especially costly.

INSTALL BARRIERS WHERE FREEZE VENTS OCCUR

5½" FIBERGLASS BATT INSULATION OR EQUAL (R-19) VAPOR BARRIER TOWARD HEATED WALL (ROOM SIDE)

3½" FULL THICK FIBERGLASS BATT OR EQUAL (R-11) VAPOR BARRIER TOWARD HEATED WALL (ROOM SIDE)

RECOMMENDED (NOT REQ'D) FLOOR INSULATION 4" FIBERGLASS BATT

BAFFLE FOR AIRFLOW OF BASEMENT VENTS

Figuring Areas for Insulation

A shortcut calculation when buying insulation is to multiply the gross area by 90 percent, if framing is on 16-inch centers. Use 94 percent if framing members are on 24-inch centers.

To put it another way, knock off 10 percent for the closer spacing, 6 percent for the 24-inch.

This formula applies to any ceiling, wall, or floor area.

So: with 1,000 square feet of ceiling to insulate, joists spaced 16 inches, you'd buy 900 square feet of insulation.

How Much Insulation?

In general terms, it's not hard to find out how much insulation your house should have—a lot! Magazines dealing with home ownership tell you this regularly, and a phone call to your utility company in our energy-conscious era will also evoke a figure.

Instead of the old standard of R9, you're now likely to be told to try for R19. As heating and cooling costs escalate further, you're going to hear more about that as a minimum, with something up towards R-25 to R-30 for ceilings. Not that such a mass of insulation pays for itself in most climates with most fuels—just that it soon will do so. And we might as well start getting the benefits now. After all, insulation is not going to wear out.

What does all this R-stuff mean? R-value is the figure you get when you add up all the components of a section of wall or ceiling or floor or roof.

To figure the heat resistance of any part of your structure you need only find the R-value of each layer—and add 'em all together.

This directly proportional cumulative effect simplifies calculations—but it can be deceptive, too. Although when you double the thickness of insulation you double the insulating value, you by no means double the savings. Look at it this way. If the first layer of insulation cuts heat loss 60 percent, there's no way the second similar layer can shave off more than 40 percent of the original loss, since that's all there is left to work on.

What it will do, of course, is reduce the loss by 60 percent of that remaining 40 percent, for an actual figure of 24 percent. Each increment of insulation is bound to save you fuel money, but it is also bound to save you less than the preceding one. The worse the wall, the greater the result!

Here are some typical R-values for the structural parts of an ordinary frame wall:

Film of air on outside of wall	0.2
Film of air on inside wall surface	0.7
Wood siding or wood sheathing	0.8
Half-inch-thick insulating sheathing	1.0
Half-inch-thick gypsum wallboard	0.5
Airspace within the wall	1.0

A typical combination of these, forming an uninsulated frame wall, would use one layer of each. Add those six values together and you get a total of R4.2, which is not very good. So consider the following table.

Fiberglass rolls or batts	3.5
Mineral wool rolls or batts	3.0
White granule plastic insulation board	4.0
Blue closed-cell insulation board	5.5
Polyurethane foam or board	6.5
Vermiculite or perlite loose fill	2.5
Most types of cellulose fibers	4.5
Urea-formaldehyde spray foam	5.5

A typical two-by-four stud wall has 3-1/2 inches of air space.

As you can see, substituting 3-1/2 times one of those figures (since they are for each inch of thickness) for the mere 1 produced by an empty airspace will give your wall a far more respectable resistance rating.

To make the actual calculation, first multiply the R value (3.5 for fiberglass, for example) by the thickness in inches (presumably 3-1/2 again, since that's the thickness of the cavity you are filling) and add the figures for the other components. The total you'll get is 15.45. (If you come up with 16.45 instead you probably forgot to subtract or omit the value of the airspace, which you've now lost by stuffing it full of fiberglass.)

If 15.45 isn't good enough, and in many climates it truly is not, you may consider going to a more efficient insulation—comparing costs and not overlooking flammability. In new construction, you might go to a wall framed with two-by-six instead of two-by-four lumber. By simply making room for these two additional inches of fiberglass, you can add a potent 7 to the R value at relatively small cost.

Another excellent possibility in new construction, and sometimes in a major remodeling, is adding a layer of additional insulation on the outside of the wall. Since this layer of insulation board covers not only the wall cavities but the framing lumber (which can transmit significant heat) as well, it is especially worth while.

Shading a House

AT MIDDAY in midsummer in mid-America (40 degrees north latitude) the sun hurls an average of 230 Btus of heat against each square foot of a south-facing wall or window.

In the instance of the window you can estimate that 80 percent of this comes through the glass and into the house to increase discomfort or add to the air-conditioning load.

Figure that fiberglass solar screening (which is not the same thing as plain old-fashioned insect screen) installed one inch outside the window will reduce this intake by about 80 percent, to 185 Btus.

To Ventilate a House

FOR COMFORTABLE summer cooling by natural ventilation, you might choose as your goal one change of air every three minutes.

This will call for a rather generous collection of openings—windows, doors, louvers, vents.

How generous? There is a formidable formula for finding out, but happily it can be simplified to the following routine.

First, determine the wind-direction factor. For breezes that customarily blow directly into the opening, this is 0.5.

For diagonal winds, it is 0.3.

From these numbers, you can estimate factors for other angles—0.4 if the wind on a south wall is from south-southwest, for example, or 0.1 if it comes along almost parallel to the wall containing the opening.

Multiply this factor by the wind speed in feet per minute. Lacking an official figure for wind speed, estimate it in the more familiar form of miles per hour, as you hear it given in weather bulletins. Multiply by 88 to convert it to feet per minute.

Now multiply by the number of minutes you're allowing for each air change, perhaps 3.

CUBIC FOOTAGE OF HOUSE
÷ BY TOTAL OF
WIND SPEED/FT PER MINUTE
X MINUTES FOR
AIR CHANGE =

Find the cubic footage of your house (length times width times 8 for each story, unless the height is unusual). Divide by the previous figure. This is the total square feet of openings needed.

Here's a typical calculation, for a one-story building 25 by 40 feet with 8-foot ceilings, for a volume of 8000 cubic feet. Ventilation is through south windows, by prevailing breezes from south-southeast averaging 2 miles per hour.

Multiplying this 2 by 88, by the direction factor of 0.4, and by 3 gives 211. Dividing this into 8000 gives almost 38. So that is how many square feet of windows and other openings you'll need in that wall to ventilate by breezes alone—helped, naturally, by outlets in other walls.

If those outlets are high and the intake openings are low, you can get another physical principle working for you—again free of charge. Temperature difference between hot air near the ceiling and cooler stuff in what should be a shaded spot just outside the ventilating openings will move air even with little breeze.

To figure what this will produce, begin with the height difference in feet between centers of your low air inlets and high outlets. Multiply by the temperature difference between the air just outside and the hot air under the ceiling. Find the square root of this product, and multiply that by 9.4 and by the number of minutes in which you wish to make a complete change of air.

If this is 3 again and there's a height difference of 7 feet and a temperature difference of 10 degrees, you should get a figure of about 236. Dividing a house cubage of 8000 by this gives 34 square feet that the inlet openings should add up to.

Perhaps yours is a house that has some of both things going for it. Then you should get by nicely with openings less than either calculation alone would call for.

If 38 square feet will do the job by breeze alone and 34 by temperature differential, 25 should suffice when both are working for you.

What's a Square Foot of Ventilation?

BUILDING CODES and manufacturers' instructions often warn you to allow a specified area of opening to ventilate a crawl space under a house, or an attic, or to provide combustion air for a furnace.

A typical prescription for crawl space is 1 square foot of opening for each 25 linear feet of wall.

So a house that is 32 by 40 feet, with 144 feet of perimeter, will need openings totaling nearly 6 square feet. But there's a catch to this you'd better take account of: such openings usually are screened or louvered to keep creatures out.

A rough rule is that louvered openings count only half. So double the area of ventilating openings. For our example that means 12 square feet.

If an opening is both louvered and screened (with 16-mesh screen) multiply by 3 instead of 2.

Better—since screen may be clogged by paint, cobwebs, and dirt—is to use 1/2-inch mesh instead. Then you don't have to allow anything extra.

How Much Can the Sun Save You?

THE SUN'S heat is free—but it costs money to use it. That's the dilemma if you're contemplating—as you probably should be—solar heating for the hot water for your house.

Before putting in a solar device, you naturally want to know if it will pay its way, and how many years it will take to do so.

One installer of such things has worked out this calculation:

A typical family in Hawaii has an electric bill of about $200 every two months. Let's say this family uses 2349 kilowatt hours per billing period, or 39 kwh per day, of which 50 percent are estimated to be for hot water: 19.5 kwh times 365 days times 7.76 cents per kwh equals an estimated savings of $552.30 per year, $46.02 per month. A system for this typical family costs $4600, installed. The payback period is between eight and one-third and nine years—or so the installer we're quoting concludes.

But is it?

(Note first that since a couple of his base figures are necessarily approximations, he has couched his conclusion similarly. That's very much to his credit. Especially in making sales pitches, savings estimates are often presented as though they could possibly be known to the last cent.)

But that's not where the trouble lies. As the contest puzzles in magazines used to say, what's wrong with this picture?

Nothing's wrong with the method up to the final four words.

That's why I've quoted it as a useful example for anyone giving sober consideration to solar water heating or to other devices having a long but not unlimited life.

But—the payoff period here is a faulty figure. True, if you divide $2300 by $276.15, you will indeed get 8-1/3 for the number of years. But will the maker of the equipment install it on your roof and let you pay for it over that period—without interest? Not likely.

So, to find a realistic answer, you need a more elaborate calculation, ready-made or made by you.

For the ready-made answer, consult any time-payment table for mortgages and installment loans, to be found in the top drawer of any banker or money-lender's desk.

Assume any reasonable rate of interest. Say 8 percent. That's about as much as you've been able to get for your money in a guaranteed deposit in recent times and is somewhat less than you could borrow at if you wished to finance a home improvement.

Turn $276.15 a year into monthly payments of $23.01.

Look up $2000 and $300 in the book and you'll find $19.83 and $2.98, respectively, as the monthly payments for a 14-year loan. So (doing a little adding) $22.81 a month will pay off $2,300 in 14 years. That's close enough to $23.01 for any reasonable purpose.

So it seems the true payback period is 14 years—close to double what the man said.

If Hawaii seems remote, here's a somewhat different calculation, from a solar-hot-water man in New York:

"A family of four using 80 gallons of hot water per day will require about 20 kwh to heat the water electrically. At Lilco (Long Island Lighting Company) rates (five cents per kwh then) it costs that family $1 per day for hot water—$365 a year, not including fuel-cost adjustments, and assuming a constant utility rate."

This time, the system would cost $1600 and provide 80 percent of the family's hot water. The payback period is given as between five and six years.

Again, this is a method you may find useful. But, also again, don't accept the figure given in the example for payback period and don't estimate your own by dividing the cost of the equipment by the annual savings, as obviously was done. Instead of five to six years it would really take more than seven.

Capacity of a Solar Water Heater

THE FIGURE to start from, in sizing tanks and collectors for solar water heating, is daily consumption.

Per person this can be as high as 50 gallons, or so the utility companies used to like to tell us. In grimmer times it perhaps had better be more like 5 or 10 gallons except in households making prudent use of the free heat from the sun.

With solar assistance a good figure to work from is 20 gallons of hot water for each person each day. Unless there will be an auxiliary tank in the system, double that figure. Multiply by 1.1 to allow for heat loss. If you will have a heat exchanger, again multiply by 1.1 to allow for its losses.

This will tell you the approximate size of tank to obtain.

This result will also help you estimate the area of solar collectors you will need.

To find this in square feet, divide the tank size (which you have found in gallons) by 2 if you live in the south or southwest. This assumes you have just one layer of glass or plastic over the heat-absorbing element in your collector.

If you live in Canada or the northern U.S., multiply the tank size by 1.5 and also use two transparent layers.

If you live somewhere between, use either one or two transparent layers and make the collector size in square feet equal to the tank capacity in gallons.

What Will a Circuit Take?

To FIND OUT what you can load onto a single household lighting or appliance circuit, first examine the fuse box or breaker panel. You'll probably find the circuit is fused for either 15 or 20 amperes, this capacity being marked on the fuse or on the breaker handle.

Multiply this by the voltage, usually 110 to 120, that your power people deliver. If you don't know this figure, you can call the electric company and ask for it. If they're not sure what it is at your house, they'll come out and measure it for you. Or you can just assume 115 and be close enough for most purposes.

At 115 volts, a 20-amp circuit will give you 2,300 watts to work with.

Name plates on appliances usually tell you the wattages they require. If you have a coffee maker rated at 600 watts and a small electric heater rated at 1,000, already in use on such a circuit as just described, there goes 1,600 watts. With only 700 watts of capacity remaining, you know what will happen if you plug in a 1000-watt toaster as well. The fuse will blow or the breaker will trip.

Heavy appliances, such as permanent heating devices, ranges, ovens, or large air conditioners, commonly require more voltage. One of these will have a 220- to 240-volt circuit to itself. The calculation is made in the same manner, so a 20-amp heater wired for 230 volts will consume 4600 amperes.

What's a Switch Worth?

I WAS SHAVING under a couple of hundred watts of lighting that illuminated a whole bathroom. It occurred to me that a lot of electrical energy is wasted in such arrangements. If directed at my face, a fraction of the light I was using would have served as well.

But in wiring such a bathroom would it have paid to install two switches instead of one? Would the power saving have been more than balanced by the cost of the extra switch and wire?

With my next house in mind, I ran through these figures.

Extra switch box, 75 cents; extra switch, ditto; additional 12 feet of cable, $1.50. Total outlay, $3.00. (If I weren't doing the work myself I should double this total, to allow for the several extra minutes of the electrician's time.)

Continuing to ruminate while shaving, I guessed that the bathroom light is in use two hours a day, including the time it's burning because someone neglected to shut it off. Even if both bulbs would still be used half the time, the additional switch would save a 100-watt bulb one hour each day, or one-tenth of a kilowatt hour each day.

That comes to 36 kilowatt hours a year.

At a power rate of about 11 cents per kwh, that's $4 a year. The extra switch would pay for itself in a year or two in most localities. And saving electrical power is a good deed in itself.

Working Out an Island Sink

KNOWING that sinks and other plumbing contrivances require vents that go up through the roof, I've occasionally wondered how the problem is solved when the sink is in a counter in the middle of a room.

Well, now I know. And you do too. The trick includes a long sweep fitting below the floor with a pipe connection running horizontally to the nearest partition, where 45-degree fittings you see here make for easy cleaning of any stoppage that might develop.

The return vent you see beside the sink should be as high under the drain board as possible. The loop is made as gradual as possible with a pair of 45-degree fittings and one long-radius elbow.

Loading a Waterbed

A WATERBED can weigh a ton or more, enough to break through some floors. If you're putting in one or renting to a tenant who might, a little arithmetic may be in order. So says an expert on landlording.

First you should figure the area of the bed. For queen size, which is most usual, it's 60 by 80 inches. You can figure the area directly in feet or multiply those two dimensions together and divide by 144. Either way you do it, you get 33-1/3.

Multiply the water capacity in gallons by 8.33. Then, to find the total weight you're dealing with, add about 4 percent to this for the weight of the bed parts if it's an all-water bed. Make that about 20 percent if it's a hybrid that uses urethane foam or something of that sort as well as water.

For an all-water queen, 9-inch fill, capacity may be given as 196 gallons. Multiplied by 8.33, and by 1.04 to add that 4 percent, gives a total weight of 1698 pounds. With a couple of occupants, that's just about a ton. King size will be even heavier and some of the big round beds heavier yet.

Divide your weight figure by the area of the base, calculated in feet as above.

If the weight of the bed plus sleepers is a ton, you'll find the loading on the bedroom floor to be about 50 pounds to the square foot.

A result like this might be taken as a caution warning, since residential floors are most often designed to accept a loading of 40 pounds to the square foot, in some areas 30.

Consultation with the building department in your town might be the next step.

Aiming a Satellite Dish

THE ANGLES you need to know when aiming a communications-satellite dish can be found in four routines. They're all based on this trio of basic figures:

A. *Longitude of the site* of your dish. If it's west longitude, put a minus sign in front of the figure.

B. *Longitude of the satellite.* Again, if it's west longitude make the figure negative.

C. *Latitude of the site* of the dish. In the Southern Hemisphere, make this negative.

(If these figures are in degrees, fine. But if any include minutes as well, change those minutes to decimal degrees by dividing them by 60. That way 40 degrees, 15 minutes becomes 40.25 degrees.)

From these three location points, you can find the great-circle angle between your site and the point on the Equator directly below the satellite. You can also find the site azimuth from due north.

Then you can use that great-circle angle to find the range from earth-station site to satellite. Use the range, in turn, to find the elevation-from-horizon of the satellite.

To find great-circle angle:

1. From A, subtract B. If the result is not between minus and plus 180 degrees, add or subtract 360 to make it so.

2. Find the cosine.

3. Multiply by the cosine of C.

4. Take the arc-cosine. (If this is more than 81.3 degrees, you're in trouble; the satellite will not be visible from that location.)

To find azimuth:

1. This step is the same as Step 1 above.

2. Find the tangent.

3. Divide by the sine of C.

4. Find the arc-tangent.

5. If the site is in the Northern Hemisphere, add 180.

6. If the result is negative, add 360.

To find the range in kilometers:

1. Multiply cosine of great-circle angle by 536,950,000.

2. Subtract this result from 1,818,590,000.

3. Find the square root.

To find the degree of elevation from the horizon:

1. Square the range and subtract 1,737,517,200.

2. Divide by the range and also by 12,734.

3. Take the arc-cosine.

4. Subtract 90.

Applying these procedures to aiming an earth-station dish at Orlando, Florida (longitude 81 degrees 20 minutes west, latitude 28 degrees 32 minutes north) for Satcom 1, with west longitude of 135 degrees, first change minutes to decimal degrees. Your working values then are: A = -81.33, B = -135, C = 28.53.

Great-circle angle: A minus B is 53.67, of which the cosine is 0.59. Cosine of C (28.53) is 0.88. Multiplying the two cosines together gives 0.52, of which the arc-cosine is 58.67 degrees. (If you do all this with a calculator or computer, not rounding off anything, you'll get the more precise 58.6346 degrees.)

Azimuth: The tangent of 53.67 is 1.36, which divided by the sine of C is 2185. Arc-tangent of 2185 is 70.65. Adding 180 degrees for the Northern Hemisphere location makes it 250.65 degrees, or more precisely 250.64.

Range: The cosine of the great-circle angle 58.63 degrees is 0.52. Multiplying by 536,950,000 and subtracting from 1,818,590,000 gives 1,539,097,476, of which the square root is 39,231 (kilometers).

Elevation: Beginning with the square of the range and subtracting 1,737,517,200 and then dividing by the range (39,231) and by 12,734, taking the arc-cosine and then subtracting 90 gives 23.40 degrees elevation.

Since this is all a little involved, you might like to check the method against the results calculated for a second location. It's a spot in the Carmel Valley in central California. A friend of mine has been installing and aiming an earth-station dish there. It's at 121.8 degrees west longitude, 36.5 north latitude. The calculations are for the same Satcom 1, at 135 degrees west.

My four results are 36.50 degrees, 210.75 degrees, 37,394 kilometers, and 45.42 degrees.

The basic routines for these calculations came to me from Stephen Reed of Maitland, Florida, by way of David Nelsen of Lethbridge, Alberta, who mentioned another formula that could come in handy in conjunction with these others. It's one you can use to figure how far a dish must be set back from a structure or tree to avoid interference.

Having obtained the degree of elevation by the last of the four calculations above, subtract an arbitrary 5 degrees as a margin for avoiding noise. Find the tangent of the result. Divide the height of the object by that tangent.

For instance, using the 23.4-degree elevation found for Orlando, you'd subtract the 5-degree margin to get 18.4. Dividing 50 by the tangent of 18.4 (which is 0.33) warns you to keep your dish at least 150 feet back from that 50-foot tree.

Chapter N
  OPERATING YOUR SHELTER

To Build a House
Managing a Mortgage
Shrink a Mortgage?
Your Depreciating House
How to Sell a House
Property-Tax Strategy
How Much Home Insurance?
Cost-Effective Energy Saver?
Fueling the Water Heater
How Much Heat in Firewood?
What Does Fuel Cost?
Mobile Home—Does It Figure?
Free Color TV?
Cork in the Eye

To Build a House

As many a family has discovered to its sorrow, becoming the owner of the kind of house you dream about can be a shockingly expensive proposition. But a custom-built house is still what most of us want, so it's worth taking a hard look at how to figure the cost—and how to reduce it.

Making decisions about such major matters as this is, after all, one of the things that figuring is for.

Cal and Midge Mason are about to build a new home for themselves and young Mark, who is seven. They know what they want—pretty much—and they have already searched their town in vain for anything like it.

"No wonder!" Midge sighs. "If I had a house like the one I'm dreaming about, I wouldn't part with it either."

Having consulted one architect, several real-estate salesmen, and a couple of custom builders, the Masons have learned that a house of the size and quality they want will cost about $110,000 in their small city.

At that, they're lucky. In applying this and the cost estimates to follow in many larger towns, or any city in most areas, the dollar figures would have to be doubled or worse. On the thesis that it is easier to multiply than divide, a near-minimum example may be the most useful.

The arithmetic behind that $110,000 price is simple. The least-expensive suitable lots the Masons have seen are listed at $22,000. They want a house of about 1600 square feet, and the going rate for custom construction is a minimum $50 a square foot. That comes to $80,000.

Few architects will design a small house and supervise its construction for a fee of less than 10 percent, $8000 in this case. And to get a house that fits both site and occupants, there has got to be a designer in the picture.

Before coming to any decision, Cal and Midge list the parts that add up to $110,000, using round numbers:

Land: $22,000.

Design: $4800.

Supervision: Architect, $3200; time, overhead, and profit for contractor and sub-contractors, $20,000. Total: $23,200.

On-site labor (carpenters, plumbers, etc.):$30,000.

Materials: $30,000.

For that amount of money, $110,000, they can have their special house. They need only choose their architect well and make certain she understands their needs and desires. She will help them choose a lot, draw a detailed sheaf of plans and specifications, first submitting rough sketches. They can be reasonably sure of getting exactly what they want and can afford, provided they don't add too many extras along the way.

They can also see that, although the down payment is within possibility, the monthly mortgage payments will be excessive.

The first place to save is on the land.

"If we can buy a lot directly from the owner, we can avoid a real-estate commission, and that's a flat 10 percent on vacant land," Midge points out.

It's a good idea, though it means some work. It means touring the area they're interested in, looking for vacant lots that fit their specifications, noting location so they can identify them on tax maps at city hall or county court house. And then obtaining names and addresses of owners from tax records, and phoning or writing to find out if the owners are prepared to sell.

For the Masons, this procedure produced two level lots and a third that sloped sharply down from the street.

"Looks like we can buy either level lot for about $19,000 or the steep one for a couple thousand less," Cal says.

"From what I've been reading, the saving on the steep lot is about what it will cost us extra to build on it."

Either way, the Masons' do-it-yourself land-buying approach should save them about $3000, mostly from bypassing commission at the cost of some shoe leather.

Now how can they save on the next item, design? They can skip the architect and go instead to a student or a recent graduate who is unemployed or moonlighting or find a designer who is less than a full-fledged architect.

Midge and Cal plump for an alternate economy solution: ready-drawn plans. Home magazines offer such plans for $50 or $100 or so. Many have been drawn in the first place for clients who paid a lot of money for them.

A ready-made plan may be excellent, but this solution is not without its risks. For one thing, it takes good judgment to choose a plan, especially if it must be adapted to an unusual building site with special engineering requirements or code problems.

Whether Midge and Cal can safely save more of the planning and supervising costs by acting as their own general contractor will depend upon their experience and how much they are willing to learn in a hurry. They will have to hire people to do concrete work, framing, finish carpentry, plumbing, wiring—instead of paying a general contractor to do this or to subcontract the work.

If the Masons go this far—finding their lot, bargaining for it, drawing their own plan or buying a stock plan, then supervising construction—they stand to save as much as $30,000. Their dream, hopelessly expensive at $110,000, is within reach if they can hold to this plan.

You, or anyone capable and willing to go this far, might consider going even further. You just might tackle some or most—or conceivably all—of the construction work itself, as I have done more than once and with highly satisfying consequences.

What will you save? Well, for a house like the one just described for the Masons, labor was estimated at $30,000. Providing it yourself can knock anything up to that off the $80,000. For a very rough estimate, the house you build yourself should cost half what it would sell for.

I can back up that generalization. Thirty-some years ago, when my family was in somewhat the position of the Masons, we slowly built a house for ourselves—big one, five bedrooms, 3,000 square feet. When finished, it was valued at just about three times the amount we had in it. With a modest boost from inflation, it sold a dozen years later for four times our investment. Loss of interest over that period on our invested money was more than balanced by the rent we weren't paying.

Meanwhile, as our kids began going off to college, we built something better but smaller. On a rather costly lot, it ran us some $26,000 and was worth about $60,000. We lived in it for 10 years, rented it out for 3, then sold it to our oldest daughter and her husband. Of course, the appraisal of $120,000 had 13 years of inflation in it.

Some people don't know when to stop, so by this time we were living in another homemade home. This time we used a good bit of hired handyman help, but contracted out nothing except the tar-and-gravel roof. Materials came to $15,000, help to $10,300. Appraisal by the tax man, two years after completion, was $51,000, not counting land.

Here's a final interesting figure to toy with.

You own a lot that is acceptable as a down payment. You want to build a $104,000 house on it...a $52,000 house if you build it yourself.

You can afford payments of $800 a month for principal and interest. The going rate is 8-1/2 percent. Consulting a table of mortgage payments, you find that your $800 a month will pay off $104,000 in just 30 years. But it will pay off one of half that size in a remarkably brief seven years plus a couple of months. By going the sweat-equity route and slicing your cost of building in half, you've cut your total outlay over the years to one-fourth.

Managing a Mortgage

If you're now carrying a mortgage that was tailored for you two or ten years ago, it is like a suit of clothes of the same age. It may be uncomfortably tight in its payments or too short in its amortization period.

The mortgage may need alterations because your house does. Rates that have dropped appreciably may translate into lower monthly payments or an earlier payoff.

Two other situations might help you.

You may have had to accept a high rate of interest because your equity was small. By now, that equity has grown through your payments to principal, improvements you've made, development of the neighborhood, wider acceptance of an unconventional design.

Small changes in interest rates are more significant than they may sound. Over the life of a 30-year $25,000 mortgage, reducing a 9-1/2 percent rate to 8-1/2 means a healthy $5,397 in your pocket. It cuts monthly payments about $18.

In refinancing, you need to offset potential savings by the cost of the change. These possible expenses include appraisal and loan fees, title-insurance costs, and any penalty provided for in your present mortgage. With the help of your loan source, you can quickly discover whether a change is worthwhile now or may be in the future.

If so, you'll have some pleasant, though possibly difficult, choices. You can reduce the monthly payments. You can shorten the payoff time. You can turn some of your equity into cash.

Example 1. When interest on your mortgage runs 10 percent and the going rate has dropped to 9, is it time to refinance? If associated costs, which may include new title insurance, survey, appraisal, legal fees, and taxes, eat up the advantage, you have little to lose by waiting in hope of a larger drop.

Example 2. When your mortgage can be refinanced at a lower rate, reducing payments even after closing costs are added, are there additional hidden advantages or drawbacks? Yes, if you need a further break, you may be able at the same time to reduce monthly payments by extending the life of the loan. And this time you may be able to include an open-end clause, which is handy if you want another loan in the future for home improvements.

Example 3. If your present mortgage is high-interest but it wouldn't pay you to change just now, are there any steps it would be prudent to take? Prepay it as fast as you can afford to. Making extra payments is really a way of investing money safely at a favorable rate.

Example 4. What are the rules about prepayments? Read the fine print and consult the lender. Sometimes you're allowed to pay extra only on the principal, and you may be required to give written notice. The lender may be willing to agree that the pre-payments can be used later on (in an emergency) to count as payments on the principal—especially if you're unemployed or ill for a period.

Example 5. Having come into or accumulated some extra cash, should you prepay part or all of a mortgage—or invest it elsewhere? Paying off a loan is the safest kind of investment, but by no means the most liquid. If you can safely park the cash elsewhere at a rate higher than the loan bears, you might as well let the mortgage stand and reap the difference. Moreover, if the mortgage is large and assumable, it might be an asset if you should sell.

Example 6. For some years, you've been paying on first and second mortgages, as well as making improvements on your home. The second runs out in a few months, when you must make a final payment of $2,000. Where can you find the money? You've probably built up a hefty equity. Refinance with a present or new lender if you can get a cheaper first that covers the balloon and closing.

Example 7. You suspect your loan payments, manageable now, may become a burden when the kids reach college age. What to do? Prepayment on your mortgage is one possible move, provided the interest rate is as high as you could get by stashing the same amounts in a savings account. Or maybe you can work out a prepayment schedule with the lender that will make it feasible to retire the mortgage before those college years arrive.

If you hope to retire before the mortgage is paid off, look for a prepayment schedule that will permit you and the mortgage to retire at the same time. For more about this, take a look at the last item in this book.

Example 8. You've lost your job and you're falling behind on mortgage payments. How can you keep your house?

Get together with the lender. He doesn't want a foreclosure either. Together you can probably work out a plan to catch up when you go back to work. If yours is an FHA loan, the lender may wait a year before foreclosing, provided the house is being kept up. VA also has helpful rules.

Example 9. You've kept up your payments so far, but now you see stormy times ahead. Your lender is not prepared to carry you if you fall behind.

He might be amenable to an entirely new loan, extending the time while reducing the payments. You may have better luck with a new source, although you might encounter heavier closing costs. Either way, your good record to date and your prudent anticipation of trouble should work for you.

Shrink a Mortgage?

ONE OF THE SILLIER of the notions that keep cropping up in books on budgeting, and in newspaper and magazine columns of financial advice has to do with the wisdom of keeping your home mortgaged because the interest is a big tax deduction.

See the fallacy?

This option arises only if you have spare money for making a bigger down payment than required. Or reducing or paying off an existing mortgage. Your alternatives are to put the money into the house or into some interest-paying investment.

Say there's $20,000 remaining on your 8 percent mortgage. Under federal tax law the interest on this reduces the income on which you must pay tax at a rate of $1,600 a year. Pay it off and you lose that benefit (and perhaps one under your state's tax laws as well). But if instead of paying off the mortgage, you invest the $20,000 in bonds or something at 8 percent, you immediately increase your income subject to taxes by an offsetting $1600 a year, gaining nothing.

So in figuring out whether to put your extra money into your own mortgage, you can forget the advice about hanging onto the tax deduction. You can decide purely on other factors.

It's pretty clear that if you can safely invest the money at a rate of return higher than you're paying on your mortgage, it makes sense to let the mortgage ride and gain the difference. It's equally apparent that if you can't find a safe investment at an equal rate, you'll be ahead in dollars if you invest in your own mortgage instead, prepaying to the extent that is permitted free of penalty.

Don't forget that a comparatively large existing mortgage can be a help in selling your house if the buyer is allowed to take it over. However, a mortgage that is small relative to the value of the house is usually no help at all. Most buyers won't have enough cash to make up the difference.

Your Depreciating House

IF YOU OWN a house and take care of it, it probably gets better every year. The landscaping matures, you update its energy efficiency and its mechanical parts as they wear out.

All the same, from the income-tax point of view it's depreciating. Which means that if it—or any part of it—is producing income, it is also generating an income-tax saving for you. This applies if some or all of it is rented, full- or part-time, or used as an office or shop or garage or photographic darkroom that produces taxable income.

Since you are allowed to depreciate an income-producing room or dwelling over precisely 27-1/2 years, the first step in figuring how this works is division by that cumbersome number. What you divide by it is the cost to you of that structure (land doesn't count, mostly because it doesn't ordinarily wear out).

Dividing something by 27-1/2 is equivalent to multiplying it by 0.03636, or taking 3.636 percent of it. That operation gives you the depreciation for each year after the first...until 27-1/2 years have passed.

Which brings us back to that first year in which you are entitled to take depreciation, the year that produces an interesting calculation.

Depreciation begins in the month when the property is put into service. But no matter which day of the month this is, tax law assumes it was halfway through the month. Not surprisingly, this assumption is dubbed the mid-month convention. It means that regardless of whether you are figuring from the first of January or the 31st, you still are credited with the same 11-1/2 months. And if you start any time in December, you get half a month.

Divide that mystic 0.03636 by 12 and multiply by the number of months in which the use occurred—a partial month counting as one half. Multiply by the cost of the building. That's the first-year depreciation.

Let's say you've spent $100,000 (not including land cost) to build a house. You rent it out, beginning sometime between March 2 and 30th. Multiplying $100,000 by 0.03636 will give you the depreciation you may take each year after the first: $3636.

In the first year, however, you must multiply that result by 9-1/2 divided by 12, coming to $2878.79.

If, by the way, you're renting out only part of the property, keep in mind that only that fraction counts. If you're occupying one-third of the property yourself, the $100,000 in our example becomes $66,667 instead. You would probably be allowed to choose the fraction in the way most helpful to you—by area, or by number of rooms, for example.

For something other than real estate, the time period, and hence the numbers, will be different. And while even the details of the real-estate computation may change from year to year, understanding the basics gone into here should still help.

How to Sell a House

WHAT KIND of figuring will persuade a prospect that he can afford the payments involved in buying the house you want to sell? In his syndicated real-estate column, Robert J. Bruss answers this very nicely.

> Dear Bob: I am a new real estate saleswoman. Although I've been working for four months now, I've only closed one sale. I could have sold many more, but the buyers balked at the high monthly payments. Yet I see other agents selling homes with high payments. Do you have any suggestions for overcoming the sales objection of high monthly payments? Bettina W., Washington, D.C.

> Dear Bettina: You've got one of the best jobs in the world, real-estate sales. You need to show your clients why they can't afford not to buy from you. Once you find a home that your client likes, the key to getting a sale is to show the buyer's true net housing cost. For example, suppose a client is considering buying a $60,000 home and he has $15,000 for the down payment. You can arrange a $45,000 mortgage for 25 years at 9 percent interest. The monthly mortgage payment will be $377.65, plus $100 per month for property taxes and $25 for insurance. But the client doesn't like the $502.65 monthly total, although he can afford it.

> At this point, ask the client what tax bracket he is in. Suppose he says 40 percent fed-state-local, including his wife's earnings. Then show him his first-year monthly tax deductions: $373.50 interest, $100 taxes, for a $473.50 total. He saves income tax on 40 percent, for $189.40 tax savings. That reduces the monthly housing cost to $313.25, probably less than the cost of a comparable rental. In addition, the mortgage principal is increasing while the value of the house is probably appreciating. Show your clients why they can't afford *not* to buy a home from you.

(In figuring something like this for your own purposes, you should include one factor overlooked here. From now on, the buyer will lose the interest he could be getting on the $15,000 he put down. Call it 8 percent. So the loss is $1200 a year, $100 a month, leaving a net after taxes of $60. Realistically, that must be added to the $313.25. And in dealing with a $240,000 house, that $60 becomes a more impressive $240. Bettina will still have a winning argument in most cases, though.)

Property-Tax Strategy

I GO ALONG, and no doubt you do too, with Oliver Wendell Holmes, Jr.: Taxes are what we pay for civilized society.

But that doesn't imply any eagerness to pay more than our fair share, even if it takes a certain amount of figuring to insure that this doesn't happen.

Given neighbors, with houses comparable to yours, who are willing to discuss their tax bills, you can quickly estimate whether the valuation placed on yours is out of line. Evaluating property is far from an exact science, and an error or simple misjudgment is not at all unlikely. Essentially, it is up to the homeowner to spot an instance of this and get it rectified.

If you seem to have a case worth pursuing, consult city-hall or courthouse records, open to the public. You get a precise comparison on as large a scale as you choose.

Property taxes are usually in proportion to estimated market value, with land and improvements often calculated separately.

Full value, or some arbitrary percentage of it, becomes the "assessed valuation." Your mathematical approach begins with dividing this assessed valuation by that percentage to see exactly how much the appraiser has said your place is worth.

If you don't agree, it's up to you to say so. If you can show, by actual instances (and a lot of exact numbers down on paper—always effective) that comparable homes aren't selling for that much today, you have a much more telling argument to present.

In the end, you may have to appeal formally to get a reduction, but an informal appeal is less of a nuisance for everyone and often gets more hospitable treatment.

Over the years, I've followed the latter course four times and four times had my taxes reduced. Since my complaints were legitimate and typical of those of many householders, you may find some suggestive value in the details.

Forty-five years ago, in a rural area, I built my first house, brashly using simpler methods than some in general use at the time. My framing of two-by-fours spaced 24 inches and my masonry of hollow blocks not plastered or stuccoed on either side, both commonplace today, were radical (and economical) short-cuts at the time. Having produced a house at far less than normal costs of materials, I was flattered—but not particularly pleased—to find it valued by the tax people at the same level as houses of similar footage but far more elaborate construction.

I addressed a mild note of protest to the office of the county assessor. One of his assistants turned up a few days later, examined my case and my house, and ruled that my structure should be reclassified—not at all flatteringly this time—as "partially substandard." This cut my tax bill by about 10 percent.

The house, by the way, is still standing sturdy and would be regarded, if anyone questioned it today, as conventional in construction.

My second brush with tax people came after a move to where taxes are assessed in three parts: land, improvement, personal property. There I built a rather elaborate large house. I received a proportionately huge tax bill, but found unwarranted only the very heavy levy on the personal-property portion. The $900 valuation, implying nearly $5,000 market value, didn't square with our scanty array of home-made and makeshift furniture that was all we needed in a house loaded with built-ins. It wouldn't have fetched $500 at auction.

I asked how come, and the county sent a man to explain. "To save time, we assess personal property in proportion to house value," he explained. "That way we don't have to poke through your closets and everything. A house as big as this, we figure you'd have maybe a grand piano and other expensive furnishings. Or you're going to buy them."

"Well, I don't," I said. "And I'm not. And so I shouldn't pay tax on something I don't own."

The man eventually agreed that there was nothing in the law to justify taxing expectations—even his. When I received my final tax bill, the $900 was crossed out and $300 written in—still ample for our modest possessions but quite a tax saving for us.

A little later when we built another house, a similar informal protest brought agreement that full valuation was premature on an unfinished structure, even though occupied. (Occupancy before completion was permitted in our town, which was small, still friendly, and not overfond of red tape.)

My companion argument, that our lot should be valued at less than its neighbors because an extreme slope added to foundation cost, fared less well.

"Right you are," said the assessor. "So we didn't charge you for the cost of the foundation in valuing the structure."

He had me there.

A few years back, the assessment and taxes nearly doubled in a single year on five rocky, waterless, unimproved acres we own in northern California. To buttress our opinion that the new valuation was more than the land was worth, we listed it for sale at that figure with a reputable real-estate office. It failed even to get an offer, and we used this fact in our letter of complaint. The only reply was a form acknowledgment, but the next tax bill showed a drop in "full value" from $15,000 to a reasonable $10,000.

To me, these experiences say that where there is an avenue for an informal request, one well-based and carefully worded letter may produce a substantial benefit that goes on year after year. If you are not confident of your grounds, avoid formal protest if possible. There's always a risk, however slight, of what happened to our dentist. He told me about it over the whir of the drill.

"I demanded a hearing on my tax bill," he said. "The county went over all the figures and jumped my taxes nearly $100 a year. And there's nothing I can do about it."

How Much Home Insurance?

BALANCING costs against benefits to decide how much insurance to carry on your home starts with estimating its value.

The first thing to keep in mind is that you are valuing the structure—the part that can burn or otherwise be destroyed—not the house-and-lot combination that you would buy or sell. And for insurance purposes, you will ordinarily count foundation and pipes that are below the surface of the ground as part of the land, not as part of the house.

Here is how I went about putting a value on my house.

I began by measuring the number of square feet of floor space in the house. I did not count open porches or patios. I did count the garage, but only half its area since it is generally assumed that garage space costs about half as much to build as living space. If there were any complicated, roofed porches, I'd have counted half their area as well.

I found my house to have 1576 square feet of living space, plus an attached single garage of the usual 12-by-20-foot size. Adding 1576 to half of 240 gave me roughly 1700 square feet.

A phone call to my insurance broker told me that their current figures suggested replacement costs in the range of $65 to $75 a square foot. An architect I was playing tennis with a couple of weeks earlier told me he figured local building costs at $70 on up. (Never pass up a chance to get free professional advice—except possibly from M.D.s, who are more inclined to resent your maneuvers than to be flattered by them.)

Multiplying 1700 by $70 gave me a valuation of $119,000.

Since my house had been reassessed this year for tax purposes, I consulted my county tax bill to see what the assessor's person thinks my house would fetch if put up for sale. I found the assessment for "improvements" corresponded so remarkably to my own estimate that I couldn't help suspecting it may have been reached in precisely the same way.

Although I know that assessments tend to run just a trifle on the low side (perhaps to minimize ratepayer squawks) and also lag a year behind, I decided to go for the $70 figure. I'd have made it more, except for three things:

1. Mathematically speaking, insurance is equivalent to gambling, so the same rules apply. Insurance is a relatively poor buy for those with less than average chance of needing it. These are people known in the jargon as good risks and are beloved of insurance companies. I counted myself as one of these, since my house is solidly constructed, lived in by two fearful people, and is very close to our fire department.

2. Since I built the house myself I could rebuild it more economically than a contractor could. Hence, my apparent loss would be somewhat greater than my real loss.

3. I needn't concern myself with the effects of inflation on the value of my house during the period of the policy, since it contains an "inflation guard endorsement". This ups my coverage every three months.

Putting all this together, I can figure my house as safely insured at that $70 figure.

But now I have another question to face. Do I want to pay a premium big enough to keep my house fully insured? The chance that it will be totally destroyed is pretty tiny. I'll be paying a lot to cover a very unlikely eventuality—smart only if the loss would be crippling. I review my financial state and conclude that I can afford to cover a part of this minor risk myself.

The question is: how much of it? Perhaps I'll decide to insure my house for just half its value.

Clever of me? No. It's good economy from my point of view, but not from that of the insurance people. They've figured out a way to forestall this move. It's the 80 percent rule.

This stipulation says that if you insure your property for less than 80 percent of value, you can wind up collecting less than you expect. You can be limited to the same proportion of replacement cost as the insurance you're carrying is of 80 percent of the value.

In other words, insure at 50 percent of value and you may collect only 50/80 of your loss. Have yourself a little $1000 fire and 5/8 of $1000 is what you get; the other $375 of the repair bills is on you.

For most people, this makes it sensible to keep coverage up to 80 percent of real value.

On that basis, I multiply the full value of $119,000, arrived at a few paragraphs back, by 80 percent, to get $95,200. I round that off and insure for $95,000. Or maybe an even rounder $100,000.

Cost-Effective Energy Saver?

INNUMERABLE formulas have been devised to estimate the cost-effectiveness of an energy-saving home improvement. But here's one that is both simpler and more realistic than most.

It applies to any home you plan to live in for quite a while, and assumes that if you sell your house the buyer will recognize the value of the improvement—even if, as with such vital but prosaic contributions as insulation, it may not be readily seen.

Begin by estimating how many years the new item will last. For most energy-savers, such as double window glass or attic insulation, this may be the life of the house, perhaps 35 or 50 years. For a mechanical device, it may be nearer 10 or 15. To find the annual percentage of depreciation, divide 100 by that number of years.

Now figure the interest on your investment. It's better not to use an interest rate as incomplete and as temporary as the rate you'll pay if you borrow or the return you're sacrificing by spending the money instead of keeping it invested. True long-haul return may be better estimated by subtracting from the going rate both the tax you'd have to pay on the interest and that part of the return that is going down the drain because of inflation.

Add that cost-of-money figure to the depreciation percentage calculated according to the second paragraph above. Multiply by the cash price of the project. Subtract the annual cost of upkeep, if any. This will tell you how much the improvement will really cost you each year. To find how cost-effective it will be, compare it with the dollar saving that will be produced by the improvement.

Let's try the procedure, for a $2500 insulation job in a house you think should be standing for 50 years to come. Dividing 100 by 50 gives you 2 percent annual depreciation.

If you can get 12 percent on your money, with a quarter of that being taxed away, you're sacrificing a net of 9 percent. If inflation looks to average 4 percent, the difference is 5 percent. And, in fact, experience in the '70s, '80s, and early '90s suggests that 5 percent is as good a guess as any at the amount by which return on invested money can hope to exceed the erosion by inflation.

To that 5 percent, you should add the 2 percent from depreciation. Since 7 percent of $2500 is $175 (and there's no maintenance cost to add in) you now know the annual cost of the improvement.

Compare that amount to what you can expect to save in fuel and you'll know whether your energy-saving measure is likely to prove cost-effective.

If the price of gas, oil, or electricity goes up faster than inflation in the future, your picture will become increasingly favorable. And, of course, many improvements yield an intangible bonus in comfort or convenience as long as you're in the house.

Fueling the Water Heater

T HERE IS an easy formula for comparing the cost of gas and electricity for use in heating water:

Electricity is cheaper than gas only when the cost per therm (100,000 British thermal units, or Btus) of gas is more than 22 times the price of a kilowatt hour of electricity.

If you cannot tell from your bill how much you are paying for a therm of gas, a phone call to the gas company will get the information.

Similarly, the power company can tell you about your electricity cost.

These costs may differ according to quantity consumed, so keep in mind that the cost you need for this purpose is the highest rate you pay. The lower-cost units of power aren't involved in this, so you look only at the top level of cost.

Example: If the cost of the last increment of power on your bill is

10 cents per kilowatt hour, electricity is cheaper only if gas is costing you $2.20 or more per therm.

The same figures apply to other comparisons of gas and electricity, but you may have to allow for how the fuel is used. Electric room heaters, for instance, may not be as costly as the formula says because it may be easier to leave some rooms unheated part of the time than it is with a gas furnace.

How Much Heat in Firewood?

CONTRARY to what many people think, you can figure on just about as much heat from softwood as hardwood, pound for pound. The thing is, oak and hickory weigh twice as much to the cord as pine does. So there's twice the heat to the cord.

Kind of wood (air dried)	Weight in pounds per cord	Equivalent in gallons of fuel oil
hickory	3900	191
white oak	3600	175
beech	3500	165
hard maple	3500	160
yellow birch	3500	160
ash	3300	154
elm	3000	135
soft maple	2800	132
tamarack	2500	132
cherry	2550	129
spruce	2100	99
hemlock	2100	98
aspen	1900	97
basswood	1900	93
white pine	1800	94

I don't know how it is where you live, but in my central California coastal country, when oak runs $200 a cord (green, unstacked) to $225 (dry and put where you want it) pine costs at least $110. You can see that it's only the unknowing among us, deluded by the numbers and the impressive bulk, that buy the pine. Anytime softwood costs more than half what hardwood does, it's a bad buy so far as heat is concerned. It's not very good for your chimney either; use much pithy wood and a chimney that's not frequently cleaned can become a fire hazard.

Having figured the comparative values of kinds of firewood, you may wonder about cost of wood compared with other fuel. The table offers a reasonable approximation of how many gallons of fuel oil it takes to give you the heat to be found in a cord of firewood. Multiply 191, for example, by the gallon price of oil, and you'll learn how much you can pay for a cord of hickory and come out even—until you take efficiency into account.

What Does Fuel Cost?

T HIS IS an alternative way to compare true costs of using such fuels as coal, wood, electricity, natural gas, liquid-petroleum gas, and oil.

To use it, you need to know three things: heating value of the fuel, cost per unit of the fuel, and the level of efficiency at which the heating device operates.

Coal: Heating values are 8000 to 11,000 British thermal units per pound, which is 16 to 22 million per ton.

Wood averages 7000 to the pound, or 14 (for some softwoods) to 31 million (for the densest hardwoods) to the cord.

Electricity: One kilowatt hour produces 3413 Btus.

Natural gas: A cubic foot has a heating value of 850 to 1000. Your public utility company can no doubt provide a more precise Btu figure for your area. However, even for a given location, it's a so-called constant that in fact will vary from time to time.

L-P Gas: Figure on 95,500 Btus to the gallon.

Oil yields 140,000 to the gallon.

Efficiency of electricity used to heat by resistance is virtually 100 percent.

A good airtight stove will be 40 to 60 percent efficient. For a circulator fireplace, the figure is 20 to 30 percent, but count on only 5 to 15 percent for an ordinary fireplace. Furnaces usually burn their fuel with efficiency of 50 to 85 percent.

If yours is one of the high-efficiency kind, you probably know it and have the actual test figure to use.

With these figures, or more precise ones if you can obtain them for your own fuel and equipment, you can predict relative costs.

Multiply the number of Btus per unit by the efficiency, expressed as a decimal.

(Remember that this means that 60 percent becomes 0.60.)

Divide by the dollar cost per unit. This tells you the number of Btus you can expect for each dollar spent.

Thus the best firewood, used in a good stove and costing $150 a cord gives 31 million times 0.60 divided by 150, or 124,000 Btus for each dollar of cost.

And electricity at 10 cents a kilowatt hour gives 34,130 of those cozy British units for a dollar (3413 times 1 divided by 0.10). At last, here's a calculation for which you don't need even paper and pencil since dividing by 0.10, or 1/10, is the same as multiplying by 10.

Mobile Home—Does It Figure?

THERE ARE a lot of personal considerations in buying a mobile home rather than a conventional one plunked down permanently on a lot by itself. Nobody can help you much with those decisions.

But the financial aspects are something else again. They can be clarified by a little paper-and-pencil work.

To begin with the plus side, the typical mobile home costs perhaps one-fourth as much as the average conventional house. That is partly because it is smaller. But even when you compare homes of the same size, the mobile comes in at half the price.

There is more to the story, beginning with three financial draw-backs.

• Mobile homes depreciate fast. Just as they are commonly licensed much like motor vehicles, they depreciate almost as fast as cars do. Even in our worst inflationary times, mobiles were losing half their market value in the first 6-1/2 years. After that, the value drops more slowly until, by the time about 15 years have gone by, they are worth one-fifth to one-third their original cost. Regular houses, as you know, have tended in our time to appreciate in value instead of losing; and, despite short-term setbacks, this trend has great staying power in regions of population growth.

• Most mobiles are financed at automobile interest rates—10 to 12 percent when home mortgages are going at 7 or 8.

• When you buy a house you are also investing in land, which can be expected to increase in value year after year. As an owner in a mobile park, you rent the use of land and cannot expect to share in the appreciation of its value. (But if you buy a lot for your mobile home, on your own or in a mobile-home subdivision, you eliminate this disadvantage.)

Since no general rule will tell you whether a mobile is financially advantageous to your own special situation, you'll have to add up some figures and alternatives to find out. You can do this by following the procedure worked out for one western family. Let's call them the Wrights.

Anyone using these figures will have to adapt them to their own tax bracket, current interest rates in the area, and housing level. For instance, although the Wrights are comparing a $20,000 mobile to a $48,000 conventional house, a larger family in a more expensive area might find a need to double or triple these numbers.

The Wrights began by looking at the newest mobile models, averaging 12 feet wide by 60 feet long—pretty big, but much smaller than the houses formed by putting two or even three basic units side by side. They chose a "doublewide" costing about $20,000. (Half the units sold in California were "doublewides," they learned.)

This unit could be purchased for 20 percent down and payments spread over 7 years at 10 percent interest. Their payments would be $265.62 a month, according to a standard loan table. Since there are 84 monthly payments in 7 years, and $16,000 divided by 84 is $190.48, they could find their average monthly interest payment by subtracting $190.48 from $265.62. Rent for space in the trailer park they favored would be $120 a month for the unit in question.

From residents in the park, they learned they could expect to pay a modest $400 a year vehicle and personal-property tax. There would be no ordinary county or city property tax, of course, except to the extent that one of the things their site rental was calculated to cover was a portion of the property tax paid by the owner of the land on which the trailer park sat.

They were also able to make some reasonably good estimates of maintenance, heating and other utility bills and insurance costs by talking to owners of units similar to the one they were considering.

Interest and property taxes are deductible for income-tax purposes. But this is helpful only to those who can't advantageously use the standard deduction. This being true for them, the Wrights added these together ($75.14 plus $33.34 is $108.48). They estimated 25 percent of this amount as their average monthly tax saving from purchase and ownership of the mobile home. (Actual saving would be more than this in early years, when most of each monthly payment would be for interest, and less in the later years of the loan.)

Then they made a similar list of costs for owning a house comparable in living qualities to the mobile home. They did this for the 7-year period in which the mobile would be paid off. Although it was only slightly larger than the mobile, the selling price of this house was more than twice as much, because it included land and involved a construction cost per square foot about double that of the factory-built mobile. Its price came to $48,000, of which they estimated $8,000 represented value of the building site.

MONTHLY COSTS	$20,000 MOBILE HOME	48,000 HOUSE
loan repayment and interest	265.62	306.98
park rent	120.00	
taxes	33.34	100.00
maintenance	100.00	150.00
utilities-insurance		
TOTAL	518.96	556.98
income tax saving	27.12	90.00
NET COST	491.84	466.98

Beginning with the same amount of cash, $4,000—to keep costs as comparable as possible—they would require a mortgage of nearly $45,000 for a total of $49,000. The extra $1000 was for some closing costs not involved in a mobile-home purchase, and furnishings they would need in a new house but not in a largely prefurnished mobile.

At 7-1/4 percent, a 30-year $45,000 mortgage requires monthly payments of $306.98. Since very little goes toward principal in the early years, they estimated that about $260 of this would be for interest, on the average, during those initial 7 years.

Where as owners of a mobile home they would have monthly site rent, as house buyers they would have none. But in a typical town or city their taxes as homeowners would be much higher; in this instance about $1200, as they learned from owners of a house similar to the one they were contemplating.

For maintenance-utilities-insurance, they estimated an average monthly cost of $150 compared with $100 for the mobile home.

They added their estimated monthly property-tax cost ($100) to the figure for mortgage interest ($260) to find they'd have $360 a month to deduct from income for tax purposes—a net saving of about $90 a month for people in their bracket.

With the same initial expense, the comparison for monthly costs worked out as shown in the table.

The first thing these figures show is that for conventional housing of comparable size, monthly costs would, in this instance, be pretty much the same. Although the conventional house came out slightly ahead in the tabulation, this was true only because of the greater income tax saving. In any year in which it was no more costly to take the standard deduction in filing their federal return, this family would run somewhat ahead with a mobile home.

Like all taxpayers, the Wrights could either list and deduct from taxable income such things as mortgage interest and property taxes paid and charitable contributions and possibly some medical costs— or they could deduct a flat amount based on a percentage of their income up to a maximum. Deductible expenses produce a saving only to the extent that all of them together add up to more than the standard deduction would be.

They also noted that by dropping to a smaller mobile home, they could reduce their net cost for most items about 60 percent—the natural consequence of buying a $12,000 unit instead of a $20,000 one. This could cut their outlay per month to as little as $340. But with single-family houses, it might be hard to find anything that much smaller and cheaper and in an acceptable neighborhood.

Being forward-looking people, the Wright family considered the position in which they would find themselves after 7 years. If they bought a mobile home it would then be fully paid for, and their expenses would drop by $246.84 a month ($265.62 loan payments ended, but $18.78 tax saving lost). If they had bought a house instead, their payments might go on for 23 more years.

Since that is actually the more realistic way to look at it—the average American family moves at least that often—our people finally zeroed in on the most helpful comparison of all.

Where, they asked, will we stand 7 years from now? With mobile homes depreciating 50 percent in the first 6-1/2 years, it is clear that they would be owners of a unit worth about $10,000. With the conventional house, the Wrights would have paid $3800 on the principal in 7 years. How much that equity would be worth would depend upon the real-estate market in their area. In recent years, their equity would have grown by more than $10,000.

But that's clearly a gamble based on continuing inflation. The economic aspect remains something of a toss-up. The Wrights could make their decision on the basis of what they really preferred to do, without predictable financial indiscretion. Clearly, that is the most comfortable position to be in.

Free Color TV?

SOME BANKS have been making an exciting offer. They'll give you a color television set if you'll deposit a mere $1000 and leave it with them for 55 months. Virtually no interest is paid, of course; at the end of the four years and seven months your account will be $1003.76.

Is it a good offer?

A simple way to look at the question is by asking how much you would be willing to pay today to have this amount at that future date. Use either a table or this formula that is often useful in a "today's worth" calculation. For rate of interest, use the highest rate being offered for any equally safe account in any financial institution.

Briefly, the formula says: Express the rate of interest as a decimal and divide by the number of compounding periods in a year. Add 1. Raise this to the power of the total number of dollars you will then have in the bank.

When you subtract the result from the $1000 you deposit today, you'll learn what you are truly paying for the television set.

The example given here is an offer made by a bank in Oakland, California. Using their usual rate of interest for similar accounts at the time, 7.25 percent, it turns out that what you are paying for a 19-inch Hitachi is just under $280. If comparison shopping tells you that's a bargain, so is the offer (although you'd also want to look into any other stipulations attached, such as what happens if you need to withdraw any part of the money before those 55 months are up).

By the way, since California has a sales tax, there is in this instance, another—and simpler—way to check what you're really paying. The Oakland bank says it must collect $18.20 sales tax. Knowing that the sales tax rate is 6-1/2 percent, you cannily proceed to divide $18.20 by 0.065 to get—surprise!—$280.

Evidently, your banker knows how to do interest-bearing arithmetic—and so must you.

Cork in the Eye

Having myself known two celebrants who suffered from champagne even before drinking it, I was pleased to come upon a valuable calculation offered by Dr. Richard Gordon of "Doctor in the House" fame.

"In swinging 1965, when Britons drank 5,181,185 bottles of champagne, eight Londoners between 24 and 72 years of age were taken to Moorfields Eye Hospital with injuries from the corks. Seven were serious enough for admission, three developed cataracts. Four of the eight patients were waiters. The relevant physics should be in the mind of anyone broaching a bottle of Cliquot.

"Chilled champagne (47 degrees F) has an intra-bottle pressure of 90 pounds per square inch, rising at room temperature (65 degrees F) and with shaking about. This can shoot the cork 40 feet (a bottle of claret needs 300 pounds of force on the corkscrew to open it)."

Dr. Gordon goes on to a calculation that the cork from an ill-aimed bottle "strikes the eye with a velocity of 45 feet per second. It arrives in half the time of a blink, with a force of 100 atmospheres. This is comparable with blast injuries to the eye in mines and quarries.

"To avoid ruining the party, champagne should be opened with a linen napkin over its mouth from the first breaking of the golden foil, the cork eased out pointing at someone else."

Chapter 0
WORKSHOP NUMBERS

How Big a Box?

YOU HAVE a sheet of tin. You want to cut the corners, removing the four little squares or folding them over, so you can fold the sheet into a box open on the top. You want to construct a box with the largest volume you can get from a square sheet.

How much should you cut out?

It's not an unusual problem, actually. You might be making little drawers or open boxes to store bolts or other metal parts. The solution, however, calls for elaborate graphing or the use of calculus—a page full of equations.

You may be, as I was when the question first arose, happier just to accept the solution ready made. Which is that the little squares you cut out should have sides just one sixth the length of the side of the original square.

Maybe that's not clear. Well, given a sheet of tin a foot each way, the biggest box (the one with the most cubic inches of capacity) that you can fold from it is one that is 2 inches high and has a base 8 inches square.

That works out so neatly you might have hit it by guess. But note that with a sheet of length not evenly divisible by 6, things become less obvious. The height that will give you the maximum volume with a 10-inch square is one-sixth of 10, or approximately 1.67 inches.

How Wide Is This Board?

CALLING a hunk of wood an inch and a half thick and three and a half inches wide a two-by-four seems faintly fraudulent, but that's the way it is.

The idea, of course, is that this is what's left when you start with something two inches thick and four inches wide and smooth it up.

It's true that when you buy rough lumber, you often get the dimensions you ask for, though not precisely.

Anyway, you frequently need to know true dimensions when building or when making calculations about strength of beams.

So it is useful to memorize the simple rule that lumber called 2 to 6 inches thick or wide dresses down to half an inch less than that.

From nominal 8 inches on, the amount you pay for but don't get is three-quarters of an inch. These figures go for seasoned—dried—lumber.

If you buy your lumber unseasoned—green—it can be expected to shrink to these dimensions. But while it is green, it will measure a bit more. The 2-, 3-, and 4-inch dimensions will be shorted by 7/16 of an inch. A nominal 6-inch dimension will be off by 3/8 of an inch, making it really 5-5/8 inches. Anything beyond that will be off by half an inch.

To Divide a Board

GIVEN a plank that you wish to cut into two (or three or any other number) equal lengths, how do you measure it if you're stuck somewhere without any measuring device?

A piece of string as long as the plank (or the garden plot or whatever you're dealing with) will do fine—if you've got a piece of string.

If you have no string and if the plank seems consistent in weight and is light enough to lift, just balance it on a finger and cut at the point of balance.

If you're dealing with a heavy timber or a section of garden or anything else you can't lift, you'll have to find another way. A useful way is one called successive approximations and the beauty of it is it will work for quite a number of situations. It will serve to divide a length or distance into three or more parts just about as easily as into two. It will even solve practical mathematical problems, as you will see elsewhere in this volume in material dealing with finding square roots.

Use the principle this way.

To mark a long garden into three equal plots, break off a stick to a length that you can see is at least one third the length of the garden. Lay it down three times and then pick it up and break off and throw away what looks to be not quite one third of the part that stuck out beyond the length of the garden. Repeat the process with your shortened stick. Go on repeating as many times as seems necessary.

After a few trials, you'll have that garden marked off into precise thirds, right down to a gnat's eyelash.

How Big the Bolt or Screw?

K<small>NOWING</small> the number size of a bolt or a wood screw you can easily determine its diameter without chasing down a chart or handbook.

This is useful when you have to drill a hole for the screw or bolt, using a bit for which the diameter is given as a decimal or fraction of an inch.

The routine called for is a trifle weird but not complicated. Multiply the size number of the screw or bolt by 0.013 and add 0.060. Don't ask me why.

Anyway, you'll find that a No. 10 wood screw has a diameter of 0.190 inches (10 times 0.013 plus 0.060 is 0.19).

Confounded Compound Cuts

H<small>AVE YOU</small> ever tried to make one of those popular planter boxes with many tapered sides?

If you have, you've encountered the intricacies of compound-miter cuts and perhaps discovered how hard they are to work out by trial and error.

Actually, setting a circular saw to cut tricky compound miters for such home-workshop projects can be a cinch.

So says R. F. Gyger, Houma, La., who burrowed through a fairish bit of trigonometry to supply this straightforward routine, which he has sent to me to share with the world. Or with such part of it as hankers to cut compound miters but was afraid to ask how.

Decide first how many sides your box is to have and how many degrees from vertical you'd like the sides to slope. You might, for example, choose to build a hexagonal planter with a side slope of 10 degrees.

The geometric calculations that follow can be handled nicely with paper and pencil—and even more nicely with a simple hand-calculator used in the presence of a table of trig functions.

But there's no denying that one of the basic scientific calculators makes it all faster. And, to be honest about it, a good bit more fun.

So:

Divide 180 by the number of sides. Find the tangent of this, then multiply by the sine of the angle of slope. The arc tangent of this is the setting for your saw arm or miter. For the hex planter with 10-degree slope just mentioned, this means multiplying tan 30 by sin 10 and then getting the arc tan, which turns out to be 5.7 degrees.

After again dividing 180 by the number of sides, this time multiply the sine of this result by the cosine of the slope.

The arc sine of this result gives the tilt angle, which proves to be 29.5 degrees.

And there you are with the two saw settings you need for your power saw.

How Big a Pipe Wrench?

SINCE ANY size pipe wrench will work on any pipe within reason, you might think size doesn't make any difference.

But a wrench that's too small won't do the work, at least not without excessive effort. And—this is important—one that's too big will multiply your muscle power so effectively you may over-tighten a connection. Besides making it excessively difficult to undo later on, this overpowering may split a fitting. If you discover this you'll have the job to do over again, possibly after a trip to the plumbing-supply shop. If you don't, you could end up with a concealed leak—of water, which is bad, or gas, which is worse.

Here are recommended sizes:

10-inch wrench for pipe up to 1-inch diameter.

18-inch wrench for pipe from 1-inch to 2-inch diameter.

24-inch wrench for pipe 3-inch diameter or over.

Or use a chain wrench for pipe over 3-inch or of irregular shape.

(Note: a midget wrench 6 to 8 inches long is a good auxiliary for working with pipe under 1-inch when space is short. Note also that for taking apart old pipe that is frozen with corrosion, there's no law against using a bigger wrench than the table indicates. If that's not adequate, or you don't have a big wrench, try slipping a length of pipe or tubing over the handle to extend it and increase your leverage. Your strength will be as the strength of ten, even if your heart isn't pure.)

From RPM to SFM in One Jump

IT'S QUITE COMMON under workshop conditions to know the speed of a tool in revolutions per minute—and need to find what the surface speed is.

The surface speed is frequently expressed in surface feet per minute. Multiply the known rpm by the diameter of the pulley in inches and divide by 4 to get a quick approximation of the sfm. What you are doing actually is multiplying the diameter of the tool or work by pi to find the circumference and dividing by 12 to change inches to feet.

And that is indeed the routine to follow if you need a precise answer. Which will prove to be not quite 5 percent more than your quick approximation.

Shelf Math

Aʙᴏᴜᴛ ᴛʜᴇ ᴏɴʟʏ sensible way to determine the depth (distance from front to back) of a shelf or set of shelves is by first measuring whatever is to be stored. You may be surprised to discover that shelves only six inches deep will do nicely for most books. You can save a lot of lumber and some space by making them only that deep—if you also provide one or more deeper shelves for outsize books.

The only other rule applying specifically to book shelves is that they should be far enough apart to allow 1-1/2 inches above a typical book for finger room.

More important—and harder to determine by logic or guess—is how long a span to allow either for book or general storage. The problem is to avoid sagging.

Research at the University of Illinois has produced some useful rules of thumb for this:

When using 3/8-inch plywood, limit the spans to 26 inches for normal loads such as toys, clothing, light dishes. For books and records and other heavy loads, keep spans under 16 inches to avoid sagging.

But 3/4-inch plywood will usually handle normal loads on 46-inch spans, heavy loads on 29-inch.

If you're using solid wood 3/4 of an inch thick (nominal 1-inch lumber, that is) you can stretch these spans slightly. Figure on spans to 48 inches for normal loads, 30 inches for heavy ones.

Thicker lumber usually makes much better looking shelves as well as stouter ones. When it's 1-1/8 inches thick, you can increase spans for shelves to bear normal loads to 66 inches; 42 inches for heavy loads.

For a major shelving project, or for shelves that are to bear unusual loads, you may wish to be more scientific. Use the table for acceptable loading for shelves of any width made of the usual thicknesses of lumber (nominal 1-inch and 2-inch) or of plywood (3/8-inch, 1/2-inch, 5/8-inch, 3/4-inch).

The figures in pounds are for the loads, in pounds per square foot, that the shelf would support without sagging more than 1/240 of the span, which works out to 1/20 inch for each foot of length.

FIGURING SHELVES

Distance Between Supports	Pine and Cedar Lumber		Douglas Fir and Larch Lumber		Douglas Fir Plywood			
	$3/4$"	$1\,1/2$"	$3/4$"	$1\,1/2$"	$3/8$"	$1/2$"	$5/8$"	$3/4$"
12"	670 lb.	2,685 lb.	1,134 lb.	4,537 lb.	163 lb.	330 lb.	593 lb.	817 lb.
16"	377 lb.	1,510 lb.	638 lb.	2,552 lb.	68 lb.	138 lb.	249 lb.	374 lb.
20"	222 lb.	967 lb.	356 lb.	1,633 lb.	35 lb.	71 lb.	127 lb.	191 lb.
24"	129 lb.	667 lb.	206 lb.	1,126 lb.	20 lb.	41 lb.	75 lb.	111 lb.
32"	54 lb.	378 lb.	87 lb.	638 lb.	9 lb.	17 lb.	30 lb.	46 lb.
36"	38 lb.	306 lb.	61 lb.	489 lb.	6 lb.	12 lb.	22 lb.	33 lb.
40"	28 lb.	223 lb.	44 lb.	356 lb.	4 lb.	9 lb.	16 lb.	24 lb.
48"	16 lb.	130 lb.	26 lb.	206 lb.	2 lb.	5 lb.	10 lb.	14 lb.
60"	8 lb.	66 lb.	13 lb.	106 lb.	–	–	–	7 lb.
72"	5 lb.	38 lb.	8 lb.	61 lb.	–	–	–	–

The table tells you, for example, what to expect if you use cedar boards 3/4 of an inch thick, supported at 32-inch intervals. Without objectionable sag, they'll hold 54 pounds per square foot. So if each shelf is to have three spans, for a total length of 8 feet, and to be 2 feet wide, the calculation is 8 times 2 times 54. Evenly distributed, 864 pounds should be an acceptable load for the whole shelf.

Figuring Torque

WHEN YOU get into many mechanical things, even ordinary car repairs, you are likely to be told how energetically to do such things as tighten nuts.

An extreme example is a flywheel gland nut on some Volkswagens. Torque it to 220 foot pounds, they tell you.

"They" think you have a torque wrench that measures how hard you twist. But what if you don't?

Estimate by understanding what foot pounds of torque really is all about. This is just the number of pounds of pressure you're putting on, multiplied by the leverage.

Suppose the point at which your hand is exerting pressure on a wrench (or on the pipe sleeve you've slipped over the wrench to extend its handle) is 18 inches from the center of the nut. You're putting all your 150 pounds behind the push. Multiply the leverage distance in feet (18 inches equals 1-1/2 feet) by your 150 pounds of pull and you get 225. Which is almost exactly what that VW nut needs.

Though by no means as accurate as a torque wrench, this calculation at least brings you to a well-educated estimate.

If you'd rather be metric about it, the term is meter kilograms. It takes about 7.2 foot pounds to make one of them, so that gland nut has to be torqued to 30 mkg.

What Size Wrench?

IF YOU MUST choose a wrench to fit a nut or bolt that's not in the same place as your wrenches, you may be glad to know that there's a fairly reliable and consistent pattern to sizes.

And happily it's a relationship that's not hard to remember.

Multiply the diameter of the bolt—its thread diameter, that is— by one and one-half. For a half-inch bolt you'll want a three-quarter-inch wrench.

How Fat a Wire?

IF THAT MYSTIFYING unidentified copper electrical wire that you come upon in the course of household wiring is the thickness of a penny, it is No. 14. This is the size often used in wiring a household circuit for lighting only, and its capacity is limited to 15 amps.

Thickness equal to that of a nickel will identify No. 12 wire for you. This is the size required for appliance circuits so that they can safely be fused at 20 amps; and it is often used for 15-amp circuits as well, just to avoid having to fool with an extra size during wiring.

If the wire is as thick as two dimes, it's No. 10. Since this size can be fused to carry 30 amps, it is useful for single-purpose, fairly heavy-duty circuits on either 110-120 or 220-240 volts.

Radius of an Arc

AN EXPERIENCED woodworker says he understands there are various ways to calculate the radius of an arc. But this is the one he finds simplest.

First, draw a line connecting the ends of the arc.

(In doing this, if you place your yardstick or whatever you are measuring with so that this distance is an even number, your arithmetic will be simplified.)

Measure to the center of this line. Call the distance A.

At a right angle to the line at this point, measure the distance to the arc. Call that distance B.

Square A and square B and add those results together.

Divide by two times B.

You can now reproduce the circle whose arc you began with, because you've found its radius.

How Tall Is My Grandfather?

GRANDFATHER clock, that is.

When building a clock of the pendulum variety, one thing you'll need to know is precisely how long that pendulum must be.

You can find out from a rather simple formula. Square the period (time per swing) of the pendulum. Multiply by the acceleration due to gravity. Divide by 4 times the square of pi.

That's not so hard. You may recall pi as 3.1416 or, have it on your pocket calculator. For gravity, use 32 feet (or for a metric answer, 975 centimeters) per second per second.

If you want greater precision you'll have to find someone to tell you the figure for your own little portion of the earth. It does vary just a trifle from place to place.

So for the length of a pendulum with a period of 2 seconds, square 2, multiply by 32, divide by 4 and by the square of pi.

If you don't have a pi-bearing calculator at hand, you can save time by using a figure I've just pulled out of mine. To find the length of the pendulum in feet, multiply the square of the time period by 0.81057 (or, for most purposes, just by 0.81). For the length in centimeters, use 24.7 instead.

The formula behind this works in reverse, naturally. So you can calculate the period (swing time) of the pendulum from its length.

Just divide the length by that acceleration constant, take the square root, and multiply by 2 times pi.

Quick example: The square root of 8/32 is 1/2, which multiplied by twice pi has to be just plain pi. So an 8-foot pendulum will have a period of 3.1416 seconds.

Chapter P
   YOUR CAR

What Does a Car Cost?
How Long to Charge a Battery
What Do Skid Marks Tell?
Going Sixty
Analyzing Gas-Station Lines
Engine Displacement

What Does a Car Cost?

WHAT'S THE BIGGEST item of expense in automobile ownership? That's right—for most of us it's depreciation.

What's second? Gas? Insurance? Repairs?

Actually, for most owners it's none of the above. It's the cost of money. This refers not only to the interest you and I pay but also the hidden interest we lose on the hard-earned stuff that's tied up in our cars.

Surprisingly, this robust cost figure is missing not only from most published material—but even from that offered by statistical organizations and by the federal Department of Transportation. It's a neglect that harks back to those happy days when interest costs were half what they've become and fully deductible to boot.

All of which makes interest a good place to start in adding up what it costs you to operate your car for a year.

Interest. Financially speaking, your car is divided into two parts. There's the part that the bank or other loan agency owns; and there's your equity, the part you own.

To find the cost of the first, add together the sums owing on the car at the beginning of the year and the end of the year and divide by two. This is the average amount you pay interest on, so multiply it by the APR or average percentage rate in your contract.

Next add the market value of your car (what you could reasonably expect it to fetch if put up for sale) at the beginning and at the end of the year in question and divide by two. From this, subtract the average amount of the bank's equity, found in the previous paragraph, to determine your average equity. Multiply this by the rate of return your money could be earning if not tied up in a car.

To find your total interest cost, add this interest you lose to the amount you pay.

Depreciation. This is both a biggie and a toughie. It varies enormously from one make of car to another. So the loose generalizations tossed off for estimating this are to the real world as one-size-fits-all shirts are to actual people.

However, if yours is a car of average price and popularity and you've paid full sticker for it, these approximations may prove close enough to be of some use: 35 percent of new price the first year, then

17, 13, 11, and 8. After the fifth year how well you've maintained the car will count far more than age.

Another fairly quick way to arrive at a meaningful figure for depreciation is to browse classified ads and compare cars like yours with similar ones a year earlier and a year later.

However, there's another tactic I've found much quicker and less tedious than perusing used-car ads. By consulting an automobile blue book, you can arrive at a quite reliable estimate. Bank loan officers and many libraries have these compact and detailed guides. By consulting two issues dated a year apart, you can find out how much value a specific model has lost in a year.

Since it's not always easy to get hold of two issues a year apart, I've worked out another way to produce an excellent estimate of depreciation for the year gone by or the year to come.

All it requires is one up-to-date blue book.

At one time, a current blue-book figure told me that my daughter's aging Civic station wagon should retail at $5445. A similar car newer by a year, also a 4-door with average mileage and standard equipment, was valued at $6465. So a quick subtraction suggested depreciation of $1020 for the last 12 months.

Similarly, when I looked up a Honda of my own that was three years older, I found its retail value given as $4255, which was $510 less than the figure for an equivalent model one year newer. This confirmed my suspicion that an average 6-year-old car tends to run up depreciation only about half as fast as a 3-year-old.

To estimate depreciation for the year to come, I could subtract the book value for the previous year's model from the current figure for my car.

The blue book will also tell you how to refine your figures by adding or subtracting for miles racked up, major accessories, and so on. But if you can keep these elements consistent within the group of cars, you can safely ignore them.

Blue books give two values for each car—low and high. Either will serve for making comparisons, but an average of the low and high—wholesale and retail—will be especially useful. It will be close to what the car might go for in a private deal. The main thing is to consider what you're going to use the result for—and then be consistent.

Gas and oil. This one's easier. Even without records you may know about how often you put in a fairly consistent number of gallons and how much you pay for each on the average. You may do better by making use of the odometer. Estimate yearly gallons from what that instrument says divided by the miles-per-gallon figure you've probably been keeping track of—or occasionally sampling—all along.

Insurance. Ay, there's the rub—for some of us anyway. Only when I began totting up figures did I fully realize what a major chunk of my driving expense insurance, especially liability, had become in recent years.

The cost of keeping a car conservatively insured can be right up there with interest and depreciation. For a young urban bachelor, it can top all other costs, especially if he drives something old (so that depreciation is negligible), high-powered (scares insurers to death), and sporty (which they don't especially admire either).

At least it's easy to predict the premiums. Just look at your bills or check stubs. And it's not hard to get quotations on a car you're planning to add.

Remember, though, that if you've got more than one car on a policy, taking one off may affect the rate on any other.

You could find, as I did, that the insurance premium is Cost No.1 on an old, low-depreciation car driven sparingly. This may well be the primary motive for getting rid of a second or third car that is not really needed.

Maintenance and fees. Since these costs, like those for insurance, tend to stay high however little you drive, this is as good a place as any to list and add in the registration fees and the charges for any required periodic safety checks.

Your maintenance budget, including tires and cosmetic as well as mechanical care, might start at a couple of hundred dollars or less the first year and rise to three or four times that by the third or fourth.

Very low mileage drivers will spend less, of course. If you're one of those, you'll make a better guess by suspecting somewhere between five and fifteen cents a mile, depending mostly upon size and age of the vehicle, who made it—and how you drive it.

Your own habits, experience, and any records you've kept will tell you the most about where your costs fit into the all-purpose estimate.

Garage, parking, tolls. A nickel or dime a mile is not an unusual figure for this trio.

However, for a low-mileage driver who incurs two or all three of these costs regularly, the total may be several times that. And for the many drivers who seldom encounter any of them, it may be too small to reckon.

Adding up parking and tolls is not hard, but drivers who do not keep track often seriously underestimate this category. Don't overlook fines.

As for garage, if your car is kept in your own space on which you pay rent or taxes whether there's a car in it or not, you may have no outlay here to charge against your car expenses. But if in the absence of a car the garage could and would become a rentable storage space, there's a cost worth counting. And that's equally true if in the absence of a car the garage space could be turned into a workshop for which you would willingly pay $50 to $100 or so a month.

That bottom line. Realistic analysis of your car costs, both visible and hidden, can lead to profitable decisions.

You may find that you can easily part with a car that your figures prove is costing you more than you thought—and more than it's worth.

You may happily discover that trading up to something far more comfortable or more fun than what you're driving will cost you little more in the long run. This can most easily happen when you move to a model notable for its low maintenance or low depreciation.

If you've been considering any of the touted leasing plans you'll now have sound comparative data. You can save quite a bit by not erroneously believing you're in one of the comparatively rare situations where leasing is thrifty. (More about this toward the end of this chapter.)

If you're putting business miles on a car, you'll now know whether to use actual costs for income-tax purposes. The alternative, of course, is to go with the per-mile figure the government allows.

As car prices and operating costs creep upward and some insurance premiums double in a single year—you have the best of times and the best of reasons for calculating precisely where you stand when you sit behind the wheel.

How Long to Charge a Battery

There's a simple formula for estimating how long it will take to charge a storage battery in your car and other equipment.

First you read the battery's capacity in ampere hours—on the battery. For most batteries it's between 30 and 130.

Then you divide that by the rating of your charger—3 to 10 amperes with most inexpensive home-type chargers.

The answer is how long it should take to charge the battery after it is run down. As you can see, this will take anywhere from about three hours to a couple of days.

Another use for this calculation is in choosing a charger. If you're likely to be charging a battery of 100 ampere hours or so, and you'd like to be able to do it overnight, you're going to prefer a charger rated at 8 to 10 amperes.

(Incidental intelligence: Since a typical battery will require about 1.2 kilowatt hours of power to charge it, the cost per charging operation will be around a nickel or a dime at most utility rates.)

What Do Skid Marks Tell?

Next to the shriek of brakes, there's no more disturbing reflection of highway disaster than those long black stripes of rubber you see on the pavement. And like a sentence printed in an unfamiliar language, when translated they have a story to tell.

They do a pretty good job of revealing how fast a car was going when its driver hit his brakes.

To interpret those cryptic skid marks, you use a physical law that says that braking distance varies as the square of the velocity. It takes a potentially disastrous 9 times as long a distance to stop a car going 90 miles an hour as one going 30.

In cases of major importance, law officers may use elaborate methods for converting skid-mark measurements into estimated speeds. These may include measuring the actual coefficient of friction of the road surface. They may take into account special circumstances when the friction was not the same for all wheels, as when one side of the car skidded along a muddy shoulder.

But generally the calculation will depend on coefficients established by measurements on typical surfaces. Here are some figures in miles per hour for skid marks 10 feet long.

Ice, 6
Packed snow, 8
Wet oiled gravel or wet brick or concrete, 12
Wet asphalt or wet or dry gravel, 13
Dry brick, concrete, asphalt, or oiled gravel, 15.

To apply the table, divide the measured length of the skid mark by 10 and find the square root. Multiply this by the figure listed for the pavement and its condition.

For an 88-foot mark on wet asphalt, divide 88 by 10 and then find the square root of 8.8, which is 2.97. Multiplying by 13 gives 38.56. In recognition that all the assumptions and measurements you're working with are approximations, it would be sensible to round this off to 39.

If you'd taken the quite reasonable course of rounding 88 feet to 90 in the first place, 39 mph is the result you'd have obtained anyway, in a matter of seconds.

Going Sixty

REMEMBERING that 60 miles an hour is 88 feet a second makes various mental calculations about driving easy.

At 55 miles per hour, the figure becomes an even simpler 80 feet a second (80-2/3 to be precise).

It may be salutary sometimes to contemplate at 55 how much closer than 80 feet you are to a car or anything else ahead—and how brief a second is for responding to an emergency.

Analyzing Gas-Station Lines

WHY IS IT that at certain periods, as in 1974, motorists found it necessary to wait in lines a block or more long to buy gasoline?

Was it because of a shortage of gas, as most of us assumed at the time?

No. In fact, during much of the apparent crisis there was more gas than usual in customers' tanks. For fear of not being able to buy the stuff tomorrow or the next day, motorists were keeping their tanks nearly full at all times, to the tune of half a billion gallons in what might be called mobile storage.

The only real shortage was in service-station time.

Since it takes nearly as long to dispense a couple of gallons as a tankful, hysterical customers strained station time before they began to exhaust fuel supplies.

We·may be able to avert future crises, now that the manner in which all this works has been figured out by students of queuing theory. That's a branch of the mathematics of statistics and probability.

What follows is an example that can serve as a pattern for forecasting many comparable occurrences. Here is how it goes for a typical situation.

A small gas station able to serve two cars at a time is open 60 hours a week. It takes 5 minutes to serve a car, so the station can handle as many as 24 customers an hour.

In a typical region, it might be found that such a station serves 300 customers who drive, on the average, 200 miles a week in cars that go 300 miles on a tankful.

Normally such a customer would stop for gas once a week. To meet this demand, the station would have to serve only 5 customers an hour, which is almost ridiculously far below its capacity of 24. Allowing for occasional chance stack-ups and for the likelihood of certain times of day being busier than others, the probabilities still remain more than 9 out of 10 that when you drive into the station you could be served instantly.

But suppose those same customers respond to rumors of a gas famine by topping up every time their gauges show them down a couple of gallons. They stop 5 times a week instead of once and the station is now being asked to handle 25 customers an hour. There will be very long lines. In fact, this condition calculates to a 30-minute average wait with 1 chance in 6 that your time in line would be more than an hour.

And this without any gasoline shortage at all!

Now suppose occasional very slight shortages of fuel do develop. To ensure—in all fairness—that every customer gets at least something, a station operator may begin limiting each sale. Result: even customers who were not coming in every time they burned a couple of gallons will have to turn up more frequently than had been their custom, compounding the problem.

Of course, if people will respond more rationally, curb their gimmes, fill tanks only when necessary, and meet any shortfall by curbing use proportionately...then, as these queuing-theory calculations show, we'll all do just fine.

Engine Displacement

SINCE THERE are almost exactly 61 cubic inches in 1 liter, you can use that factor to make the leap between the two measuring systems.

If the displacement is given in liters, multiply by 61 to find it in cubic inches.

If the displacement is given in cubic centimeters, divide by 1000 first—that is, slide the decimal point three places to the left, converting a 2000cc figure, for example, to 2 liters—and then do the multiplication by 61.

To go the other way, from liters (or cubic centimeters divided by 1000) to cubic inches, divide by 61 instead of multiplying.

If all you want is a reasonable equivalent in a hurry, you can estimate it from this:

3000cc is about 183 cubic inches; 5000cc is about 305 cubic inches.

To put it another way, multiply the cc figure by 6 and divide by 100 (set the decimal point two places to the left).

In the other direction (turning cubic inches into cubic centimeters) you just divide by 6 and add on two zeros.

I don't own a car with a rotary engine. Very likely, you don't either. But some folks do, so bear with me.

To figure the displacement of a Wankel rotary engine, multiply 3 times the square root of 3 by all of these factors: rotor eccentricity, rotor radius, rotor width, number of rotors.

(For occasions when you're doing a Wankel calculation without a square-root-finding calculator or a computer, that square root is 1.732 and 3 times that is 5.196.)

For the Mazda RX-2, later RX-3, and early RX-7 automobiles, these factors are 1.5, 10.5, and 7 centimeters and there are 2 rotors. So it seems that all these models have 1146-cc engines.

For RX-4 and American Cosmo models, the 7 becomes 8. For the Japanese Cosmo Sport, the R-100, and the early RX-3, the 7 becomes 6. For the NSU Ro80, the factors are 1.4, 10, 6.7, with two rotors. Same for NSU Wankel Spider, but only one rotor.

If you want the displacement in cubic inches, divide the cubic-centimeter result by 16.387.

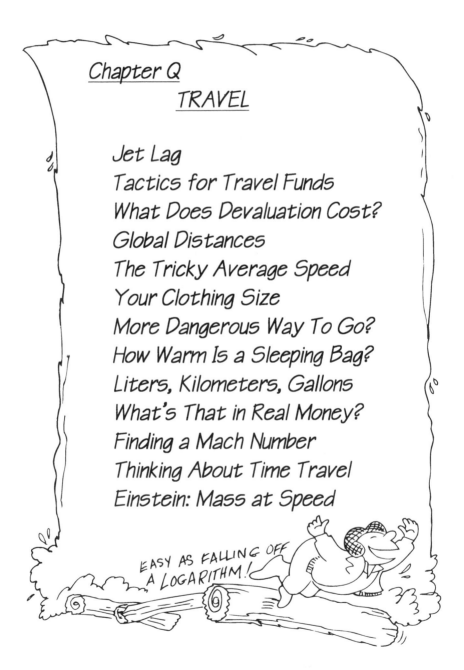

Chapter Q
    TRAVEL

Jet Lag
Tactics for Travel Funds
What Does Devaluation Cost?
Global Distances
The Tricky Average Speed
Your Clothing Size
More Dangerous Way To Go?
How Warm Is a Sleeping Bag?
Liters, Kilometers, Gallons
What's That in Real Money?
Finding a Mach Number
Thinking About Time Travel
Einstein: Mass at Speed

EASY AS FALLING OFF
A LOGARITHM!

Jet Lag

T HE MAIN thing most of us need to know about jet lag is how to minimize the effect of an air trip that drags your reluctant body across several time zones in a few hours. Some doctors at the National Institute of Mental Health have suggested a kind of fool-the-eye approach. It comes down to visually simulating the light conditions at your destination from the time you get on the plane till some days after you get off.

During the hours when it's daylight where you're going, keep yourself exposed to as much light of any kind as you can where you are. Similarly, during the hours when your destination is in darkness, simulate that condition by wearing an eye shade or dark glasses when you can.

From my own experiences over the years, I'd advocate trying to start this some days before getting on that plane.

But it isn't easy.

You may be delighted to find that science and informed observation have sprung to our aid in respect to another facet of the same problem. How many days rest do you need after a long air trip before you're fit for further serious travel or business?

Something called the International Civil Aviation Organization has worked out a formula for estimating this. These folks advise you to apply it to avoid making bad business or diplomatic or personal decisions while your body is out of tune with your new surroundings.

First, divide the number of hours of travel time by 2.

Second, subtract 3 from the number of time zones crossed.

Third, find your departure-time coefficient. If you leave between 8 a.m. and noon, it's 0; noon to 6 p.m., it's 1; for 6 to 10 p.m., it's 3; for 10 p.m. to 1 a.m., it's 4; and 1 to 8 a.m., it's 5.

Fourth, find your arrival-time coefficient. If you arrive between 8 a.m. and noon, the coefficient is 4; noon to 6 p.m. it's 2; for 6 to 10 p.m., it's 0; for 10 p.m. to 1 a.m., it's 1; and for 1 to 8 a.m., it's 3.

(Some of these seem a little too arbitrary. I'd suggest that if your timing is borderline, you split the difference; arriving at 1 a.m., for instance, use 2 as the coefficient.)

Now add your four figures together. Divide by 10. The result is the recommended number of recovery days.

To see how it works, consider a flight leaving New York at 7 in the evening and reaching Rome at 9 the next morning, crossing seven time zones in seven hours. The figures you get will be 3.5, 4, 3, and 4. Their sum divided by 10 is 1.35.

Your body will need about a day and one-third to get over this flight.

And that's not even considering the possible effects of airline cuisine and the sleep you lost if you haven't bought first-class space or learned to curl up like a kitten.

Tactics for Travel Funds

O N A VACATION or business trip you'll probably use travelers checks to pay-as-you-go or charge cards to go-now-pay-later. Cash is too risky and the traditional bank letter-of-credit is outdated as too much trouble for most circumstances.

Between checks and cards, which is the better economy?

If you carry checks with you, you'll have paid for them in advance. When you use cards you postpone the outlay until the chits drift in days or weeks later and then get billed to you—when you still have something like a month of free interest coming before you have to pay. You can count on gaining at least a couple of months interest, which could be worth $100 to you on a $5000 trip.

Using cards may also save you the fee that is sometimes still charged—though frequently waived—for travelers checks. Against this you should reckon the annual fee for the card unless you can get it free or would be paying rent on it anyway.

There's another offset for some travelers. Since not so many of the less-expensive establishments accept the cards, wanting to use yours may persuade you into a costlier place than you'd had in mind. (Now you know why pricey hotels and restaurants are willing to split their take with card issuers. On the other hand, a few of the tonier establishments, notably fine restaurants, can get all the customers they can handle without offering plastic bait, so they don't.)

For foreign travel, another factor enters—rate of exchange. It was my clear impression during the 1970s that the charge-card people generally used less favorable rates than I could have obtained by exchanging cash or travelers checks as I went. But a shift seems to have occurred by the middle 1980s and my careful records show that exchanges in more recent years have been consistently in my favor.

The last I knew, both MasterCard and Visa were using a standard international rate, but where Master was adding a 1 percent charge to it when figuring exchange, Visa was taking only 1/4 of 1 percent. If you want to save a bit by bargain hunting even before your foreign trek begins, ask your card suppliers about their policies in this respect.

The difference in rates between card and check will be even more likely to favor use of the card if the alternative is cashing a check in a hotel or restaurant. When the merchant sets the rate he is likely to round it off pretty roughly and not in your favor, either.

(How much this can hurt I found out one rainy night in a Shell station on the German Autobahn. With my gas running low I was forced to accept the offered rate for my dollars even though it was some 10 percent worse than I'd been getting at banks. I wouldn't have minded that percentage levied on the cost of the gas, but it applied to the whole $100 travelers check that unfortunately was the only kind I had left.)

When exchange rates are fluctuating, the card user won't know how much he's spending till the bill comes months later. The exchange rate is not set at the time of the purchase but days or even weeks later when the ticket turns up at the bank. You could win. You could lose.

To discover, when you receive your bill, how much you've lost or gained by using the charge card, you might try this sequence I've used to check a bill of mine.

At the time and at my bank, anyway, MasterCard (unlike Visa) bills did not show the rate of exchange being used. Fortunately, I'd retained the original charge slip, which came to 860 Portuguese escudos. I had only to divide this figure into the $20.48 that MasterCard was billing me for to discover the exchange rate used: 0.0238.

My trip diary showed that, on approximately the day of the charge, I was getting a rate of 0.0225 at Portuguese banks—a difference of 0.0013. Dividing this difference by the 0.0225 rate told me that I was paying a penalty of 0.0578—or 5.78 percent. In return for paying by card I'd had use of the money for a couple of months. That's one-sixth of a year, so I multiplied the penalty by 6 to discover I was being assessed interest at an annual rate of about 35 percent. Too much.

This experience, you'll note, was contrary to most recent ones. This time, I'd lost instead of gained by using the card.

If you snobbishly distinguish between tourists and travelers, you'd have to say that all this is becoming less important, at least for the former. The big cities and other highly developed places, where tourists throng, commonly offer access to those remarkable machines that read your bank card and thrust cash into your hand. And, amazingly, do it at lower cost to you.

Unless you have evidence to the contrary, you probably should give the nod to the iron teller—if you can find one where and when you need it. Even dedicated travelers, those who disdain the beaten paths, are now coming upon them even in many unlikely places.

What Does Devaluation Cost?

You COULD BE traveling, or importing, or investing. How can you figure what happens to you when there's a change in the value of one or the other of two currencies you're dealing with?

Let's say you were in Germany when the Deutsche mark was revalued upward 10 percent. That's simple enough. Everything you bought in Germany would still take the same number of marks—but you'd have to pay precisely 10 percent more dollars for each mark you bought.

Now suppose you were in New Zealand (as I was) on the day in 1973 when the U.S. dollar was devalued by a flat 10 percent. You might suppose that your tour down under had just become 10 percent more expensive, and indeed that's what most newspapers and bankers told people.

But they were wrong. It was worse than that. From that moment on, the New Zealand dollar began to cost Americans 11.1 percent more. And so it followed that costs for people traveling with American dollars instantly increased by that 11.1 percent.

How come? Well, from that point on, the tourist was using a 90-cent dollar. Add 10 percent of 90 to that and you're up to 99. It takes 11.1-plus percent more than 90 cents to make 100.

If that's a little hard to swallow—and it is—try an exaggerated case.

When your currency drops to half its former value you will need twice as much of it to do the same job. So a devaluation of 50 percent (half) means a cost increase of 100 percent (double). For calculations, you'll need to figure out how much more the old dollar was worth than the new one—instead of the other way around. When devaluation is 10 percent the calculation you want is this: What percentage of 90 is 10?

That calls for the usual formula—multiply 10 by 100 and divide by 9.

Which gives 111.1 percent, or 11.1 percent more than before.

Global Distances

YOU CAN FIND the distance between any two points on the earth's surface by using either of two methods—bearing-and-range or rectangular-coordinates. Either way, the procedure is unavoidably pretty complex.

You won't need either of those methods, and you might prefer to avoid them, when the two points happen to lie on the same meridian. Here's one for that situation. It involves no tools more difficult or unfamiliar than the value of pi.

If pi is not built into your calculator, you'll probably recall that it is commonly expressed as 3.1416. You can easily type it in.

If both of your points lie in the same direction, subtract one from the other to find the number of degrees they are apart. If they're in different directions—one north of the equator and one south, for instance—add them together. Change minutes and seconds to decimal degrees.

Multiply this distance in degrees by 7917.6 and by pi and divide by 360. For example, if you've started with points 6 degrees and 30 minutes north and 18 degrees south of the equator on the same meridian, you'll find they're just over 1692 miles apart.

If, by the way, you don't recognize that odd number, 7917.6, it's the diameter of the earth in miles.

The Tricky Average Speed

IF YOU DRIVE from here to anywhere at a steady 20 miles an hour and return at an equally steady 30, what's your average speed?

Sounds simple, but the problem's difficult enough to have found its way into college entrance examinations. As you might guess from this, the answer is not 25. That's the average of the speeds, true enough, but it's not the average speed.

Since you have spent more of the trip driving 20 than 30, you must find a way to give the slower speed more weight in computing the average.

How it works is easier to understand if you assign some arbitrary distance to the trip.

It's neatest if you make it a distance that works easily with the figures you have.

So call the one-way distance 60 miles.

Then you must have driven at 20 mph for 3 hours and at 30 mph for 2 hours. It took you 5 hours to go 120 miles. That's 24 mph.

Since there are many work problems ("If John can build a fence in 5 hours and Bill in 3, how long will it take the two of them together?") and pump problems ("If one pump, or pipe, can fill a tank in 40 minutes and another…") that follow the same principle, it may be worth while to look at a quicker, though less instantly clear, way of solving such problems.

To begin with, remember that the reciprocal of a number is what you get when you divide 1 by that number. Many hand calculators will find this value for you at a single punch of the key marked 1/x.

The formula is brief: Divide the number of trips by the sum of the reciprocals of the speeds.

In our example, divide 1 by 20 and 1 by 30 and add the two fractions together. Divide that result into the number of trips—two—and you'll again have 24.

Your Clothing Size

NOT ALL Continental and British clothing sizes can be transmuted into their American counterparts by simple formulas, but many can. As follows:

Men's suits, overcoats, sweaters, pajamas: English and American sizes are the same. Add 10 to get the Continental size. (Example:

English 40 is U.S. 40 is Continental 50.) For U.S. small, medium, large, extra large, try 38, 40, 42, 44 in Britain.

Men's hats: Add the fraction one-eighth to the English size to find the American. (That is, if you wear a size-7 hat in the U.S., buy a 6-7/8 hat when in London. But consider it risky to buy for anyone not there to try it on.) Continental sizes are 8 times U.S., our 7 being a European 56.

A shirt in London and a shirt in New York have the same collar size. To find your Continental size, just multiply by 2-1/2, and round off upwards if necessary. At home, when I buy a decently marked shirt—not one of the small-medium-large abominations—I choose something with a 15 collar. In Europe, I multiply by 2-1/2 the easy way, getting 37-1/2. That rounds up to 38, which has proved to be just right.

(The easy way? Multiply by 5 and divide by 2.)

Shoes: Subtract 1 from the U.S. size to find the British equivalent. To discover the Continental size, try adding 30 to 33 to the U.S. size. (But this won't work with some of the smaller sizes and is not entirely consistent from one country to another.)

Women's blouses, dresses, sweaters: Try adding 2 to U.S. sizes in Britain, 8 on the Continent. Where that produces ridiculous results, make it 22 and 28 instead.

More Dangerous Way To Go?

"IF YOU ARE worried about your chances of being killed on a coast-to-coast trip, you won't get much relevant information by asking whether trains, planes, or cars killed the greatest number of people last year. Get the rate, by inquiring into the number of fatalities for each million passenger miles. That will come closest to telling you where your greatest risk lies."

This advice, published in a book of mine about how to lie with statistics, fetched a letter from an engineering student in Atlanta: "I strongly disagree on rating fatalities in passenger miles. It seems to me they should be plotted with respect to time exposed to the danger...."

This is by no means hairsplitting. You'll come to quite a different conclusion (you should forgive the expression) if you use one method than you will if you use the other. It is evident that airline people are going to prefer to figure by the mile and train or bus or car folks by the hour.

I think this is a case where you can make sense only by applying real-world, or human standards.

Perhaps your inner dialogue runs like this. "I'm going to fly or drive for the next eight hours. Which is more dangerous?" If that's what you want to know, seek figures using passenger hours.

I think it far more likely, however, that you will be deciding between air and surface for a trip in which the constant is the distance—here to Chicago, say, or across the country, as in my example.

Then it's passenger miles that count, not hours.

How Warm Is a Sleeping Bag?

WARMTH of a sleeping bag is pretty tricky to measure objectively. It depends on bag design and size and stuffing material—and of course on who is sleeping in it.

I can testify to this after hundreds of nights in down- and Dacron-stuffed bags of various shapes and designs, in climates from those of Fiji, New Zealand, and Baja California to Romania and Finland.

The common method of comparison for down bags—by total weight of stuffing used—doesn't really tell the whole story. Four pounds of down in an ineptly designed rectangular bag may not keep you as warm as two pounds in a mummy bag of overlapping-tube design.

In comparing bags of similar construction, figure goose down as the warmest insulation, duck down next, and the more costly of the synthetic fibers as an excellent third.

To compare bags of varied construction (mummy, rectangular, and semi-rectangular shapes; different methods of arranging the down) the best criteria I've found are quality of down—and loft.

WIND CHILL CHART—EFFECTIVE TEMPERATURE

Actual Temperature °F								
Wind Speed MPH	40	30	20	10	0	-10	-20	-30
Effective Temperature °F								
10	28	16	4	-9	-21	-33	-46	-58
20	18	4	-10	-25	-39	-53	-67	-82
30	13	-2	-18	-33	-48	-63	-79	-94
40	10	-6	-21	-37	-53	-69	-85	-100

U.S. ARMY—THICKNESS-COMFORT DATA

Effective Temperature °F	Thickness of Insulation Required for Comfort (in inches)		
	Sleeping	Light Work	Heavy Work
40°F	1.5 "	.8 "	.20 "
20	2.0	1.0	.27
0	2.5	1.3	.35
-20	3.0	1.6	.40
-40	3.5	1.9	.48

Loft is how thick the wall of the bag is when you've fluffed it up—as you should when getting into it on a cold night.

A bag that lofts to 3 inches (for each wall, not the total height of the bag) should keep you comfortable down to 20 degrees.

Add enough down to increase the loft by half an inch and you should be able to take a temperature drop to zero.

These calculations should hold for average wind conditions. If you're well protected or there's no wind at all, you'll be comfortable at even lower temperatures.

To be more scientific about it while allowing for extreme conditions, the U.S. Army has come up with a thickness-comfort chart, to be used along with a simplified wind-chill calculator. It takes into account the familiar fact that when the wind blows, it can turn chilly weather into freezing.

(As a kid in the American middle west I used to enjoy February tennis occasionally—after brooming snow off the court—on still days well below freezing. But more recently a gale wind helped frostbite my ears painfully in the mere minute or so it took me to dash from my camping bus into a Turkish border station. At 9 in the morning, the guards were—wisely—still sacked out.)

To use this pair of charts, read across from wind speed to the column under the actual temperature in degrees Fahrenheit.

This figure for effective temperature will tell you how thick the army thinks your insulation should be. You will note that the chart is for clothing as well as sleeping equipment.

Example: Opposite wind speed of 20 mph and under the 0 degrees temperature column, you will find that the effective temperature for this combination is 39 degrees below zero. The nearest temperature to this in the thickness-comfort chart is minus 40, which calls for 3-1/2 inches of insulation for sleeping.

Liters, Kilometers, Gallons

Americans touring abroad find gasoline quantities reasonably easy to understand. Distances in kilometers, too. But fuel consumption is something else again, since it depends on what you might call a double or simultaneous translation—between liters and gallons and between kilometers and miles.

Figures of the kind you see here in the third column are often listed in car comparisons in motor magazines. That's why a traveler put together this little table.

I've emphasized one handy set of figures that would be easy to memorize and use to compare one car's appetite with the hunger of another.

KM per liter	KM per gallon	Liters per 100 KM
3	7.1	33.3
4	9.4	25.0
5	11.8	20.0
6	14.1	16.7
7	16.5	14.3
8	18.8	12.5
9	21.2	11.2
10	23.5	10.0
11	25.9	9.1
12	28.2	8.3
13	30.6	7.7
14	33.0	7.2
15	35.3	6.7
16	37.7	6.3
17	40.0	5.9
18	42.4	5.5

What's That in Real Money?

THE TRAVELER, real or armchair, has frequent need to figure the value of foreign currencies in American dollars. Doing this with a paper and pencil is not difficult. With a pocket calculator, even a $2 model, it's downright easy...even when extreme precision is wanted.

So let's look at the various procedures that may be called for when there's time to make a calculation. Spur-of-the-moment estimates that must be made during actual shopping in a foreign country are a bit trickier, but we'll get around to those in a moment.

Dollar exchange for the Danish crown may be listed as 0.1759. This gives you the decimal fraction of a dollar that a crown is worth: in other words, about 17-1/2 U.S. cents. So that Kr 10,000 fur coat in the window will set you back by a mere $1759. You have found this figure by multiplying the price in Danish currency by the value of the dollar today.

Sometimes you have to go the other way. Your Danish cousin has unblushingly asked you how much the Volkswagen you're driving cost you back home. (Danes, even more than most Europeans, are delightfully unencumbered by those feelings of delicacy about prices that afflict Yanks.) About $9,500 you tell him. "Okay, okay," he says, "but how much is that in real money?"

To calculate in this direction, you divide the price by the rate: 9,500 divided by 0.1759 is 54,008 *kroner* and never mind the few odd Danish cents.

The wistful glint that comes into his eye reminds you that the car would run him far, far more than that, with Danish taxes.

Sometimes you'll find currency quotations in a different form: the number of units of the foreign currency required to make a dollar. That crown will be listed not as about 17-1/2 cents but as about 5.7 to the dollar. Then the route from U.S. price to Danish is by multiplication.

Quite often you'll find a quotation in one form when you'd rather have it in the other. To make the swap, just divide 1 by the number. To get technical about it, the two rates are reciprocals of one another.

With a pocket calculator having a constant function built in (most do) you just punch in the number and then hit the division key three times. More elaborate calculators have a reciprocal key to save you half a second; it's the key marked 1/x.

A budget-conscious tourist abroad will usually need speed more than precision. What he really wants to know is whether that pizza or *croque monsieur* on the menu is in the two-dollar neighborhood or more like five or ten. Never mind exactness.

So his first fiscal move upon reaching a new country will be to learn the rate in whichever form is more manipulable by mental arithmetic. His second will be to reduce it to a simple approximation formula.

Sometimes it's easy to do this. If you find the French franc at about 20 cents (where it is as I edit these lines) you need only divide by 5. (Or multiply by 2 and divide by 10. Take your choice.)

When German marks, Swiss francs and Dutch guilders were hovering close to 40 cents, they were pretty easy to figure. Multiply by 4 and divide by 10. That 60-D-Mark tripod you were considering became $24 worth almost instantly in your mind. As you multiply by 4, your brain shoves the decimal over by a point without your consciously thinking about it.

But what do you do with a franc that's quoted at 22 or 23 cents or some other unfriendly figure? At 22 cents or thereabouts, I'd say divide by 5 for a rough approximation, getting more precision (if called for) by then increasing that result by 10 percent or one-tenth.

At an exchange rate of anywhere around 0.225 dollars, it's more accurate, and almost as easy, to divide by 4 and then reduce the result by one-tenth. When you're dealing with a rate around 24 to 26 cents, you'll come close if you divide by 4 and—on major amounts—then reduce or increase by half of one-tenth.

Finding a Mach Number

Since the Mach number for the speed of an aircraft is simply that speed divided by the speed of sound, the idea is simple enough. Mach 2 merely means twice as fast as sound. This is convenient, since aviation buffs love Mach numbers.

The apparent simplicity, however, can mislead. It overlooks the fact that the velocity of sound changes with altitude, making the actual computation more complex than it would at first appear to be. But tackled a step at a time it is quite straightforward and, with the aid of a pocket calculator having the powers key found on even the most modest scientific models, not difficult.

Given instrument readings of pressure altitude (PALT) of 25,500 feet and calibrated airspeed (CAS) of 350 knots, here's how it goes.

1. Divide the CAS by 661.5 and square the result (getting, in our example, 0.2799).

2. Divide by 5, add 1, raise to the 3.5 power, and subtract 1 (getting 0.2101). Save this result for later use.

3. Multiply the PALT by 6.875 and divide by 1 million by moving the decimal point 6 places to the left (getting 0.1753).

4. Subtract this from 1 and then raise the result to the power of -5.2656. (In our example this should give you 2.7592.)

5. Multiply by the number stored in Step 2. Add 1. Raise to the 0.286 power. Subtract 1. Multiply by 5 (getting 0.6984).

6. The square root of this (0.8357 in our example) is the Mach number. The plane is flying at a little more than eight-tenths of the speed of sound.

Thinking About Time Travel

WHAT WOULD HAPPEN if you were to mount a time machine, as in the famous H.G. Wells story and ride a few years into your past? Paradoxes would fly like confetti.

You might meet your younger former self and start a disturbing conversation. You would understand the situation but he would not. About the healthiest conclusion he could come to, everything considered, is that although you know an amazing lot about him, you are probably crazy.

A way to look at such a situation from the outside was devised by Richard Feynman, a Nobel Prize winning physicist. To follow the life of a time traveler, just place a ruler or even a piece of paper horizontally on the drawing.

Move it slowly upwards. You begin with the man's birth, following his life until he's about twenty. Suddenly, then, he appears simultaneously on another track at the age of forty. Now as the ruler slides upward you are following both his lives until he reaches forty on the left-hand track—and vanishes. By this time, he's sixty on the right-hand track, and he goes on from there.

Our man's time travel is represented by just a line on the diagram, since on this he's traveling backward in time and so is invisible.

Interesting.

Einstein: Mass at Speed

ACCORDING TO relativity, the mass of an object changes with the speed of its motion. This is described numerically by what I believe is called the Lorenz-Fitzgerald contraction.

The phenomenon is not especially important on the highway, or even at supersonic air speeds, but it is for spaceships as they approach the speed of light (186,000 miles per second) so perhaps it behooves us to have this formula ready. You never know when you're going to need something.

Step 1. Square the speed of light. From this, subtract the square of the velocity of the object. Find the square root.

Step 2. Multiply the speed of light by the mass of the object (at rest).

Step 3. Divide that result by the result of Step 1. This answer tells you the perceived mass relative to the given velocity. In the calculations, you naturally must be consistent in the measures used. If you use the speed of light in miles per second, then you must express the velocity of the object in miles per second as well. If you feed in mass in pounds, the result will be in pounds.

Check the procedure against these results: A 5-kilogram object, traveling at 93,150 miles per second (just half the speed of light) will have a perceived mass of 5.7735 kilograms, which is rather a significant weight gain. It suggests what would happen as a spaceship and its passengers came closer and closer to the speed of light.

If you find any of this confusing or doubtful, recall the fine old anonymous limerick (of which the version here is probably as authentic as any) from the days when the writer Gertrude and the sculptor Jacob were almost as much in the public eye as Albert was:

There's a wonderful family called Stein.
There's Gert and there's Ep and there's Ein;
Gert's poems are bunk,
Ep's statues are junk,
And no one can understand Ein.

Chapter R   OUTDOORS

I REALLY THINK YOU'VE GOT IT!

Weigh a Mountain, Measure a Pond

To ESTIMATE the weight of a great mountain—or a mere pile of dirt—think of it as a series of cylinders piled one on the other like a stack of pancakes of varying sizes. Then apply to each cylinder the standard formula of height times pi times the square of the radius to get an approximation of the volume.

To turn your volume figure into weight, you need to have some idea of what the stuff it's made of weighs per cubic foot or meter. In pounds per cubic foot, that's chalk 143, hard coal 94, concrete 137, earth 75 if loose and 100 if rammed, granite 168, gravel 109, gypsum 150, ice 56, iron slag or marble 168, limestone or quartz, 162, sand 100 dry and 125 wet, sandstone 143, slate 175.

Sometimes you can get a sample of the material and weigh a cubic foot or other measure of it.

If your mass is more or less in the shape of a cone, you'll save time and possibly come closer by treating it as such. Handiest version of the formula for the volume of a cone: Multiply the height by the square of the diameter and by pi (3.1416 or 22/7 will do for this) and divide by 3.

Same approach works well, naturally, for finding the area of any irregular surface, such as a body of water. Draw it on square-ruled paper. To the full squares, add estimates of each of the partial ones.

In anything like this, it's important to round off your final figure. A precise conclusion founded on estimated data is an illusion and a snare.

Water Over the Dam

FOR MANY purposes, including construction of a water-powered electrical generating system, it is necessary to measure the flow of a stream with some accuracy.

(If you don't need precision, it's easier to use the estimating method described in the next item.)

WEIR GATE

DAM PERPENDICULAR TO FLOW OF STREAM

RULER

5'

TOP OF STAKE LEVEL WITH BOTTOM OF WEIR

HEIGHT OF WATER ABOVE STAKE

STAKE

PACK MUD BEHIND DAM FOR WATERPROOFING

WIDTH SHOULD ALLOW ENTIRE FLOW OF STREAM TO PASS THROUGH WEIR

⅛" MAXIMUM

BEVEL PLANK

90°

90°

THIS EDGE TO BE LEVELED

The method giving the more accurate result begins with construction of a kind of simple dam with what is called a weir gate.

The weir is a rectangular opening in the top of the dam, at its center. This opening should be at least several times as wide as it is high. It should be level and its bottom must be beveled to 45 degrees so that the edge the water passes over is no more than 1/8-inch wide. Width and depth must be such that the whole flow of the stream can pass through the opening.

Of course your dam must be directly across the stream—that is, perpendicular to the flow of the water.

WEIR MEASUREMENT TABLE

Table shows water flow (in cu. ft./min.) that will flow over a weir one inch wide and from 1/8 to 20 7/8 inches deep.

Inches	$\frac{1}{8}$	$\frac{1}{4}$	$\frac{3}{8}$	$\frac{1}{2}$	$\frac{5}{8}$	$\frac{3}{4}$	$\frac{7}{8}$
0	0.00	0.01	0.05	0.09	0.14	0.19	0.26	0.32
1	0.40	0.47	0.55	0.64	0.73	0.82	0.92	1.02
2	1.13	1.23	1.35	1.46	1.58	1.70	1.82	1.95
3	2.07	2.21	2.34	2.48	2.61	2.76	2.90	3.05
4	3.20	3.35	3.50	3.66	3.81	3.97	4.14	4.30
5	4.47	4.64	4.81	4.98	5.15	5.33	5.51	5.69
6	5.87	6.06	6.25	6.44	6.62	6.82	7.01	7.21
7	7.40	7.60	7.80	8.01	8.21	8.42	8.63	8.83
8	9.05	9.26	9.47	9.69	9.91	10.13	10.35	10.57
9	10.80	11.02	11.25	11.48	11.71	11.94	12.17	12.41
10	12.64	12.88	13.12	13.36	13.60	13.85	14.09	14.34
11	14.59	14.84	15.09	15.34	15.59	15.85	16.11	16.36
12	16.62	16.88	17.15	17.41	17.67	17.94	18.21	18.47
13	18.74	19.01	19.29	19.56	19.84	20.11	20.39	20.67
14	20.95	21.23	21.51	21.80	22.08	22.37	22.65	22.94
15	23.23	23.52	23.82	24.11	24.40	24.70	25.00	25.30
16	25.60	25.90	26.20	25.50	26.80	27.11	27.42	27.72
17	28.03	28.34	28.65	28.97	29.28	29.59	29.91	30.22
18	30.54	30.86	31.18	31.50	31.82	32.15	32.47	32.80
19	33.12	33.45	33.78	34.11	34.44	34.77	35.10	35.44
20	35.77	36.11	36.45	36.78	37.12	37.46	37.80	38.15

With this done, drive a stake into the stream bed at least 5 feet upstream from the weir. Make its top just level with the bottom edge of the weir. With the steam flowing freely, use a ruler to measure the depth of water over the top of the stake.

The table figures water flow over a weir one inch wide with depths up to 20-7/8 inches.

Choose the number from the table that represents this depth of water. You'll notice that the first column is in exact inches, while each other column is for that number of inches plus the fraction given at the top of the table.

Multiply the number you've found by the width of the weir, the opening, in inches. The result will be the flow of your stream in cubic feet per minute.

Here's an example. You've built a weir 4 feet wide. You find that the water flow establishes itself at a depth 5-1/2 inches above the top of the stake. Opposite 5 inches in the column headed by 1/2 you find the number 5.15. Multiplying this by the weir width of 48 inches gives you 247. Your stream is flowing 247 cubic feet per minute.

Water Power Calculated Quickly

With a dam and a water turbine, how much power can you develop from a small stream?

How can you do the measuring and calculating quickly—in short, without first building one of those temporary mini-dams called weirs?

Take these steps:

1. Measure the width of the stream in feet.
2. Measure the depth of the stream in feet.
3. Find the average velocity.

(One way is by stretching strings across the stream, one 20 feet downstream from t'other and counting the seconds a cork takes to float from the first string to the second. Divide 20 by this number. Multiply by 0.7 to reflect the fact that the bottom and edges of a stream move more slowly.)

4. Measure the depth in feet of the fall at the dam.
5. Multiply these four figures together and divide by 23. (This assumes your turbine will have the usual efficiency of about 50 percent. If you know otherwise, divide this answer by 50 and multiply by the actual percent figure to compensate.) Result is how many kilowatt hours of power you can expect to develop.

For the answer in horsepower, divide the kilowatt result by 0.75. For a monthly total, multiply either figure by 720.

Try it for a small stream about 3 feet wide and 6 inches deep with a flow of 2.5 feet per second (that's 1.7 miles per hour) and a 3-foot drop. Multiplying together 3 and 0.5 and 2.5 and 3 and dividing by 23 gives the figure of 0.65 horsepower. In kilowatt hours per month it comes to about 360.

With a very tiny stream, by the way, it may be easier to measure the flow with a bucket. Divide the bucket capacity in gallons by the number of seconds required to fill it. Multiply by 0.14 to convert from gallons to cubic feet. Use this result instead of the three figures from the first three steps above.

Horsepower in a Stream

HAVING DETERMINED the flow of a stream that you're planning to dam in order to generate electric power, you'll very likely want to calculate how much usable power is there.

In addition to the flow (in cubic feet per minute) you'll need to know the head—the vertical distance in feet that the water falls.

Multiply these together.

Now multiply that product by the weight in pounds of a cubic foot of water (62.4) and divide by the number of foot pounds per minute that equals one horsepower (33,000).

(Instead of those last two operations, you can get almost the same result more quickly if you just multiply by 190 and move the decimal point in your result five places to the left.)

Either way, your answer is in available horsepower.

But what does that mean in terms of the amount of horsepower you can actually generate? Since a reasonable assumption is that your water wheel will be at least 75 percent efficient and your generator 80 percent, multiply the available horsepower by the product of these, 0.6, to find what might be called real, or switchboard, power.

Water and Kilowatts

THERE IS more than one way to estimate the power to be had from running water, and this may be just the formula you're looking for.

How much electric power you can develop from a stream depends upon two things. One is how fast the water is flowing. The other is how far the water drops, which is called the head.

Having measured the rate of flow in cubic feet per minute, express in feet the distance the water drops.

To find the power available, in kilowatts, multiply those two figures together and divide the result by 709.

So if your stream flows at 20 cubic feet each minute and has a head of 50 feet, it could generate just over 1.4 kilowatts (20 times 50 divided by 709).

To this information you might want to apply a rule of thumb that says a combination of flow rate and head that produces no more than half a kilowatt (500 watts) probably isn't worth developing.

But you'll decide that for yourself. It is, after all, enough power to keep a pretty bright light going indefinitely.

How Far Can You See at Sea?

THE RULE for figuring out how many miles you can see at sea (which naturally depends upon how high up you are) goes like this:

Multiply the height of your eye above sea level in feet by 3, divide by 2, then take the square root.

Say you're on a big ship or up in the rigging, or whatever, of a lesser one, so your eye is 96 feet above the water line. Divide 96 by 2, multiply by 3, and you get 144—of which the square root is an easy 12. From 96 feet in the air, you can see about 12 miles. Of course, you won't often come out with a perfect square like 144, but your trial-and-error estimate of the square root will be close enough for most purposes.

The original formula created for a slightly different purpose, the one from which I derived the above, is equally useful. It says you can find how high you must climb, in feet, by squaring the number of miles you wish to see and then multiplying by two-thirds.

The Accurate Cricket

THE PALE green snowy-tree cricket, common in the Americas, is often called the thermometer or temperature cricket. He (or, for all I know, she) is so sensitive to temperature that you need only count chirps for 15 seconds and add 39 to determine the temperature in degrees Fahrenheit with remarkable precision.

If you think better in Celsius, you might want to count some chirps and work out a formula of your own. Or simply add 7 instead of 39 to that 15-second count, then multiply by 5 and divide by 9.

Or take up the matter with your local cricket or entomologist.

Camera's Hyperfocal Distance

IN THE GREATER part of the photos most of us take, we want the distant background sharp and as much as possible of the foreground as well. Even an autofocus camera won't always do that for us.

A way to insure that result when we want it is to take over command from the camera's workings and set the lens at its hyperfocal distance.

This refers to the nearest distance for which you can set your camera and still keep far objects sharp. With this setting, everything from infinity down to half the hyperfocal distance will be sharp.

This fact makes hyperfocal distance a sort of multi-purpose setting, a good one on which to leave your camera between uses. If you then must take a picture in a hurry—or if you forget to check the focus—there's a good chance all will still turn out well.

Hyperfocal distance depends on two things: focal length of the lens and the f-stop you're using. To find it, choose any f-stop and set your distance at infinity. Read the near limit of the depth of field. This, the closest point that is sharp when the camera is set for far distance, is the hyperfocal distance.

Now set your camera so it is focused at the hyperfocal distance. Everything from half that distance to infinity will be in focus.

Here are a couple of examples.

With the distance scale on my old twin-lens reflex set to the farthest point possible, the depth scale on the side tells me that at f/22 everything will be sharp from 15 feet on. So I reset the focus to 15 feet, knowing that everything should be sharp from half of 15 feet to infinity. The depth-of-field scale confirms this.

Applying this to a 35-millimeter camera with a 50mm lens, I find that at f/8 with focus at infinity, the nearest point of sharpness is 10 meters. If I set the camera at 10 meters, I can be confident of sharpness from about 5 meters to infinity.

With the same lens set on f/16 to get greater depth, I find that the hyperfocal distance is only 5 meters. Everything should then be sharp—from 2.5 meters to the moon.

Sometimes it's more convenient to find hyperfocal distance by calculation.

The focal length of a lens is engraved on its barrel. Multiply this quantity by itself and then divide by the f-stop you're planning to use.

Divide also by what you consider to be an acceptable circle of confusion, or blur. A commonly used figure is 0.03 millimeters.

So for a typical small-camera 50-mm lens, to be used at f/8, the calculation proceeds like this:

The square of 50 divided by 8 and by 0.03 is 10,416. Changed to meters, that's 10.4, which confirms the 10 meters found by the direct experiment. And you know that if you use a 10-meter setting while working at f/8, everything from half that distance to infinity will be in focus.

Need a Ped Xing?

THE SUBJECT of pedestrian crossings, those marked rights of way for foot traffic the British call zebras (they like to stripe them) reminds me of something I heard about a foreign visitor to the U.S. The only American phrase he encountered repeatedly and couldn't find in his phrase book, he said later, was ped xing.

If you were to come into contention with traffic authorities about whether a pedestrian crossing is needed somewhere that you go, you might have use for a formula. It could also serve you if you should find yourself on a town planning committee of some kind. Or in a P.T.A. where safety of children was involved.

A formula has been arrived at by the Department of Environment, in England. It is called PV2.

It says that to justify establishing a marked pedestrian crossing, you must first count the number of pedestrians crossing the road in one hour. Then count the number of vehicles, square that number, and multiply it by the first number. If the product exceeds one hundred million, the ped xing is justified.

If your count shows 100 people crossing in an hour at a point used by one vehicle every four seconds...well, that isn't quite enough. (Fifteen cars a minute is 900 in an hour; 900 squared times 100 is only 81 million.). But make that a car or truck every third second and PV2 is more than satisfied.

By the way, you're not permitted to count organized kids in this. That is, children on the way to school are presumed to be protected by special traffic wardens with those little round signs on sticks. (I am delighted to discover that the British call these guards lollipop men.)

So much for zebra crossings.

As for what are known in England as pelican crossings—traffic lights used on straight stretches of road to stop traffic when a pedestrian pushes a button—there PV2 has to reach 500 million.

Acceleration from Gravity

IGNORING air resistance (which, of course, you can't with relatively light objects unless they are in a vacuum) you can figure that a freely falling object increases its velocity by about 32 feet per second during each second of its fall.

Using this figure you can find out just about how fast an object will be falling after a given number of seconds. Or you can find out how far it has fallen if you know long the fall has taken.

Such a useful figure may be worth knowing more precisely. It is generally given as 32.173 feet per second per second, for those who want it more exactly than 32. However, because the earth is not a perfect sphere, even this may not be precise enough to please you, since acceleration from gravity varies according to latitude by as much as one half of one percent. The figure of 32.173 is close only for latitudes of about 45 degrees.

For 0 (the Equator) to 5 degrees, add 0.09 to the working approximation of 32. For 10 or 15 degrees, make that 0.1; for 20 degrees, 0.11; for 25, 0.12; for 30, 0.13; and for 35, 0.15.

For 40 degrees, use 0.16; for 45, 0.17, as mentioned; for 50, 0.19; 55, 0.2; 60, 0.21; 65, 0.23; 70 to 75, 0.24; 80 to 85, 0.25. For 90 degrees, use 0.26. You are at one of the poles.

With a table of trig functions, or even more quickly with a scientific calculator or a computer, you can dig out yet more precise acceleration figures. All you have to do is subtract the cosine of twice the latitude from 32.173.

How To Bank a Road

Hᴏᴡ ᴍᴜᴄʜ slope should be built into a curve in a road to counteract the tendency (that we call centrifugal force) for the car to run off the road?

Though you may not build many roads yourself, the calculation might interest you the next time this happens to your car. Sloppy road engineering—or was your speed different from what the road builder had in mind?

The gradient in degrees can be found by multiplying the square of the velocity in miles per hour by 0.067 and dividing by the radius of the curve in feet.

Thus if you are planning a curve of radius 2500 feet to handle traffic at 60 miles an hour maximum, you would multiply 60 by 60 and by 0.067, and divide by 2500.

Result is the information that you'll need a lateral gradient, as it's called, of 0.09648. Which times 100 is 9.648 percent.

If the road is to be used by heavy vehicles as well, the builder may have to compromise. With slope more than 6 percent, a big, slow-moving truck might slide off the road when it's icy.

With a road that is already designed or built you can work backwards to find the ideal, or balancing, speed for a vehicle taking a curve of known radius and gradient.

Say the radius is 2400 feet and the gradient 6 percent. The balancing velocity will be 3.86 times the square root of the product of the gradient and the radius of the curve.

Multiplying 2400 by 0.06 gives 144, of which the square root is 12. And 12 times 3.86 is 46 miles per hour.

Underpass for a Bicycle

Wɪᴛʜᴏᴜᴛ going into just how many of us are likely ever to be called upon to design an underpass, it is worth noting that one of the things the world needs is more and better bike paths. They're major ammunition in our continuing, and I like to think, accelerating battle to wrest the world back from the automobile. Well, anyway, here's how Kenneth Claxton—architect of some of the best cycleways anywhere—at the English new town of Stevenage—sums up the formula:

"Two up, one down."

That means elevate the roadway for cars two yards and drop the bike path one yard.

Make that meters and it should still be a close enough approximation of the ideal, in our metric future.

I hope quite a few people will have occasion to use the formula soon. *Vive la bicyclette!*

When the Tide?

F̶OR BOATING or other salubrious purposes it is often useful to know at just what minute the tide will reach some specified height.

There is a procedure for finding this out, starting from the information offered in widely available tide tables. The calculation is straightforward arithmetic, except for one step that calls for the use of a trigonometric ratio called the arc cosine or reverse cosine.

A cosine, as most of us can recall from high school, is simply the number you get when you divide the base of a right triangle by the hypotenuse (longest of the three sides). The arc cosine of a number is the reverse. It's simply the angle of which your number is the cosine.

The cosine of 30 degrees is 0.866; so the arc cosine of 0.866 is 30 degrees. A scientific calculator will get it for you at the touch of a key or two. So will a computer. Or you can read it from a table. You don't actually need to know all that stuff in the previous paragraph to use it. They do say, though, that the unobserved life is not worth living.

For navigational or other reasons, you might like to know at what time on a morning shortly before Christmas the tide will be at 4 feet at Sandy Hook, N.J. It is a day for which a tide table gives high-water mark as 4.8 feet at 6:21 and low water of 0.3 feet at 12:40.

If those heights were given in feet and inches, you'd have to change them to decimal feet or express them as feet plus twelfths of a foot (5 feet 4 inches becoming 5 plus 4/12 feet). If times are not given on a 24-hour clock, it would be necessary to add 12 to any p.m. hour. Any minutes given should be treated as fractions of an hour, for example by expressing the time of 6:21 as 6 plus 21/60 and 12:40 as 12 plus 40/60.

With that out of the way, here's the procedure.

From the chosen height—the one for which you want to know the time—subtract the low-water height. (In our example that means taking a minus 0.3 from 4, giving the result 4.3.) Next, divide that by the result of subtracting from the high-water height the low-water height. (Divide 4.3 by 5.1, getting 0.843.)

Now multiply by 2 and subtract the result from 1 (1 minus 1.686 is -0.686).

That's the number for which you must find the arc cosine by calculator or table. (Arc cos -0.686 is 133.336.)

Divide by 180 (getting, in our example, 0.7408).

Multiply by the difference between the time of high water and the time of low water (which here means taking 12 plus 40/60 from 6 plus 21/60 to get -6.317, and multiplying by the previous result to come up with -4.679).

Add the time of low water to get the answer in decimal form. You can change the decimal part to clock time in hour and minutes by dropping any decimals after the first two and multiplying those remaining by 60. (In this example, adding together 12 plus 40/60 and -4.679 gives, when rounded to two decimal places, 7.99, or almost 8 o'clock. Multiplying the decimal part of that by 60 gives 59 minutes. So the 4-foot tide we've been running through all these steps to predict can be expected at 7:59 in the morning.)

How High the Tide?

WHEN YOU'RE about to sail under a bridge, or go fishing, you might want to predict the height of a tide at some specific hour.

To do so you could begin by taking, from a tide table, the times and heights of low and high water. You would convert any of these times that were expressed in hours and minutes into hours and fractions (common or decimals as you prefer) of hours on a 24-hour clock. You would do the same thing with the time for which you wish the information. (So 8:15 p.m. would become 20 and 15/60 hours. Or 20.25.)

Heights should be in feet and fractions or decimals, not in feet and inches. Of course if you're working with meters there won't be any problem. Good for you.

So here we go to find the height of the tide in Boston harbor at 8 o'clock on a December evening.

From the published table, we know that high tide will be 10.3 feet at 9:13 p.m., and low water will be -2.1 feet at 5:02 p.m. (Don't overlook the minus sign.) The table also tells us the time of high water, 23 hours 13 minutes, and the time of low water, 17 hours 2 minutes.

Now proceed.

From the time for which you need to know the height, subtract the time of low water (20 minus 17 and 2/60 equals 2.97).

Divide that result by the difference between high-water time and low-water time (23 and 13/60 hours minus 17 and 2/60 is 6.19), getting the result 0.48.

Multiply by 180 (getting 86.40).

Using a computer, a scientific calculator, or a trig table (even a slide rule if you really want to) find the cosine of that result. (You'll find the cosine of 86.40 to be 0.06.)

Next multiply by the difference between the high-water height and the low-water height (10.3 feet minus a minus 2.1 feet comes to 12.4, so now you have 0.78).

Divide by 2 (bringing your result so far to 0.39).

Subtract this from the average height of the tide, found by adding high-water height to low-water height and then dividing by 2.

(In our example, you're adding 10.3 to a negative 2.1, and dividing by 2 to get 4.1. Subtracting the previously obtained 0.39 from this gives you about 3.7 feet. And that's the height you may expect the tide to reach at the hour you've chosen to fish or to sail your boat under a bridge.)

Ship's Propulsion Power

How much power does a ship need?

To figure that out, you naturally have to know some things about the ship in question. You must know its displacement tons and the required normal speed in knots—as well as a constant, called the Admiralty Coefficient, for ships of its particular class.

Then you can find the horsepower needed by squaring the displacement tons and taking the cube root of that result, then multiplying by the cube of the speed in knots and dividing by the Admiralty Coefficient.

It's not as difficult as it sounds, if you have a table of cube roots and an ordinary calculator. You don't even need a table if you have an ordinary calculator and employ the method given in this book for calculating cube roots.

With a scientific calculator, this formidable little operation is actually a cinch; you do the first two steps in one by asking for the 2/3 power.

You might like to give the procedure a trial run. See how it goes for the U.S.S. Constitution, displacing 30,090 tons, having a normal speed of 22.5 knots and an Admiralty Coefficient of 370.

Square of 30,090 is 905,408,100, for which the cube root is found to be 967.419. Cube root of 22.5 is 2.823.

Multiplying these two results together, and dividing by the coefficient of 370, tells you that this 1951 ship requires 29,782 horsepower.

How Far the Projectile?

You fire a shot.... The gun crew of a cruiser shoots a cannon....

How far will these or any other projectiles travel when launched at a given speed and a known angle with the surface of the water or flat terrain? At what angle will the projectile travel the farthest?

To answer either question, begin by squaring the velocity, in feet per second, and dividing by 32.

With the velocity expressed in meters per second, square as above but divide by 9.8 instead of 32.

Either way, call this Answer 1.

(Those divisors, by the way, are the acceleration per second, per second, with which a body falls—in a vacuum—at a point near the earth's surface.) The figures are approximations, but don't worry about that. They are as good as any for most calculating purposes, since ignoring air resistance has already introduced a modest error.

(If you'd like a more precise average value for the earth's surface generally, use 32.174 feet instead of 32; divide by 9.80665 instead of 9.8 meters. If you know you're dealing with a point near the poles, use 32.2575 feet or 9.8321 meters. At the equator, use 32.086 feet or 9.7799 meters.)

Next step:

Multiply the launching angle by 2 and find the sine of this result (in a trig table or, more easily, by pressing that handy SIN button on a scientific calculator). Multiply this sine by that Answer 1 you obtained previously, and you will know how far the projectile will travel.

If you want to find out what angle to choose for the longest possible flight, you can use a trial-and-error technique. It is not at all tedious with a calculator.

Double each of several possible launching angles, multiplying the sine of each by Answer 1. You'll soon find the maximum result comes when you use 90 degrees—not surprising if you've noted that the sine of 90 is 1 and that they don't get any bigger than that. So the elevation for maximum flight is 45 degrees.

Which you may not find too amazing, either.

Quick Height Measurement

THIS IS ONE WEIRD way to determine the height of a distant object, such as a tree. But it works, so it must be anatomically correct.

Hold a foot ruler vertically at arm's length as you approach the tree or other object to be measured. Stop when one end of the ruler lines up with the top of the tree and the other with its base. Now pace off the distance to the tree, taking normal strides. The number of paces will be the approximate height of the tree in feet.

This trick obviously can be adapted to various circumstances. Clearly your measuring stick can be longer or shorter as long as you multiply its length in feet by the number of strides.

You can refine the method for greater precision by tuning your stride until you get consistent precise results for an object of known height.

What if you can't conveniently get far enough away from the tree to sight it in this manner? Well, there's a trick for this situation too.

Stand a yardstick or other object of known height against the trunk. Back off just far enough to sight on this stick with your ruler. Raise the ruler up the tree in jumps. The number of increments it takes to bring you to the top will tell you how many times as tall the tree is as the length of the stick.

JUST A LITTLE FURTHER...

How High the Shot?

A PLANE is going by up there and you are on the deck of a cruiser with a cannon. You determine that the plane is at an angle of elevation of 70 degrees. Given a muzzle velocity of 700 feet per second, can a shell from your cannon reach the plane if it is flying at 4000 feet? How about 6000 feet? Or 8000?

To find the maximum height of a projectile, you must square the velocity, multiply by the square of the sine of the angle of elevation, then divide by twice the acceleration of gravity.

That acceleration constant is 32 if you're working in feet, 9.8 if you're working in meters.

Whether you find the sine from a trig table or by the touch of a key on a scientific calculator, you can use either the square of the sine or the sine of the square of the angle. They come to the same thing.

So in this example you multiply 490,000 by -0.8830 and divide by 64 to get 6761 feet as the maximum height your shell can reach under the conditions given.

How Long to Fall?

IF YOU DROP something down a well, or fall off a mountain, how many seconds will it take the object, or you, to hit bottom?

Answer: one-fourth the square root of the distance in feet.

Drop a rock off a 400-foot cliff and it should arrive in 5 seconds, the square root of 400 being 20, one-fourth of which is 5. A fall from a mile-high airplane should take one-fourth of the square root of 5280, or just over 18 seconds.

Actually, it will take longer, since our formula neglects the resistance of the air.

A large, dense object will fall almost as fast as the formula says it will, especially until it reaches quite high speeds and the resistance increases. Even when both are made of the same stony stuff, a small pebble will be slowed more than a huge boulder will. A pillow will not fall so fast as a heavy suitcase of similar size and shape.

J.B.S. Haldane describes the difference vividly in a classic essay, On Being the Right Size.

"You can drop a mouse down a thousand-yard mine shaft; and, on arriving at the bottom, it gets a slight shock and walks away. A rat would probably be killed, though it can fall safely from the eleventh story of a building; a man is killed, a horse splashes."

The formula we're using here is based on an acceleration of 32 feet per second per second, which is a slightly imperfect approximation. The true figure, in fact, varies from one place on the face of the earth to another.

Would you rather do your figuring in metric? Start with the square root of the distance in meters. Divide by 2.214.

Where on Earth Is It?

THERE ARE TWO common ways of describing where one point on the earth, whether on land or sea, is in respect to another. It is often necessary to translate one of these descriptions into the other, especially in navigation.

One of the methods is called bearing and range, which means simply in what direction and how far away. An example is "14.14 miles northeast of Podunk Post Office."

The other familiar method is by rectangular coordinates: "10 miles east of the post office and 10 miles north."

Both describe the same point. But how do you get from one to the other? Here's an example, courtesy of the U.S. Navy.

"A contact is at bearing 075 degrees, range 20,000 yards. What are the rectangular coordinates?"

The north (also called the x-axis) coordinate is found by multiplying the cosine of the angle (75 degrees in this example) by the range (20,000 yards here) to find that the object is 5176 yards to the north. Repeating, but with sine instead of cosine, tells you it is 19,319 yards east—on the y-axis, that is.

If, on the other hand, you know the coordinates and wish to find the range and bearing, you naturally reverse the procedure. But perhaps the simplest approach is to start by squaring each of the coordinates, adding these squares together, then taking the square root. That's the range. The bearing, in degrees from north, is found by dividing the east coordinate by the north coordinate and then deriving the arc tangent.

Some of this may sound formidable, but it can be done with any calculator, or a home computer, or even paper and pencil. The easiest way is on a scientific calculator having rectangular-polar capability built in, making for a fast five-step solution.

Your calculator may be one (very likely from Hewlett-Packard) that uses what is called reverse-Polish logic, and is quickly identifiable by the absence of an equals sign.

To translate from bearing and range with this breed of calculator, start by entering the angle. Putting in the range and pressing the rectangular-conversion key produces the first coordinate. Pressing the x-y exchange key gives the other.

To go the opposite direction, enter one coordinate, then the other, press the polar key for range and the exchange key for the bearing.

If you're using the more common type of scientific calculator, one with what's called algebraic logic, enter the range and then press the x-exchange-t key and enter the bearing. Press the polar-rectangular conversion key for one coordinate and that same exchange key to find the other. For the reverse routine, successively enter x-coordinate, exchange, y-coordinate, rectangular-polar, exchange.

Speed and Power

WHEN YOU know the propulsion power a ship requires at one speed, how can you determine how much that power must be beefed up to permit some higher speed?

Or, knowing the normal power and speed, you wonder how much that speed will change when available power is increased or decreased.

The rule for this is one that says the power is proportional to the velocities cubed.

So what will happen to the normal speed of the vessel when there's a change in horsepower from, say, 29,782 to 35,000? To find out, divide the proposed horsepower of 35,000 by the present horsepower of 29,782, find the cube root, and multiply by the present speed of 22.5 knots.

Turns out the new velocity is 23.7 knots.

Not much improvement.

What if the horsepower were doubled from the present 29,782? Instead of applying that cube rule to the speed itself, it's quicker and easier to apply the rule to that simple ratio of 2:1. First, find the cube root of 2 and multiply that by the present speed of 22.5 knots. Available speed goes up respectably to 28.348 knots.

Suppose you know the speed available at present (22.5 knots, say) and the speed you need to power for (let's make that 25 knots). If you're the U.S.S. Constitution with a present horsepower of 29,782, to what must that be increased? To find out, divide the present velocity of 22.5 by the 25 knots you want, cube it, and divide the present horsepower by that result.

To get this modest increase in speed, up your horsepower must go—all the way to 40,853.

Figuring Air Pressure

IF YOU KNOW air pressure in pounds per square inch and need it in kilograms per square centimeter, what do you do?

You can consult a log table, if you can find one.

You can convert each of the factors separately and then divide.

Or you can make use of a remarkably simple formula:

Multiply by 7 and then move the decimal point two places to the left.

Thus 16 pounds per square inch becomes 1.12 kilograms per square centimeter.

Siting a Windmill

IN CHOOSING a site for a windmill, you'll presumably wish to avoid the turbulence that would interfere with its smooth operation. Proximity to buildings or other obstructions is a frequent cause of this condition.

Here are some rather crude general formulas that have been found to deal with this problem.

If the windmill is upwind, raise it above the ground by twice the height of the obstruction. You should have little interference when placing a windmill downwind if its distance is at least ten times the height of the obstruction. If you must place the windmill very close, go up to three times the building height.

How Much Wind Is Enough?

TO MAKE a windmill a cost-effective power source, it has been found by experience that you'll need an annual average wind speed of at least 10 miles an hour.

If this information leads you to a difficult decision, however, it would be prudent to check it against the latest research in novel windmill designs.

You will probably find that it could take you a long time to gather data for your own particular location.

However, if there is a weather station anywhere near you in surroundings at all like yours, data from it that shows 12 miles an hour or more is a pretty good guarantee that you have suitable conditions.

If it's a little less—10 to 12, say—you will probably want to make some tests.

To avoid making your tests all year round, you can run several over a short period and use them to establish the relationship between your wind speeds and those at the nearest airport or other weather station.

If you divide the average figure measured at your site by the one reported at the same time by the station, you'll find a factor by which you can multiply the official records from other times of year—to estimate seasonal conditions at your site.

Thus you might make three measurements over a short period in the spring and find 11, 13, and 14 miles an hour, for an average of 12.67. (The kind of average we're after here is the simple one, called an arithmetic mean, that you get by adding together all the quantities and dividing by the number of them, in this case 3.)

If observations made at the same time at your nearest airport are 12, 13.5, and 15, your average divided by theirs turns out to be 0.94.

Multiplying by this factor will tell you that if the airport averages 15-mile winds in December, yours should be about 14.

That Wind-Chill Factor

THE WEATHER sometimes feels a lot colder than the thermometer says it is. This is especially true when the wind is blowing, for the same reason that food solidifies faster in your freezer or locker plant if a fan is blowing on it.

You will recognize this as the phenomenon familiarly known as the wind-chill factor.

To get a good idea of the effective temperature in Fahrenheit, multiply the wind speed in miles per hour by one and one-half and subtract that result from what the mercury is trying to make you believe.

That's why 10 degrees above zero with a 50-mile gale can freeze you as fast as a quiet 65 below.

That's why it may be prudent to believe what our ears, as long as they are unmuffed, are trying to tell us.

How Many Stars?

SOMEDAY, you might—for all I know—want a trifle of information to support an argument favoring the existence of extra-terrestrial life.

Most useful for this, perhaps, would be knowing how many stars there are. Here are some handy—if immense—figures to work from.

Sir Arthur Eddington's old rule still holds up well enough in light of modern findings: The number of stars in a galaxy is about 10 to the 11th power, and—coincidentally, I suppose—so is the number of galaxies in the universe.

Since this easily written and easily remembered number, the 11th power of 10, is shorthand for 1 followed by 11 zeros, it signifies that there are quite a few of whatever it is that you apply it to.

Taking the whole thing a step further, we can multiply the 11th power of 10 by itself to find that the number of stars in the universe is on the order of 10 to the 22nd power.

This, if you've forgotten what the teacher told you more than once, follows from the rule for multiplying together the powers of any quantity. Just add those superscripts that indicate powers. Thus 10 to the 11th times 10 to the 11th equals 10 to the 22nd.

Which comes to 1 followed by 22 zeros or, in plain American English, ten thousand billion billion.

In plain British English, where billion is U.S. trillion, I guess it comes to ten thousand million billion or perhaps ten thousand milliard milliards. In any case I am clearly getting out of my depth, and I doubt that a billion or two will affect the argument anyway.

To Rate a Quake

WHEN THE BIG ONE comes and half of California drops into the sea, some of us living along the Coast may want to rate the magnitude of the earthquake.

Here is a recipe for converting comparative intensities into magnitudes on the widely used Richter scale. By knowing how many times more powerful one quake was than another and also having the Richter rating of one, you are in a position to determine the Richter figure for the other.

Divide the known or estimated intensity of the second quake by that of the first. Find the common logarithm of this figure, by using a scientific calculator (that's the easiest way) or a table. Subtract this from the Richter rating of the second quake.

It goes like this. The famous San Francisco event that people there prefer to think of as the 1906 fire (but that nevertheless has been rated 8.25 Richter) was 105 times stronger than the Nicaraguan earthquake of 1972. That statement in effect puts the relative intensities at 1:105, so it is 105 of which the common log (not the natural log, please note) is needed. That log is 2.02. Subtracting from 8.25 puts the Nicaraguan quake at 6.23 on the Richter scale.

Figuring Binoculars

SOME CALCULATIONS are useful in selecting binoculars.

To begin with, binoculars are usually identified by two performance figures: 7x35. This tells you first the magnification (objects seen through it will appear to be only one seventh as far away as they actually are) and, second, the diameter of the objective lens—the one farthest from your eye—in millimeters.

You can also use these two figures to find a third one, called exit pupil. This is the diameter of the image or light-disc as it leaves the eyepiece. Since the pupil of the human eye ordinarily has a diameter between 4 and 5 millimeters, binoculars with exit pupils under 4 millimeters don't take full advantage of the seeing power of the eye. On the other hand, binoculars with exit pupils of more than 5 millimeters are providing more light than the eye may be able to make use of. Note, though, that after you've been in the dark for a while your pupils get larger and can make use of a somewhat greater exit pupil.

Since the exit pupil is found by dividing the second of our figures by the first, it comes to 5 for the common 7x35 binoculars. This corresponds well to the capability of the typical eye.

Something called "relative brightness" is often used to compare binoculars. It is found by squaring the figure for exit pupil. Do this for 7x35 binoculars and you are first dividing 35 by 7 to get 5 and then squaring that to get 25. For 7x50 glasses, it works out to 51. This tells you the 7x50 gives twice the brightness of the 7x35.

That's all very well, but the human eye doesn't see doubled brightness as doubled. It will perceive the one as no more than one-third brighter than the other. Apparently it is for this reason that some optical manufacturers, especially in Germany, prefer to use what is called the "twilight factor."

To find this factor you multiply the power by the objective diameter, then take the square root.

For the 7x35 it works out to 7 times 35 is 245, of which the square root is 15.7. But 7 times 50 is 350, with a square root of 18.7. This relationship comes closer to what your eye might tell you, so perhaps is more useful in making a decision.

(The 7x50 instrument is likely to be a good bit costlier than the 7x35, as well as bulkier and heavier. In making a decision it may be helpful to know that in exchange for the added bulk, weight, and cost you are getting an improvement that your eye will see as being in the ratio of 15.7 to 18.7—roughly one-fifth brighter.)

Another thing you may have need to calculate in order to compare binoculars is field. You may find this printed on binoculars, along with magnification and diameter of the objective lens. It is usually expressed in a rather eccentric fashion: as the width in feet of the field you can see at a distance of 1,000 yards. Some makers, however, prefer to express it in degrees, so in comparing instruments you may need a conversion factor.

To convert degrees to field in feet at 1,000 yards, multiply by 52.5. Thus the fairly common 7-degree field becomes 367 feet. Binoculars with fields of more than about 9 degrees fall within the loose designation of wide angle.

Chapter 5
FUN AND GAMES

How Far Can a Gamble Miss?
Will You Win a Trifecta?
Balancing a Tournament
The Cool, Calculating Tennist
Basic Blackjack
Doing Better at 21
Actually Winning at Blackjack
Horseshoe Math

How Far Can a Gamble Miss?

Aₗₜₕₒᵤɢₕ millions are eager to try, no one can predict what a roulette wheel or pack of cards or pair of dice or any other instrument of chance will do today or tomorrow.

Good thing, too, else where's the fun?

There are, however, ways to predict with considerable success what each of those instruments will do in that well-known period, the long run. (But remember: we live in the short run.)

The wheel, for instance, will produce, within a very small range of deviation, an equal quantity of reds and blacks, of odds and evens, and of each of the various numbers.

As for the short run, well, there is a kind of predictability there, too, that has been reduced to numbers. It is considered reasonable and proper to say that when a result turns up that is further off than would be expected to occur 99 percent of the time, that is unreasonable, and we should strongly suspect that something other than chance is at work.

By this test, what would you say of a roulette wheel that has been found to have produced red 6578 and black 6442 times over a period of observation? That's fairly close to the expected fifty-fifty, but is it close enough?

The deviation from even-steven can easily be calculated to be half the difference between the two figures, or 68. (Put in another way, 6578 minus 6442 is 136, and half of that is 68.) Does this mean anything? Can it be called significant?

Try this simple test.

Multiply the total number of trials by the product of the chance of winning and the chance of losing and then take the square root. Any deviation from fifty-fifty that is larger than the result given by this calculation can be expected to occur no more than 1 percent of the time.

Let's apply that general statement to the problem at hand.

Multiplying the number of trials (6578 plus 6442) by the product of the chance of winning and the chance of losing (one-half times one-half, or one-fourth, here) and then taking the square root gives an answer of 57.

But the deviation of 68 found earlier is somewhat greater than 57, so we have to call it highly unusual and perhaps even suspicious. Such a thing would not occur even once in a hundred times over the long run.

Will You Win a Trifecta?

A TRIFECTA is something that happens at race tracks. You and I don't hang around such places. But they do offer instructive mathematical models that can apply to many things. Anyway....

To win a trifecta, I'm told, you must name the nags that will come in first, second, and third—and name them in the right order, to boot. In an eight-horse race what would you say would be your chances of doing so?

Assuming you know nothing about these horses or the opinions of the odds makers on them, you'd have to give yourself 1 chance in 8 of predicting the winner. Then you'd have 1 chance in 7 of choosing the horse that will finish second among the remaining 7 horses. And 1 chance in 6 of predicting the third finisher.

That gives you 1 chance in 8 times 7 times 6—or 1 chance in 336, of hitting all three.

Expressed another way, the odds against you are 335 to 1.

When you combine things this way in a prescribed order you are dealing with what are called permutations. In this case it is the permutation of 8 things taken 3 at a time. The number of permutations is found by dividing the factorial of the number of things by the factorial of the difference between the number of things and the number of them to be taken at a time.

(A factorial of a number—for the benefit of those who may have forgotten or nodded off just long enough to have escaped learning— is simply the quantity you get when you multiply the number in question by all the smaller whole numbers. Factorial 5, which is usually written 5!, is 5 times 4 times 3 times 2 times 1, or 120.)

Thus the number of ways our 8-horse trifecta could come out is equal to factorial 8 divided by factorial 5. This is perhaps not easier to understand than the method we used first, but it is often a lot faster.

In real-life wagering you would probably have some information, or opinion at least, about the relative merits of the eight horses. You might estimate that the chance of the best of them winning was no worse than 1 in 2. You might believe the next horse of your choosing to have 1 chance in 3 of placing precisely second to your first-choice horse. And you might be able to choose another horse with what impressed you as 1 chance in 5 of coming in right after the other two. This would lead you to calculate that you have 1 chance in 2 times 3 times 5, or 30, of winning the trifecta.

You might also be wrong about the quality of your information or even your judgment. But that's the risk you assume.

If you don't like your gambles too wild, you might prefer to bet on what I am told is called a trifecta box. For this wager you choose the three horses to finish in the first three places but you do not have to say in what order.

What interests us about this, since we have agreed we're not railbirds, is this. As a mathematician would put it, we have changed a permutation problem into one dealing with combinations. The formula for finding the number of combinations calls for you to divide the factorial of the number of things by the factorial of the difference between the number of things and the number of them taken at a time (so far, just as in figuring permutations) and then also dividing by the factorial of the number taken at a time.

For an eight-horse race, that means factorial 8 (which comes to 40,320) divided by factorial 5 and factorial 3. It seems you have 1 chance in 56 of winning the trifecta box—just on chance alone— which is at least a lot better than 1 in 336.

For a shortcut in doing this kind of computation, you might note that factorial 8 is factorial 5 times 6 times 7 times 8. Canceling this factorial 5 with the one you're dividing by gets you the 336 of the previous paragraph by a simple multiplication of 6, 7, and 8. Brings the problem down to a level easily handled by mental arithmetic.

Balancing a Tournament

A SOCIABLE sort of tennis tournament with players of varied ability will be a lot more fun if you handicap it to even things up a bit. The difficulty with any ordinary form of handicapping is that you have to know your players' abilities in advance, although presumably this is what a tournament is supposed to find out.

Also you can get people sore at you unless your handicapping is done with care, and maybe even if it is.

Consider automatic handicapping. It works this way.

Suppose player A (or doubles team A) wins the first game. A then must start the second game down 0-15.

If he still wins, he starts the next game down 0-30.

If he wins again, he starts the next game down 0-40. His handicap remains 0-40 (or three points) as long as he continues to win.

Whenever he loses a game, the handicap decreases by one point. That is, when a player starts a game down 0-30 and loses it, he starts the next one down 0-15. If he loses that, the next game is played from scratch. And so on.

This system—which, by the way, can be applied to some other games than tennis—may remind you a little of vulnerability in bridge. The difference, however, is that being vulnerable at bridge increases the potential rewards as well as the penalties.

The Cool, Calculating Tennist

A<small>RE YOU</small> the conservative player who gets his first serve in practically every time? You're wasting that second serve you're allowed.

On the other hand, are you the belter who comes through with an average of a double-fault per game? You're beating yourself.

Your pocket calculator can help you work out a combination of first and second serves that will give you both a maximum first-ball attack and an acceptable minimum of double faults. A little later, we'll see how to apply the same approach to other games and gambling problems.

By keeping track for a while, you may find that you miss your first serve 33 times out of 71 and your second 10 times out of 33. Dividing the number of faults by the number of tries and multiplying by 100 gives you your failure percentage: 46 on the first serve, 30 on the second.

Multiply those percentages together and you'll find you're serving doubles nearly 14 percent of the time. No wonder you lose a lot of service games!

How can you win more often? Ask your calculator what will happen if you steady down those first serves until three out of four go in, second serves until eleven out of twelve do so. Result is equivalent to about one double fault in two sets—acceptable on any neighborhood court or even on the pro circuit.

The principle involved in this calculation, by the way, is one with many other applications. It says that the easy way to figure the chances of succeeding in at least one of two things is by multiplying together the chances of failing at each. Having done so, subtract that result from one (or from 100 percent, if you're working with percentages) to find the probability of success.

Basic Blackjack

F<small>OR ANYONE</small> interested in the strategies that follow but not too familiar with the game, here are the principal rules of blackjack as played in most casinos. Some places vary them a little.

Bets are placed first. Then each player gets two cards face down, while dealer gets one down and one up.

Each card counts its face value. Face cards count as 10s. Aces count 1 or 11, whichever gives a better total.

Winning hand is the one closest to 21—but not over 21.

A "natural" or "blackjack" is a 21 total reached with two cards. When the player receives it, he wins 1-1/2 times his bet. When the dealer gets it, the player loses his stake. When both get it, no money changes hands. (As you can see, this 1-1/2 rule gives the players an advantage over the house. They'll need it.)

The player now decides whether to accept more cards, one at a time. If he "busts" (goes over 21) he loses at once.

After the players, the dealer similarly "stands" on what he has or draws more cards. Unlike the player, he has no discretion but must keep drawing until his total is more than 16. When dealer has an ace and counting it as 11 brings his total to 17 to 21, he must stand (in many casinos it's 18 to 21).

If dealer busts he pays each player who did not bust an amount equal to his bet. If neither goes over 21, higher total wins. With ties, no money changes hands.

The house advantage, more than balancing the player's edge mentioned above, comes from one aspect of this rule: If the player busts, he pays even if the dealer later busts also. The customer has the substantial disadvantage of having to choose first.

I hope you'll forgive the digression, but I am irresistibly reminded of the story of a lone traveler who attempted to cross China during the revolution. Captured by soldiers and asked which side he favored, he answered that he was loyal to the government. The soldiers, being rebels, beat him severely before turning him loose. When captured again a little later, he hastily avowed support of the rebellion. The soldiers, this time government troops, beat him with even greater vigor. A few days later, having recovered sufficiently to resume his journey, he was taken prisoner a third time and again questioned about his loyalties.

This time his reply was, "You say first."

In addition to basic play, the player has some options he may take if he wishes.

One of these is called doubling down. After seeing his first two cards, the player may double his bet if he wishes. However, he must then turn over his two cards, accept one more card, and then stand.

The other option is splitting pairs.

If the player's first two cards are of equal value, he may turn them face up and receive an additional card on each, at the same time putting up a stake the size of his initial bet on the second hand.

He is now playing two hands. The usual rules apply to each, with one exception: If he has split aces (a good thing to do) he must then stand. A count of 21 reached after splitting is not scored as a "natural."

Another option, called "insurance," is usually unprofitable and better ignored. When dealer's up card is an ace, a player may make a side bet of anything up to half his original bet. Dealer then checks his own hole card to see if he has a natural, often called a blackjack. If yes, the player wins 1-1/2 times the amount of his side bet at once; if not, he loses it. Play of the hand then continues.

Doing Better at 21

E VEN BEFORE someone I love took a turn as a blackjack, or 21, dealer, I'd found the game the French call vingt-et-un (remarkably corrupted by the English to Van-John) especially fascinating. Something I said about blackjack in a little book I wrote years ago, on some applications of probability theory, even led me to become a footnote to an historic article in a journal of mathematics. That article launched its author, Edward O. Thorp, on a most unlikely avocation for a university professor. He began hanging around gambling casinos in Reno, Las Vegas, and elsewhere. He was bent on showing that a system had at last been evolved that would beat the house.

Unlike any past systems player who stuck around long enough to give his theory a real test, Ed Thorp didn't go broke. He made money so fast that eventually no casino management would let him anywhere near a 21 table.

How does Ed Thorp—and anyone else who is prepared to learn his methods—beat the house? Two ways: optimum strategy, and counting.

The first is what this article is about. A separate piece deals with the second.

Most of the profit the house makes at blackjack is contributed by customers who play badly, violating simple rules that any book on the subject could warn them about. Even the 21 dealers will often point out to players their obvious mistakes. They do this out of simple decency, but the casino managements do not mind; they know that the player who gambles well will enjoy himself more, stay longer—and lose as much or more in the end.

But when a player uses the very best basic strategy, something different happens.

For the first time in the history of commercial gambling, there comes a time when the customer has the edge. True, his edge is so small as to be negligible over any reasonable period of play. The point is not that the customer has gained a tiny advantage, but that the house no longer has the rather substantial advantage it holds over the ordinary player.

The result is that it is extremely unlikely that the player will, over a period of time, either lose or win much; the casino is giving him free entertainment at the table. This is quite remarkable when you consider how much it is costing the casino in overhead and wages to run a 21 table.

To realize what a triumph this new, computer-generated, strategy is, you have to compare it with previous tactics.

The popular "never bust" tactic of simply taking no more cards whenever your total has reached 12 or more (17 if your hand includes an ace) gives the house an edge of about 6 percent.

"Mimic the dealer," which means keep drawing cards till your total reaches 17 or more, shaves the house advantage to 5.7 percent.

By using a little more judgment—and studying the older books on gambling strategies—you can do even better. You can cut the house percentage to a mere 3.2 percent.

Now to Professor Thorp's "best basic strategy" for completely wiping out the house advantage. It consists of doing the following:

1. If your hand is "hard" (does not contain an ace) use the following method. When the dealer shows a seven or bigger card or an ace, keep drawing until your total reaches 17. When dealer shows

4, 5, or 6, stand when you reach 12 or more. When dealer shows 2 or 3, stand at 13.

2. If your hand contains an ace: Keep drawing until your total reaches at least 18. Exception is that if the dealer is showing 9 or 10, you draw until your total is 19. (If you go over 21 by counting your ace as 11, you automatically count it as 1—and follow the rule above for a hard hand.)

Merely following these rules cuts the house advantage down to about 2 percent. You can eliminate even this small edge by properly doubling down and splitting. (In some locations, casino rules will permit doubling down only on hands totaling 10 or 11, whittling further the player's tiny edge with the best strategy.)

3. First, the strategy for doubling down when your hand is hard: Never double down when holding 12 or more. Always double down with 11. Double down with 10 unless dealer is showing 10 or ace. Do so with 9 if dealer is showing a 6 or smaller card. If your total is 8, double down unless it consists of 6 and 2.

4. If your hand is soft: If dealer shows 2, double down only if the card with your ace is a 6. If dealer shows 3, double down if you have a 6 or 7. Always double down when dealer shows 4, unless you have a pair of aces. Always double down when dealer shows 5 or 6; never when he has 7 or more. (But if you hold a pair of aces, split them rather than doubling down, unless house rules forbid.)

5. Now about splitting pairs. Split aces or 8s always, 10s, 5s, or 4s never. Split other pairs when dealer shows 2 through 7.

This is an abbreviated version of the perfect splitting strategy, but is almost as good as the detailed form which is considerably more difficult to memorize. Thorp gives you a choice, along with a vast amount of other information in his Beat the Dealer, published in paperback by Vintage Books.

He has even calculated for you what you can expect will happen if you apply this basic strategy over a period of time. Suppose you were to play 1,000 hands at $1 a hand. This would take you from three hours up. You could lose a lot of money or win a lot, but either eventuality is very unlikely.

There is only one chance in a thousand that you'd lose or win as much as $50; and the odds are about two to one that you'd wind up within $15 of where you started.

But that's the nature of any gamble where, as when you play blackjack with the strategy outlined here, you're getting an even break. In the long run, you'll come out pretty close to where you started, which in a casino is quite an accomplishment.

Actually Winning at Blackjack

As EXPLAINED earlier in this section, you need only adopt an easy strategy to convert a casino's blackjack, or 21, table into a nonprofit institution. You can cut the house's winning percentage right down to zero.

To swing the casino's edge clear around to an advantage for you—in short, to beat the house at blackjack—you have to go a step further in your tactics. You have to "count."

A deck from which most of the high cards have been played on the first few hands builds up the casino's advantage. One that is rich in high cards favors the player. If you can remember which cards have been used (of those you had a chance to see) you can make heavy bets when the situation is favorable. When it's not, you can stick to minimum bets to keep your place at the table.

That is the essence of counting—but when it was first discovered and publicized, it was enough to throw the world of casino gambling into an uproar.

It's hardly possible to remember every card that has gone by and to remember just which strategy is then called for. There are several systems to get around the problem. Possibly the simplest of these is the point-count worked out by Edward O. Thorp. To use this system, you regard a 6 or less as a low card and 10s, face cards, and aces as high cards.

As play progresses, you count each high card you see as +1 and each low card as -1. You ignore 7s, 8s, and 9s. Thus after observing a 10, a jack, two 5s, a 9, and an ace, you'd have a running total of +1.

At all times when that running total is zero or minus, you make the smallest permissible bet, just to hold your place at the table. The rest of the time, you multiply the minimum bet by the total to determine your bet. At a $1 table when the total in your head reaches +5 you bet $5.

Elementary as that is, it's all there is to the simple point-count system.

You can increase the odds in your favor by various small modifications. One way is to take account of the fact that the fewer cards that remain in the deck, the more significant is your knowledge of their high-low proportion.

The mathematical way to take advantage of this fact is by dividing your mental point-count by how many decks are left. At a table where only half a deck was left when you reached the point-count of +5, you'd divide 5 by 1/2 to get 10. And you'd make your next bet 10 units, or 10 times the minimum.

If the casino is one where two decks are used at once and one is about half gone, you'd divide by 1-1/2 and wager only about $3 at a $1 table. (To divide 5 by 1-1/2 readily you can double both and say one-third of 10 is 3 plus.)

Horseshoe Math

YOU ARE, let's say, a horseshoe pitcher of competition class. You know, therefore, just what your season's percentage of ringers is. We'll say it's 42.75. However, you've just had a lucky string. You suddenly discover that you've tied the local record and will set a new one if you can make at least one ringer in your remaining four tosses. What's the probability that you can do it?

Applying the logic also used for tennis in this section, you can figure that your chance of a ringer is 42.75 percent on each toss—or (subtracting from 100 percent) a 57.25 percent probability of missing. Converting from percentage to decimal, you make this computation: 0.5725 times 0.5725 times 0.5725 times 0.5725 equals 0.1074. Which says there's a little more than a 10 percent chance you'll miss all four times. So there's better than 90 percent probability that your name will go down in the local record books.

Of course, if your calculator has a constant, as it most likely does, you won't have to enter .5725 four times. You'll merely punch it once, then hit the times key four times or the times key once and the equals three times. The precise procedure with your calculator is between you and your instruction booklet. If the booklet isn't helpful, try experimenting.

Chapter T
   BUSINESS DECISIONS

   Real-Estate Cash Flow
   How Cash Flow Works
   What's the Profit Rate?
   Buying Earned-Growth Rate
   Figuring Fiscal Ratios
   How Durable the Building?
   Counting Absenteeism

Real-Estate Cash Flow

MORE OF US are real-estate investors than realize it. Merely retaining ownership of a house from which for one reason or another you have departed, can make you one. This is an analysis that's useful in both choosing to invest and deciding whether to hang on or get out.

To analyze the cash flow from income property, run through four connected steps. These steps will successively calculate the potential gross income, the effective gross income, the net operating income, and the cash throw-off to equity.

Here they are, with results for a typical example, a 30-unit apartment building with rentals of $500 a month, a 5 percent vacancy rate, and an annual operation cost of $76,855. The property has just been financed with an 11 percent mortgage for $700,000 to be amortized in level monthly payments over 20 years.

First calculate the *potential gross income* by adding together all the monthly rental amounts and multiplying by 12. (Here we have $500 times 30 units times 12 months, $180,000.)

From that figure, subtract the vacancy loss and add any other income produced. (Reducing the potential income by the 5 percent vacancy loss of $9000 gives *effective gross income* of $171,000.)

Now subtract operating expenses ($76,855) to find that the *net operating income* (in this example) is $94,145.

Now subtracting the annual debt payments will tell you the cash throw-off. With these payments, taken from a table or calculated by one of the procedures given in this book, coming to $86,705, the *cash throw-off to equity* comes to $7440.

How Cash Flow Works

FINDING the amount of cash flow is a basic calculation in analyzing an income-producing property.

It is useful as the first step in determining such vital measures as rates of return. And it tells you whether an investment, however promising in the long run, is manageable in the short run.

The sample analysis that follows is founded on real figures for a 20-unit apartment building but of course the percentages used might be quite different for another time or place or structure.

The idea here is that you're considering buying an unfurnished apartment building in a suburb. Cost is $550,000 and you find that you can finance $450,000 of it with a 25-year mortgage at a fixed rate of 8.5 percent.

What percentage return can you expect to get during the first year on the $100,000 you must put up?

Gross income is the first thing to figure. With rents of $450 a month for each of eight studios, $550 for each of eight one-bedroom apartments, and $750 for each of four two-bedroom units, you find the maximum yearly rental income to be $132,000. Parking and laundry user fees promise to add $4,140 for a total gross of $136,140.

Vacancies and defaulting tenants will cost you something. Using local experience figures, you estimate this at 2.6 percent of gross. Multiplying the gross by 0.026 and subtracting expense of $3540 leaves net revenue of $132,600.

(When figuring such a discount, save time by subtracting the percentage from 100, then multiplying by the 0.974.)

Operating expenses will include maintenance (estimated at 12.2 percent of gross), utilities (12 percent), professional administration (8 percent of gross, a national average), and taxes and insurance (25.8 percent, based on a regional average).

These percentages of $136,140 ($16,610; $16,336; $10,892; $35,124) come to $78,962. With this deducted from net revenue of $132,600, what's left is net operating income of $53,638.

Mortgage payments, figured by one of the methods in this volume, come to $3623.50 a month, or $43,482 a year. Take that away from the net operating income.

Pre-tax cash flow is the result: $10,156.

And that's what you wanted to know.

Since you've gone this far, however, you might as well note also the return on investment. You can see at a glance that on the $100,000 you're putting up you'll be getting a little more than a 10 percent return. The precise figure, 10.156 percent, is found by dividing the cash flow by the investment, and then multiplying by 100 to produce a percentage.

Since you were after the immediate cash flow, you haven't added in the build-up of equity produced by that part of each mortgage payment that went toward the principal. It's of no use to you now, is tiny in the early years anyway and probably dwarfed by the appreciation—or depreciation—in value of the $550,000 worth of land and building.

What's the Profit Rate?

PR—PROFIT RATE—is an investor's or broker's term for a handy measurement of the ability of a company to make money with your money.

It can reasonably be described as measuring the profitability of a corporation. It shows the rate of return on the shareholder's equity capital at corporate book value.

To find the PR of one company for comparison with that of another, do the following:

Divide the profit per share by the book value per share as given at the start of the year.

Then shift the decimal point two places to the right to convert the result into a percentage.

Buying Earned-Growth Rate

THE WORLD of money is rife with terms that sound formidable on first encounter, but can prove to be quite simple and useful concepts.

One flourishing example is EGR, or earned-growth rate. It's defined as the annual percentage rate at which a company's equity capital behind each share of common stock is increased by net earnings after payment of dividends.

It can be important in your buying decisions. It is one of the measures used in comparing the worth of the stock of one company with that of another.

To find an EGR, first subtract the dividends paid per share from the earnings per share.

Then divide by the book value per share at the start of the year. (This can be found in all annual reports and most financial summaries.)

Multiply by 100 to convert this result to a percentage.

For example, if a stock earned $3.53 per share and paid $1.05 in dividends, while starting the year with a book value of $19.83, you'd begin with the first of these figures. Then you would subtract the second and divide by the third, and move the decimal point two places to the right.

This procedure should inform you that the value of a share grew just over 12.5 percent during the year.

Figuring Fiscal Ratios

Authorities warn us that we can deceive ourselves by judging the health of a business on its balance sheet or income statement alone. A business hovering on the verge of bankruptcy might produce a salubrious statement.

Easily calculated ratios can help us dig more deeply into the truth of things. Some of the most helpful of these have been isolated by Dennis P. Curtin, Jeffrey R. Alves, and Anne K. Briggs in their book, Controlling Financial Performance for Higher Profits. It's a good source of further information about the meaning and usefulness of the ratios that follow.

Ten ratios are used to answer seven questions.

The first question deals with liquidity: Can the company pay its bills?

Current ratio, which can be described as a general test of solvency, gives one answer to this question. To find the current ratio, divide the current assets of the company by its current liabilities. A result of 2 or more is commonly felt to indicate solid financial condition.

Quick ratio, sometimes called the acid-test ratio, is regarded as a conservative test of liquidity. Like the current ratio it can be calculated monthly from the balance sheet. There are two ways to figure it.

One is to add together cash, marketable securities, and accounts receivable and divide by current liabilities.

The other is to subtract inventory from current assets and divide by current liabilities.

An acceptable quick ratio for manufacturers is 1. Other kinds of businesses—wholesale, retail, and service companies—will usually show a ratio of less than 1.

Average collection period is a measure of how long a company waits—on the average—from the time it makes a sale until it has the cash from that sale in hand. You might wish to calculate it monthly or yearly.

Let's assume it's done monthly. Waiting till the end of a year is pretty late to find out how a business is behaving.

Begin by dividing the month's credit sales by the number of days in the month. Call this result daily credit sales and divide the amount of the receivables by it. Call that the average collection period, or ACP. A rule of thumb is that an ACP should not be greater than 1.3

times the collection period that has been established by the business. If that established period is the rather usual one of 30 days, then an ACP greater than 39 (1.3 times 30) suggests that the business is suffering from slow collections and may be extending credit un-wisely.

Inventory turnover. This is a ratio that tells a business how fast its merchandise is moving. It is an indication of how efficiently a company's assets are being used and is sometimes called stock turn or stock-to-sales. Because of seasonal variations it is well to calculate it monthly.

Look at the income statement to find cost of goods sold. This is the expense item that includes the purchase price of stock—in a retail business. Or it is total cost of goods for sale—raw materials plus labor plus overhead— in the case of a manufacturing firm. Divide by inventory, as found on the balance sheet, to find the ratio called inventory turnover.

Generally, the higher the turnover the better. However, an un-usually high rate can be a warning that the business may frequently encounter an out-of-stock problem that could lead to customer dissatisfaction. It can also be a warning that a condition of low-raw-material inventory may sometimes exist and lead to dis-ruption of production.

During a period of substantial growth a year-end figure may be misleading. It may be better to use the average inventory for the period, found by adding together the beginning and ending inventories and dividing by two. The cost of goods sold is then divided by this to produce a ratio of inventory turnover.

Safety (leverage) ratios indicate a company's ability to pay its debts. A safety ratio should be figured annually and quarterly and, for a highly seasonal business, monthly as well. The basic such ratio is debt-to-worth and is found by dividing total liabilities by total owner equity. It is often considered desirable that this ratio be no greater than 1 to 1. A stable and predictable business, however, may operate quite soundly with a larger ratio. A ratio as large as 3 to 1, or even 4 to 1, is not necessarily out of line for a utility.

A misleading safety ratio may be produced in a company's books when some assets are carried at a depreciated value for accounting or tax reasons. A hidden-assets adjustment will then produce a more revealing debt-to-worth ratio. Get this by adding the value of the hidden assets to owner equity and using this total as the figure by which to divide total liabilities.

Profitability ratios are most usefully calculated relative to sales and relative to invested capital or assets.

One of these ratios is the profit margin, which it is well to calculate monthly. To find it, divide income before taxes (excluding extraordinary or nonrecurring items) by net sales. The decimal result becomes a percentage when multiplied by 100.

Thus an income of $40,000 divided by net sales of $250,000 gives a result of 0.16, read as a profit margin of 16 percent. Which says that after all expenses are paid, except taxes, there's 16 cents left over out of each dollar of sales.

Return on assets is a ratio that measures how much profit a business is generating with each dollar of assets. Calculated monthly it tells a good deal about both profitability and asset management. Large corporations often use this ratio as a yardstick of performance.

To calculate return on assets, first find income before taxes, on the income statement. Divide that figure by total assets as found on the balance sheet. I've been told that, in a rough way, 20 percent has generally been considered good. But this should not be applied blindly to all industries.

How Durable the Building?

Most buildings could last almost forever, but they don't. The time comes when it pays someone to tear them down, often more's the pity. The truth is, far more buildings are torn down than fall down.

Anyway, the United States Treasury Department long ago decided that the figures herewith are reasonable ones to use in estimating the lives of various kinds of buildings.

BUILDING TYPE	Good Construction	Average Construction	Cheap Construction
Apartment	40 years	40 years	33 years
Bank	50	50	40
Dwelling	50	40	33
Factory	44	40	33
Farm	50	50	40
Garage	50	40	33
Grain elevator	67	50	40
Hotel	40	40	33
Loft	50	50	33
Machine shop	50	40	33
Office	50	40	33
Store	50	50	40
Theater	40	33	29
Warehouse	67	50	40

Counting Absenteeism

THE STANDARD method of looking at the absentee rate in a business is by percentage.

To find it, multiply the number of lost days by 100 and divide by the number of employees and also by the number of working days in the period.

(Of course the same method works for figuring the rate by hours as long as you use hours for both time lost and working time available.)

Suppose 8 days of absence occur during a month having 21 working days, in an office with 15 people employed. Dividing 800 by 21 and then by 15 (or dividing 800 by the product of 21 and 15 if you prefer) shows the absenteeism rate to have been just a little more than 2.5 percent.

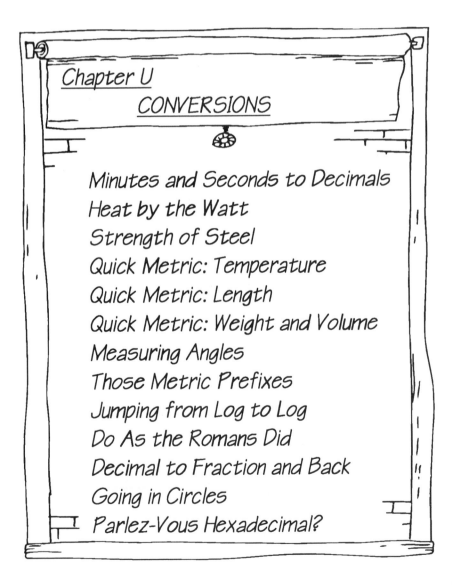

Chapter U
CONVERSIONS

Minutes and Seconds to Decimals
Heat by the Watt
Strength of Steel
Quick Metric: Temperature
Quick Metric: Length
Quick Metric: Weight and Volume
Measuring Angles
Those Metric Prefixes
Jumping from Log to Log
Do As the Romans Did
Decimal to Fraction and Back
Going in Circles
Parlez-Vous Hexadecimal?

Minutes and Seconds to Decimals

Y<small>OU MAY</small> find yourself doing calculations that involve hours (or degrees—it comes to the same thing), minutes, and seconds. Could even happen while you're navigating a boat.

It's often simplest to convert minutes and seconds into decimals and add them to the hours (or degrees). Then for each measurement you have only one quantity to deal with.

To accomplish this, divide the minutes by 60 and the seconds by 3600 and add these sums to the number of hours.

If you start with 40 degrees (or hours), 30 minutes, 32 seconds, you'll add 30 divided by 60 (which is 0.5) and 32 divided by 3600 (which comes to 0.00889) to 40 to get the result 40.50889 hours or degrees.

Heat by the Watt

H<small>OW MUCH</small> heat, in British thermal units per hour, does an electric heater produce? The factor you need, derived from something called Joule's Law, is almost exactly 3.412. Merely multiply the wattage of the heater by it.

You'll find, for example, that a heater rated at 1500 watts will turn out 5118 Btus per hour.

Strength of Steel

I<small>F YOU KNOW</small> the Brinell hardness number of the type of steel you're dealing with, you can get a very close approximation of its tensile strength by a simple calculation.

Just divide the hardness number by 2. Then move the decimal point three places to the right—or multiply by 1000, which comes to the same thing, of course.

Thus a steel with a hardness of 186 BHN will have a tensile strength of very close to 93,000 pounds to the square inch (psi).

From the Brinell hardness number you can also derive the Equivalent Hardness of a material. Divide the BHN by 10 to find the Equivalent Hardness on the Rockwell "C" hardness scale.

So 50 Rockwell "C" is about 500 Brinell; and 350 Brinell is about 35 Rockwell "C," found by dividing by 10.

Quick Metric: Temperature

As LONG AS we're still afflicted with remnants of the old English system, while trying to wean our minds off the familiar and get them more or less onto metric, there'll be occasional need for quick-and-dirty conversions.

Here are a few fast approximations.

Take half the Fahrenheit temperature and subtract 15. Be warned: This quick approximation is good enough for thinking about the weather—it's on the nose at 50 degrees F—but it's too far off in the region of 100 F. to use in deciding whether you have a serious fever.

To go the other direction, add 15 to the Celsius temperature, then double it.

Long before metrication loomed up for all of us, Americans living in Europe had these problems, and temperatures had bothered them much more than distances. For the more usual direction of conversion, C. to F., the rigmarole Americans abroad have found easiest to use is: multiply Celsius by 2, deduct 10 percent, add 32.

These suggestions, of course, are just rough ways to make the conversions in your head.

The real formulas are not very complex, either. Celsius times 9, divide by 5, add 32 yields Fahrenheit. And F. minus 32, multiplied by 5, and divided by 9 gives C.

Quick Metric: Length

Double the number of inches, then add half again, to get centimeters. (This is an approximation of the exact method, multiplying by 2.54.)

To go the other way, multiply the centimeters by 4 and divide by 10.

To convert miles to kilometers, add 60 percent. Or multiply by 16 and divide by 10. (To be more precise, you can multiply by 1.609.)

For kilometers to miles, multiply by 6, divide by 10—which means moving the decimal point one place to the left. (What you're doing here is approximating multiplication by 6.214.)

Quick Metric: Weight and Volume

To GET FROM pounds to kilograms, take half and then knock off a bit more. You'll be quite close if the amount you knock off is about 10 percent. Thus 10 pounds will become 4.5 kilos. (More exact is 4.536.)

Kilos to pounds: Double, then add 10 percent. Ten kilos will become 22 pounds, which is pretty darn close to the more precise 22.05.

For ounces to grams, multiply by 30 and deduct 5 percent. (More exact: multiply by 28.35.)

Quart and liter are nearly equal but, like the critters in George Orwell's Animal Farm, some are more equal than others. Take 5 percent off the number of quarts to get the number of liters. Add 5 percent to go the other way. (Or be quite precise by multiplying by 0.9464 or 1.0567.)

Measuring Angles

THE USUAL way of measuring an angle is in degrees, of which there are 360 in a circle. Another rather common measure is the radian, which is 180 degrees divided by pi. That comes to about 57.2958 degrees, giving a handy figure to use if you're working with paper and pencil or a simple calculator. Given an angle expressed in radians, just multiply by that figure and you have it in degrees.

Of course, if you're using a calculator and it has a key for pi, you can more simply—and precisely—do that same operation of turning radians into degrees by multiplying by 180 and dividing by pi.

To go the other direction, multiply the number of degrees by pi and divide by 180.

The one-step paper-and-pencil routine for turning degrees into radians is to multiply by 0.0174532. (This works because that ugly decimal is pi divided by 180.)

Those Metric Prefixes

Wʜᴇɴ ʏᴏᴜ'ʀᴇ working with metric measurements—as you very slowly but pretty surely will be doing more and more—you have to know your prefixes.

To keep them straight, use this silly little memory helper: Kitty, How Does My Dog Catch Mice?

And, thanks to those initial letters, K, H, D, M, D, C, M, you'll know the order of the metric prefixes as the days of fuller metrication come along.

Each denotes one-tenth of its predecessor: Kilo, Hecto, Deka, (just plain Meter fits in here) Deci, Centi, Milli.

Jumping from Log to Log

Tᴡᴏ ᴋɪɴᴅs of logarithms are often used. What is called the common logarithm takes 10 for its base. Natural logarithms use as base the number designated by the letter e, the value of which is approximately 2.718.

A problem arises if you have a table of one kind of log and need the other kind. Fortunately, there are quick fixes.

To find the natural log of any number, divide its common log by 0.4343. When it's more convenient, you can get the same result by multiplying by the reciprocal of that number, which is 2.303.

The reverse of this naturally follows. Divide a natural log by 2.303 and you have the common log. Or multiply by 0.4343.

Those key numbers are the natural log of 10 and its reciprocal. So if you are working with a calculator that has natural logs only, rather than with a table of logs, you can enter the multiplier more efficiently by asking the calculator for the natural log of 10.

Do As the Romans Did

I<small>F YOU'VE</small> gotten a little rusty on the finer points of reading Roman numerals, you can be fooled quite easily—especially when someone would be happy if you didn't read a date. Titles of movies, especially ancient ones being shown on TV, are commonly accompanied by the obligatory copyright notice—with the year briefly flashed in Roman numerals. This practically guarantees that most viewers won't discover how old the picture is.

It reminds us of the importance of reading Roman.

So did an ad I recall for a rather costly publication that offered as a premium to subscribers a "Consumer Encyclopedia." You might assume you were to receive a new and original volume unless you grasped the significance of a tiny footnote (in 4 or 5 point, I'd guess, or half the height of most newspaper or book type faces). It told you that this premium was also published as "CONSUMERS ALL" DEPT. OF AGRIC. MCMLXVIII.

Aha! It happens that each year the U.S.D.A. puts out a book of useful advice and information, topics varying from year to year, that you can purchase or, by merely writing to your Congressman, get free.

What's being offered here as a subscription incentive is evidently a reprint of a free government book—but from what year?

As we all learned in school, Roman Numeral I is 1, V is 5, X is 10, L is 50, C is 100, M is 1000. To read 'em you add 'em together...except that when a lesser number precedes a greater, you subtract it.

So the first date here is 1000 plus (100 less than 1,000) plus 50, plus twice 10, plus 5.

And the other is made up of MCM, or 1900 and LXVIII, or 68. Decoded, the message reads that the folks offering this premium back in '75 had reprinted a 1968 Federal publication, the use of which is free to all, and were not all-out eager for you to notice that detail or the vintage of the volume. Did they get the idea from those brief flashes on the TV screens?

(It doesn't apply here, but there is one more rule for reading and writing Roman that should be inserted for completeness and because it isn't widely known. When a line is drawn above a numeral, its value is multiplied by 1000.)

Maybe you've wondered how in the name of Jove ancient peoples ever found a way to add and subtract with those unlikely Roman symbols. I don't think they did, or could have. I like to believe they computed mechanically, with scratches in sand or with the beads of an abacus, and only after reaching a result expressed it in those funny alphabetic numerals. That makes me feel better. How about you?

Decimal to Fraction and Back

Have you ever tried to figure building jobs in feet and inches—and fractions of inches—using a hand calculator?

Most calculators being decimal-oriented, they can cope with the monstrosities of our traditional system only when you help them with routines like those that follow.

Decimals to fractions. Temporarily ignoring the number of whole feet, multiply the decimal part by 12 to turn it into inches. Now, temporarily ignoring the whole inches, multiply the new decimal part by the denominator you want (by 8, for example, to find the answer in eighths). String the three results together.

Say you need a plank 12.85 feet long, which you must measure off to the nearest 1/16 of an inch using an ordinary feet-and-inches tape. Multiplying the decimal part (0.85) by 12 gives 10.20. Multiplying the new decimal part (0.20) by 16, since you want the answer in six-teenths, gives 3.20. Your answer is the original 12 feet plus the whole inches (10) plus the 3/16, or 12 feet, 10-3/16 inches.

Fractions to decimals. Naturally this is the reverse of the other operation. Temporarily ignoring the feet, divide the number of whole inches by 12. Divide the upper number of the fractional inches by the lower and also by 12. Add these results together and append them to the number of feet.

Starting with 12 feet 5-7/16 inches, you'd divide 5 by 12 to get 0.4167, then 7 by 16 and by 12 to get 0.0365. Adding these two results gives 0.4531, for an answer of 12.45 feet.

Going in Circles

IF YOU work with round things you will probably have no trouble recalling that to find the area of a circle you must multiply its radius by itself, then by something called pi.

I don't know about your school, but in mine it was rated fairly hilarious to greet the formula with a well-worn witticism along the lines of, "Like heck pie are square—pie are round!"

Since the value of pi is one of those confounded decimals that go on forever and so are deservedly called irrational, it has to be rounded off somewhere for practical use. Usually it is called 3.1416. Slightly less precise, but handier if you're working with fractions, is calling it 22/7 or 3-1/7.

If what you know is not the radius but the diameter, you can divide that diameter by 2 to find the radius. Or you can use a slightly different formula, one that says the area of a circle is equal to one-fourth the product of pi and the diameter squared. You can save a step by starting with the information that one-fourth of pi is 0.7854 (which is exactly as precise as using 3.1416 for pi) or approximately 11/14.

Either of the values of pi is also useful in working with circumferences of circles. The circumference, you may recall, is equal to the diameter (or twice the radius) multiplied by pi.

Thus if your city, like some, requires a permit for removal from a building site of any tree with a diameter of, say, 6 inches at a point four feet from the ground, you can check with a flexible tape. Diameter of 6 inches means circumference of 6 times pi, which comes to about 18-7/8 inches.

Parlez-Vous Hexadecimal?

IF WE WERE listening at all, I think most of us were pretty shocked the day an arithmetic teacher first told us there was nothing sacred about our 10-based number system. If it weren't for our having ten fingers, the base might have been almost anything. When I was told of a civilization that used 20 as its base number, I guessed that it might have existed in a climate so warm that people didn't wear shoes much.

Computers work in a primitive number system called binary (bi- as in bicycle or biped), which means it is based on the number 2. Since the decimal system doesn't translate readily into binary, people who work directly with computers, in machine language, must adopt either binary or some system reasonably compatible with it. Binary is slow and cumbersome, so most computer systems have come to hexadecimal, which is 16-based. A few use octal, based on 8, or 32.

You can change a number from one system to another quite readily, because they all use the same rules.

The last, or right-hand, number has the same value in all systems; that is, just what you'd expect it to have. The value of the next to last number, however, must be multiplied by the base of the system you're using. The digit before it is multiplied by the base number squared. And so on.

That's all there is to binary or octal or decimal systems. Since binary requires two characters, you need only 0 and 1. Octal employs the eight characters 0 through 7. Decimal, as we all know, uses the 10 digits 0 through 9, combining them to create any larger number wanted.

But hexadecimal calls for 16 distinct digits. Since it is helpful for these characters to be familiar and to be available on a standard typewriter keyboard, the letters A through F have been pressed into service to follow in sequence after 7, 8, 9. So the number of eggs in a dozen becomes C.

To see how all this works, let's try putting the number of eggs in two dozen into each of these number systems.

Decimal is pretty easy for us, since we've all grown up with it. When we add 4 in the right-hand column to 2 multiplied by the decimal base, 10, we get the required 24.

As we move from right to left in binary, we must multiply the value of each digit by 2, then 2 times 2 or 4, then twice that, and so on. We write 11000, which read from right to left means no 1s, no 2s, no 4s, one 8, and one 16. Adding 8 to 16 produces the two dozen we're looking for.

The next-to-last digit in octal takes the value of the base, 8. So the number by which it must be multiplied to give 24 is 3. Two dozen eggs in octal is written 30, for three 8s plus no 1s.

Now for the important hexadecimal. Placing 1 in the next to last column will produce 16 (following the rule that anything in that column must be multiplied by the number that is the base of the system). One 16 leaves us eight short of the needed 24, so 8 goes in the right-hand column. In hex, two dozen is written 18.

For a final touch let's examine the largest quantity that can be written in binary with no more than eight digits. In decimal terms, it's 255. In binary, it's eight 1s in a row, representing the sum of 1, 2, 4, 8, 16, 32, 64, and 128. Since hexadecimal values from right to left are 1, 16, 16 times 16, and so on, we write FF, representing 15 with the F on the right and 16 times 15 with the other one. Note that the next larger quantity, 256, is written simply 100.

Chapter V
MATH IN A HURRY

Breaking Down Numbers

A<small>NY TIME</small> you can work with 10 or 100 instead of a more stubborn number, you've turned a tough calculation into an easy one. Some examples:

Add 39 and 54. Takes a little thought, does it? Try adding 1 to the first number and taking it away from the second. Now all you must do is add 40 to 53, which is no great mental stretch.

Add 98 to 49. Same trick works, except this time increase the 98 by 2 and decrease the 49 by the same amount, so you can say 147, hardly doing any addition at all.

In mental multiplication, make changeovers by multiplying and dividing.

Like this: How much is 32 times 15?

Numbers ending in zero are the easiest to work with. So double that 15 to convert it into a handy 30, balancing this (fair's fair!) by halving the first term to 16. The process becomes simply 3 times 16, tack on a zero.

When one of your numbers is close to 10, 20, and so on, or to 100, turn it into that number plus (or minus) something. Regard 9 as 10 minus 1. Treat 21 as 20 plus 1. And so on.

Then how much is 28 times 9? That's not easy at first glance, so change it to 28 times (10 minus 1) and say to yourself, 280 minus 28 is 252. (You may find yourself simplifying even this little bit of subtraction at the end, by taking 30 from 280, then adding 2.)

Any number evenly divisible into 100 (or, for that matter, 1000) can be converted into simpler form. Treat 25, for instance, as 100 divided by 4.

So to multiply 48 by 25, all you do is multiply the 48 by 100 (that is, add two zeros to make 48 into 4800) and divide by 4.

To multiply by 75, do the same thing—but include a further multiplication by 3.

In division problems, you can often simplify things by turning a single division process into several. That may not sound very probable, but take a look at this.

You are asked to divide 336 by 8. They're both even numbers so divide both by 2.

The result, 168 to be divided by 4, is still even, so divide by 2 again. And then again (84 divided by 2 is 42 and that's your answer).

Great Chief Sohcahtoa

W̲ʜᴀᴛ ꜰᴏʟʟᴏᴡꜱ is not exactly a calculation. You might call it a bit of nonsense. But it's also just about the handiest bit of mathematical help I've ever had handed to me from half a dozen directions at once.

While conducting a mathematical column in a magazine, I confessed in print to a distressing inability to recall which of the common trigonometric ratios was which. That, I said, was why I'd taped the information to the back of my favorite calculator. Several dozen readers quickly responded with word about the great Chief Sohcahtoa, the letters of whose weird but rememberable name to this day keep me reminded that in any right triangle:

The Sine is equal to the Opposite side over the Hypotenuse. The Cosine is equal to the Adjacent side over the Hypotenuse. And the Tangent is equal to the Opposite over the Adjacent.

Of course you may prefer an alternative that one reader suggested: Simple Old Herman Carries A Huge Tub Of Apples.

The Squaring Trick

T̲ʜᴇʀᴇ'ꜱ often need to multiply a number by itself. Comes up so frequently that even medium-priced pocket calculators often have a special button just for squaring.

In a way, so does your brain.

Begin with a rather specialized trick. You can use it only with numbers that end in 5 or 0.5 (or one-half, which comes to the same thing).

Among photographers, however, those who eschew automation can learn to love it. Many lens stops end in .5, and squaring a stop value is a common need, since the amount of light admitted by a lens varies as the square of the stop.

To square a whole number ending in 5, first drop the 5. Multiply what remains by itself plus one. Then append 25. (Thus squaring 95 means multiplying 9 by 10 to get 90 and tacking on 25 to get 9025.)

Similarly, to square a mixed number ending in one-half, just multiply the whole number by 1 more than itself and add on one-fourth (or .25). The square of 10.5 (or 10-1/2) is 110.25 (or 110-1/4).

Another breed of number easy to square mentally is anything ending in 1. To the number in question add the next smaller number and the square of that smaller number. That's easy to do, since that next lower number always ends in zero.

So to square 31 you just add 31 to 30 to 900.

You can apply to any number the general rule from which these special rules have sprung. Maybe you remember this squaring rule from algebra: square of the first times twice the first by the second, plus square of the second.

To use it you must first turn the number you're working with into the sum of two numbers, making the first of these end in zero. Thus 86 becomes 80 plus 6. And 122 becomes 120 plus 2.

Applying the rule to the latter gives the square of 120 (14,400, as you know from the twelves times table) plus twice the product of 120 and 2 (480) plus the square of 2. If you write it out it looks like this: 14,400 plus 480 plus 4 equals 14,884.

For numbers like 38 or 129 with a last digit closer to 10 than to zero, you may prefer another version of the rule. Express your original number as a difference instead of a sum. See 38 as 40 minus 2. The rule then asks for the square of the first MINUS twice the first by the second plus the square of the second.

Halving a Mixed Fraction

Mᴇɴᴛᴀʟʟʏ dividing in half a number plus a fraction can be confusing if the number is odd. It's a problem that comes up frequently, as in woodworking where a nominal 4-inch board may have an actual measured width of 3-1/2 or 3-5/8 or 3-9/16 inches.

There's a shortcut that will save time and make an error less likely. It's a good bit faster and easier to do than it is to explain.

Subtract 1 from the whole number and divide by 2 to get the new whole number.

Add the parts of the fraction together to get the new numerator.

Double the old denominator to get the new one.

So this is how you find half of 9-13/16:

Nine minus 1 is 8, of which half is 4.

Adding 13 and 16 gives 29.

Twice 16 is 32.

And the answer is 4-29/32.

With a little practice, you'll find that you can perform this minor operation almost instantly.

(Adding mixed numbers has its perils, too. Following the standard routine of reducing everything to a common denominator is tedious. It often leads to errors when the list is long. A more efficient approach starts by seeking out combinations that add up to 1 and crossing those out.)

Pint's a Pound

Just remember "a pint's a pound the world around" and you'll never have to wonder how many fluid ounces there are in a pint (16) or a quart (32) or a gallon (128).

Actually, if you want to be fussy about it, a pint of water weighs just a trifle more than a pound, and such fluids as oil, gasoline, turpentine, and kerosene weigh a bit less.

Illusion of the Shifting Base

Let's suppose you had income one year of $19,444 and the next year of $23,111. Subtracting the lesser amount from the greater, then dividing by the earlier figure, you find the increase to be at the rate of 18.86 percent.

Percentage increases being a little confusing to calculate, you might want to check that result in a way that is easy to understand.

You could do that check by multiplying last year's figure by 1.1886.

Or you could multiply $19,444 by 0.1886 and then add on the original $19,444. Whichever makes more sense to you.

What goes around comes around, they say—perhaps the only observation on which dedicated optimists and pessimists agree.

So perhaps at the end of the next year you will find that your income has declined by 18.86 percent or thereabouts. You'll be right back right where you started, won't you?

I'm afraid not. If you subtract 18.86 percent of $23,111 from $23,111 you'll find yourself way down to $18,607.

You're up against the illusion of the shifting base. It is the reason that a 10 percent pay cut followed by a 10 percent raise constitutes a net loss of 1 percent.

And, just to rub it all in, if the 10 percent raise comes first and the 10 percent cut afterwards, you're still the loser by 1 percent.

(Here, by the way, is another kind of calculation of percentage of increase that may prove useful in other contexts. Suppose that knowing only that this year's income of $23,111 is an increase of 18.86 percent over last year's, how would you calculate how much you earned last year? Expressing the percentage in decimal form, adding 1 (0.1886 plus 1 equals 1.1886) and then dividing the $23,111 by this gives you $19,444.)

The Catch-22 Deduction

Y OU MAY think you're in Wonderland with Alice the first time you encounter a strange rule that turns up now and then in such likely places as U.S. income-tax regulations. It has been found in, for example, the limits on Keogh Plan pension contributions by self-employed people and in deductions for contributions to 501(b) and 401(k) tax-deferred savings plans and in the limitations on charitable contributions by certain small businesses classified as "C" corporations.

The rule may tell you that you are allowed to deduct no more than 5 percent of net income. And then proceed to define net income as what you have after subtracting not only your other expenses but the contribution as well. Ay, there's the rub: how can you calculate 5 percent of a figure you haven't figured out yet?

As you will see, it's tough even when you're working with round numbers. Suppose net income before contributions is $10,000 and you're allowed a 5 percent deduction. Take 5 percent off $10,000 and there's $9500 left.

So you're allowed 5 percent of $9500, or $475.

But when you take that from $10,000, you discover that.... Seems there's no end to it.

But, fortunately, there is a formula to guide you through this thicket.

Just state the allowed percentage as a number (5 if you're permitted 5 percent) and divide by 1 plus the percentage stated as a decimal (1 plus 0.05, or 1.05). This tells you what percentage of the net income you may deduct. In our instance this comes out 4.7619 percent. This times the $10,000 you started with comes to $476.19, so that's how much you may deduct.

Totaling Fibonaccis

T HE NUMBER series named for Leonardo of Pisa, known as Fibonacci, has inspired not only whole books but clubs and magazines as well. Just one of its fascinating aspects is that it precisely describes the course of rabbit breeding if the frequent, if harsh, custom is followed of eating the parent bunnies after two litters have been produced.

Start with a pair that has a litter after one month and a second after another month. The third month there are two litters, one from the old pair, one from the young. The fourth month there are three litters, the fifth month five, and so it goes.

The number series then is one in which each term is the sum of the two preceding it and that's the Fibonacci: 0 1 2 3 5 8 13 21 34 55 89....

(If you happen to have a programmable calculator, it will probably allow you to generate this series for as many terms as you like. With Texas Instruments' TI-59, try the program Sum, 0, Esc, 0, Pause, RST. Enter 1, press RST and R/S, and watch the Fibonacci numbers fly. For several other calculators, including some from Hewlett-Packard, you can find programs for this series in accompanying handbooks.)

There may be no news in this so far for people who habitually fool around with numbers. Not so widely known, though, is a quick trick for finding the total of any ten consecutive numbers in the series. Just multiply the seventh number in any Fibonacci series by 11. Try it with the part of the series given above and you'll indeed find that 11 times 21 is equal to the sum of the ten Fibonacci numbers that end with 89.

Multiply Like a Peasant?

Do you find it as surprising as I do to learn that you can multiply together large numbers by a process of halving, doubling, and adding? That means getting by with a multiplication table ending with the twos. None of that 7 times 9 stuff that made us so much trouble in grade school.

And that, I suppose, is precisely why this method was used by European peasants three centuries ago. Presumably both illiterate and virtually innumerate, they'd have found it pretty hard to use the routine familiar to us.

As you will see, the peasant way is slow—but wonderfully undemanding.

To use it, first place the numbers to be multiplied at what will become the tops of two columns.

Divide the first number by 2 and that result by 2 and so on until you reach 1, putting the results in a column under the first number. Ignore fractions; as far as you're concerned half of 9 is 4.

Under the second number that you're working with, put the results of successively doubling it, stopping when you have placed a number opposite the 1 in the first column. Select from the second column every number that is opposite an odd number in the first column. Adding these numbers will give you your answer.

No kidding.

Here's how all this looks when you multiply 36 by 47.

36 x	47	
18	94	
9	188	188
4	376	
2	752	
1	1504	1504

1692 is the answer

When you calculate in this manner, you're actually using binary, or 2-based, arithmetic. That's the kind computers use today, so don't elevate your nose at it.

Would we have computers at the dawn of the Twentieth Century if the great mathematician Gottfried von Leibnitz had not learned this trick from watching those dumb peasants?

Easy Iteration

FANATICAL puzzle-hounds have been known to muddle away at problems for weeks, or even years, because the solution boiled down to one of those miserable third-order or fourth-order equations. Meaning they contain x to the third or fourth power.

Such equations are almost hopelessly tedious to solve by ordinary and proper mathematical methods. Indeed, a peculiarly formidable math handbook that I use mostly to intimidate myself gives only limited treatment to the solution of fourth-order (quartic) equations because the formula for the general solution is so complicated it is "hardly used."

Yet by the power of the repetitive process called iteration, which is really just trial and error, applied by way of an ordinary hand calculator, you can tame one of these beasts in minutes.

I can offer as an example a puzzle I once published that tormented readers by leading them into a nasty fourth-order equation.

Expressed as it would sound if read aloud, here's the (correct) equation most came up with:

Four times x to the fourth power, plus 12 times x cubed, plus 9 times x squared, minus 486 times x, minus 729 equals zero.

You can see how this problem can be solved by trial and error. Guess a value of x that makes sense in this problem. Substitute this value in the equation. Then try another number. As you find the range of the guesses that bring the value of the left side close to zero, you'll close in on an exact solution.

I don't think you'd want to try it without a calculator. Maybe not even with one.

Fortunately, there is a marvelous variation on this iteration method. Tolerable without a hand calculator, with one it's a piece of cake. Simply go through the equation from left to right, totally ignoring the exponents (the business about fourth power, cubed, squared) as you replace x with your best guess of what it might be.

Even if you have to make your first guess pretty much at random, further trials will rapidly bring you closer and closer until you have a solution that is correct to as many decimal places as you choose to go for.

When this procedure produces a left side of the equation that is equal to zero, you've solved the equation.

Clues in this particular problem indicated that the answer should fall in the region of 4 or 5. If you were to substitute for x those two numbers in turn, you'd see that the answer must fall close to midway between them. And, in fact, it does turn out to be just 4.5 when you make that your guess.

On the chance that I've muddied this manipulation too badly, let me run over it step by step: guess times 4; add 12; times guess; add 9; times guess; subtract 486; times guess; subtract 729. (If any power of x is absent from the equation you're working with, supply it with zero as coefficient. Thus if the equation had no term between x cubed and x, you'd fill the gap with 0 times x squared.)

Smidgens to Oodles

Though books of arithmetic note it not, there is a system of measurement that may well survive even after metrics have knocked out feet and pounds altogether.

Its units are handfuls and splashes, bits and pieces, jots and tittles, bunches and oodles.

Its informal coinages include the glug, passed along to me by one of my machetunim.

(Forgive the interruption, but this valuable word demands instant adoption into everyday English to fill a gap. It's Yiddish for those numerous relatives by marriage for which not every tongue provides a precise name.)

He got "glug" from a servant in the Southern household in which he grew up who, in orally passing along her best recipes (receipts, she called them) used the term to denote the quantity of molasses poured between each pair of glugging sounds from the jug.

There's little use offering a scale for such measurements as long as no two of us has bunches, hunks, passels, or tittles of the same size. All the same, a laborer in the field, Joe Ecclesine, has done his best for us.

PAPRIKA

PAPRIKA MEASURE

2 dashes = 1 smidgen
2 smidgens = 1 pinch
3 pinches = 1 soupçon
2 soupçons = too much paprika

OLDE CORNFIELD

ALCOHOLIC BEVERAGE MEASURE

2 fingers = 1 tot
2 tots = 1 shot
2 shots = 1 slug
4 slugs = 1 snootful
2 snootfuls = 1 night in jail

A sex-linked differential has been noted in at least one popular measure. The dollop, which is how much sour cream you add to the soup, has been found to turn into the double-dollop when it's a man who is doing the kitchen duty.

"Too good to miss," said the editor, of the contribution of a reader of *Consumer Reports* who sent in a Cheez Doodles package and "with deft analytical thrusts," pointed out that the cheese appears after the salt in the list of ingredients, which, under the law, means there's less cheese than salt. The package also indicates that the product contains about 0.002 ounces of salt. Ergo, Cheez Doodles contains less than 0.002 ounces of cheese.

"Since 'oodles' is plural, the product must contain at least two oodles. So one oodle of cheese would equal less than 0.001 ounces." With inexorable logic, the reader concluded: "For those interested in obscure measurements, an oodle would be about two-tenths of a scruple, so what we seem to have is a case of almost no scruples."

What, Me Solve Quadratics?

YOU MAY think solving quadratic equations (you know, the kind that start with x squared) is something done only in algebra classes.

But these equations turn up frequently in practical work, to say nothing of mathematical recreation. I found that out the other day, when I tackled one of those fiendish little mathematical puzzles that look so simple.

I found out that the formulas you meet for solving quadratics look pretty awful, especially the sequential version offered for use with an electronic calculator.

But actually here's all you have to do.

Arrange the equation in the normal form—something times the square of x, plus something times x, plus something. Give these somethings convenient names: first number, second number, third number.

Multiply 4 times the first number by the third number. Subtract that from the square of the second number.

Take the square root of the result.

Add the square root to the second number.

Divide by twice the first number.

Change the sign in front of the result.

Now from that same square root subtract the second number and then divide by twice the first number.

Each of these last two operations will give you one answer to the problem, since all quadratic equations have two roots. Both answers will be mathematically correct, although you may have to discard one on common-sense grounds. When a problem deals with a real distance, for instance, you'll naturally toss aside any solution that is a negative number.

Add-On Taxes and Take-Off Discounts

CALCULATING the amount of a sales tax and then adding it on, or of a discount and then subtracting it, is the slow way—if all you want to know is the final price.

In either case, it's simpler when done in a single step.

With a sales tax such as 6 percent, merely multiply the price by 1.06.

With a discount such as 10 percent, just multiply by 0.9.

Depreciation by Sum of Years' Digits

THIS ROUTINE may be more or less a historical note for most of us. It appears that the income-tax people no longer encourage its use for new purchases. But it is so wonderfully weird that it seems worth preserving for the benefit of those who appreciate an odd way to depreciate—if that makes sense. Other than that, I can only say that anyone who does need it will need it badly.

One long-accepted method of writing off the cost of a car or piece of business equipment uses a method called the sum of the years' digits. It may seem a strange way to figure depreciation, but in practice it often works out rather well. To use it, begin by determining the number of years over which the depreciation is to be taken. Add to that number all the smaller whole numbers.

Use that total as the lower part of a set of fractions. For the first year the upper part is the number of years over which you've chosen to take the depreciation. For the second year it is the next smaller number. And so on. Multiply the original value by each of these fractions to find the amount of write-off for each year.

If you're depreciating a $9000 computer over 5 years, the digits will be 5, 4, 3, 2, 1, for a total of 15. First-year depreciation is 5/15 of $9000, or $3000; second is 4/15 of $9000, or $2400. It will be fully depreciated after 5 years.

Addition Helper

OF ALL arithmetic shortcuts, this one is perhaps the biggest time saver, because you'll use it the most often.

Adding 10 is easy, so watch for 10-pairs when adding a column of figures. These are the number combinations that add up to 10, such as 3 and 7 or 5 and 5. You'll soon learn to spot and use them even when there's another digit or two in between.

You may eventually come to handling trios that total 10, like 2, 3, and 5, in the same way.

Done efficiently, mental addition is often faster than keying a list into a calculator—even assuming you have one of those lovable little machines right at hand.

Add Without Carrying

Do you get confused by carrying when adding long columns of multi-digit numbers? Maybe you've lost the knack, what with using a calculator whenever there's one around.

```
   12,344
   54,323
   98,765
   59,789
      21
      20
      20
      23
      20
  225,221
```

By totaling columns separately, you can avoid carrying.

As you can see, you could equally well start with the left-hand column. Advantage of doing it that way is that when an approximate answer will serve, you can stop as soon as you reach one that you can see will meet your need.

Subtract Without Subtracting

If you'd like an alternative to the usual way of subtracting several-digit numbers, try this.

Write the number from which you are to subtract.

Under it, instead of writing the number you are taking away, write what is called the nines complement of each digit.

(The nines complement is the number which, added to the number you have, will make 9. For instance, the nines complement of 6 is 3.)

Then add.

Now separate the left-hand digit of that answer and add it to the rest of the answer.

Suppose that from 2372 you wish to subtract 1894. The nines complements of the four digits 1, 8, 9, and 4 are 8, 1, 0, and 5.

So to 2372 you add 8105, getting 10,477. When the first digit, 1, is separated from the rest of the result and then added to it, you get your final answer, 478.

This may not be easier than the good old way, except maybe for those folks who had such a dreadful time when they first met subtraction in grammar school. But at least it could say, as Jimmy Durante said of himself, "Me, I'm a novelty!"

As a point of further interest, I'm told that this is how a computer would handle the problem, since computers are pretty simple-minded and not at all good at subtracting. A consoling thought for all of us.

Mix It the Midpoint Way

THE TRADITIONAL mixing problem is a nice exercise in elementary algebra. It's not especially difficult to handle, once you've formed the simultaneous equations and recalled how to solve them.

But not all of us have retained our algebraic skills in rust-free condition.

So a kindly soul has worked out something that might be called the midpoint method.

It's a method for solving mixture problems without recourse to algebra, and who can hate that?

Here it is, applied to a typical question.

With peanuts worth 65 cents a pound and cashews 90 cents, how much of each do you need if you wish to produce a 40-pound sack of mixed nuts worth 75 cents a pound?

1. Find the midpoint or average of the two values (adding 65 to 90 and dividing by 2 gives 77.5).

2. From this result, subtract the wanted value (77.5 minus 75 is 2.5).

3. From the midpoint result, subtract the smaller value (77.5 minus 65 is 12.5).

4. Multiply half the wanted amount by the result of step 2. Divide by the result of step 3. (Forty divided by 2, multiplied by 2.5, and divided by 12.5 is 4.)

5. To this, add one-half the quantity wanted (4 plus half of 40 is 24).

This is the amount to use of the lower-valued ingredient—24 pounds of peanuts.

6. Subtracting this from the total amount gives the figure for the higher-valued ingredient.

Forty less 24 says you should use 16 pounds of cashews.

Better Way to Mix It?

AFTER TRYING both the traditional algebraic and the novel "midpoint" approaches to mixture problems, I wondered if there might not be a simpler approach than either. Maybe you'll think this one that I've worked out fills the bill.

As an example I'm using the same nutty problem: With peanuts at 65 cents a pound and cashews at 90, how much of each kind will you require to produce a 40-pound sack of mixed nuts costing 75 cents a pound?

1. From wanted value, subtract smaller value (75 minus 65 is 10).

2. Multiply this by quantity wanted (10 times 40 is 400).

3. Now divide by the difference between the two values (400 divided by the difference between 90 and 65 is 16).

So there you have the quantity needed of the higher-valued ingredient. The remaining 24 of the required 40 pounds must be the other kind of nut.

This seems so simple it might be worth trying on another practical problem. Given an 18 percent solution and a 24 percent solution, how much should you use of each to make up 90 ounces of a 22 percent solution?

From 22, subtract 18. Multiply by 90. Divide by 6, which is the difference between 24 and 18. Result: Use 60 ounces of the higher valued (24 percent) solution, making up the rest of the required 90 ounces with 30 ounces of the 18 percent stuff.

Triangle a Problem

A SURPRISING number of measurements and calculations can be made and problems solved by converting what doesn't sound at all like a triangle into just that familiar geometric figure.

One example is estimating the area of an odd-shaped piece of land. Subdivide a map of it into rectangles and right triangles. The areas of these are easily calculated.

Here's another example. I'm putting it in partly because it illustrates a basic method...but mostly because it produces an answer I find quite surprising.

The Golden Gate bridge that takes you northward out of San Francisco is about a mile long. On a warm day (they don't hardly have hot ones there) it expands about two feet. If there were no expansion strips to compensate, how high a bulge would this make? Assume the whole bridge will bulge upward in a uniform curve.

Have you made your guess?

An easy way to calculate this is by taking the height of the bulge as one side of a right triangle that has half a mile as another side and 1 foot plus half a mile as the hypotenuse.

Going back to Pythagoras, as you usually do with right triangles, you remember that the square of the hypotenuse is equal to the sum of the squares of the other two sides. To find any side, then, you square the hypotenuse and the known side and take the square root of the difference between them.

This is quickly done with a calculator (or slide rule, for that matter, if you happen to be the last of the die-hard engineers). To spare you the agony of pencil arithmetic if you have neither of those tools at hand, I'll tell you that 2,641 squared is 6,974,881 and 2,640 squared is 6,969,600. The difference is 5,281, of which the square root is 72 plus.

So the extra two feet of length would produce a bulge more than 72 feet high.

I find that remarkable. It must be a problem to engineers.

How Big Is a Triangle?

WHETHER doing a little land measuring or cutting a sail, you may have reason to calculate the area of a triangle. If it's a right triangle—the kind that contains a right angle—it's easy enough. Just multiply together the lengths of the legs and divide by 2.

There are formulas that work for triangles of any shape. Here is one that doesn't require a trig table, though it does demand finding a square root. Your calculator probably has a key for this function. If it doesn't, you can always chase it down through a table of square roots, the good old paper-and-pencil method from school days, or through the trial-and-error method that the simplest hand calculators have turned into a cinch.

Add together the lengths of the three sides of the triangle to find its perimeter. Divide the result by 2. Since this is one half of the perimeter, let's call it half-p.

From half-p, subtract the length of one side of the triangle. Also from half-p, subtract the second side. And from half-p, subtract the third side.

Multiply those three results together. Then multiply that product by half-p.

Find the square root. That's the area of the triangle.

Give it a try. If you're sewing a sail having sides 8, 10, and 12 feet, the perimeter will be 30 and half-p will be 15. Subtracting each side from 15 will give you 7, 5, and 3 to multiply together and by 15 to produce 1575, of which the square root is 39.7 square feet or (when divided by 9) about 4.4 square yards.

Approximate a Time Payment

LOOK, MA! No algebra. No calculating device. No fat book of tables to refer to.

No logarithms, either.

Without any of those helps, you can figure out—with surprising accuracy that's more than good enough for most decision-making—how much your monthly payments will be for a home mortgage or a car installment loan.

A rather weird approximation formula lets you immediately, with reasonable use of pencil and paper, get this useful result for any preliminary purpose. Other methods given in this book can be used at leisure to find payment amounts to the penny.

All you need to supply in order to use this approximation method is the number of monthly payments to be made, the annual interest rate, and the amount of the loan. Juggling these will produce three quantities to add together.

1. First divide 1000 by the number of payments.
2. Now multiply the interest rate by 0.4286
3. Multiply the interest rate by itself and by half the number of payments. Move the decimal point four places to the left.

Add these amounts together. The result is the estimated monthly payment for a $1000 loan.

Multiply by the number of thousands of dollars of installment loan. The answer will be a close approximation of the monthly payment.

Does it work?

Try it for a $100,000 mortgage at 9 percent to run for 20 years.

Divide 1000 by the 240 payments to get 4.17.

Multiply 0.4286 by 9 to get 3.86.

Squaring 9, multiplying by half of 240, and dividing by 10,000 gives 0.97.

Adding up these three quantities and multiplying by the number of thousands of dollars of the loan (100) says the payments will approximate $900 a month. You're within a few dimes of what an actual calculation or the fat book in your banker's desk says it should be—$899.72.

Tested for a 25-year, 6 percent $100,000 loan, this method gives $644. Not bad, since the precise answer is $644.31.

It seems to work equally well for a short-term personal or auto loan, too. The book says $1000 at 12 percent for 4 years will cost you $26.34 a month. Our method estimates $26.31. How about that?

You may wonder where such unfathomable approximations as this come from. Beats me. Someone must have doodled a long time for this one. In somewhat different form, it was passed along to me by James Booher who was my first-year college-math instructor lo! these many years ago.

What Percent of...?

WHAT PERCENT of 16 is 4?

What percent is 4 of 16?

What percent of 4 is 16?

It's easy to be confused by these "percent of" questions.

Maybe it will help if you simply remember to start with the number after "is" and multiply it by 100, then divide by the other number.

Approached that way, the first two questions prove to be identical at 25 percent. And the third one comes to 400 percent.

You may not need the rule for those familiar fractions—but without it might you not fumble a bit before replying to "What percent is 73 of 37?"

Percents, Decimals, Fractions

CONVERTING percentages and decimals to fractions and the other way around is child's play. That's because children, the older ones anyway, have probably just been studying that kind of stuff in school. For the rest of us it may not be so easy.

It helps to keep in mind that all these things are fractions. A percent is just a fraction in which the lower number, or denominator, is 100. A decimal is a fraction in which that denominator is 10 or some power of 10. And all fractions can be regarded as merely uncompleted divisions.

If you want to turn 3/40 into a decimal, just do the division indicated by that slash mark. Divide 3 by 40 to get 0.075, which is 75/1000.

If you want the result as a percent, turn that fraction into hundredths by dividing each part by 10. The 75/1000 that started as 3/40 now can be seen as 7.5 percent.

Making the trip the other way, from a decimal to a fraction, requires a decision. What do you want the lower number to be? You might choose 64, since 1/64 of an inch is a useful and quite precise measurement for many things.

To convert a decimal such as 0.35 to 64ths, just multiply the two numbers together. Comes out about 22/64.

Similarly, you can turn one fraction into another. What is 3/40 in 64ths? Divide 64 by 40 and multiply by 3. The closest whole number to your 4.8 result is 5. Your answer rounds to 5/64.

Quick Fractions

W HAT YOU'RE most usually called upon to do with a pair of fractions is to divide, multiply, add, or subtract them. You can do any of these things in a hurry and almost without thinking, even if you're out of practice.

To multiply, just write the product of the numbers above the line and then, below the line, write the product of the numbers below the line.

The same goes for dividing, except that first you must lay hold of the fraction you are to divide by—and turn it upside down.

Quick way to add a pair of fractions begins with cross-multiplying and adding the products to get your new numerator. Then, below it put the new denominator, which you obtain by multiplying the old denominators together. (Save time by canceling first wherever you can.)

Subtracting one fraction from another proceeds much like adding. But this time you subtract one product from the other instead of adding the products together.

That Tricky Percentage Increase

Y OUR INCOME from some source in one year was $19,444. For the next year it was $23,111. This being during a period of notably high inflation, you wonder what the trend of your real return was. Checking back, you learn that the published rate of inflation for the year in question was 13.4 percent.

How did you fare by comparison? Try this for size.

Subtract the earlier year's income from the later year's larger one. Divide that difference by the earlier year's amount. Push the decimal point two notches to the right, thus multiplying by 100 and converting the fraction into a percentage. Well, it seems you had an 18.86 percent increase in the number of dollars coming in.

Not bad, although inflation gobbled up more than two thirds of it. To be more precise, it ate some 71 percent (13.4 divided by 18.86 and multiplied by 100).

To look at this event another way, subtract the inflation percentage from your rate of return. The effective rate of return, then, was 5.46. A rule of thumb used by some investment seers says this is pretty much what you might expect over the years.

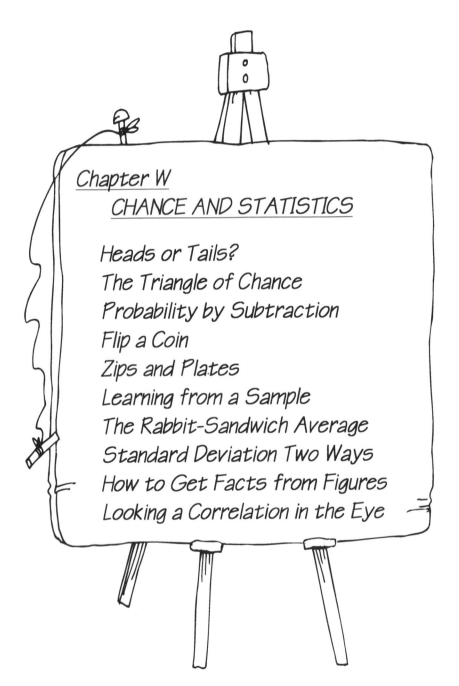

Chapter W
   CHANCE AND STATISTICS

Heads or Tails?

T HIS LITTLE item gets into a how-to-figure-it book under the heading of how not to figure it.

For background, let's begin by establishing that you're too sophisticated to fall for what is called the maturity of the chances, or the gambler's fallacy.

That error is simply a belief that if a coin has come up heads most of the time in a series of tosses, it must thereafter favor tails until things even out. Or some similar nonsense to the effect that a run of red at a roulette table indicates that black is "overdue."

The error is based on a misunderstanding of what the law of averages really means when it says that in the long run heads and tails will tend to appear with equal frequency. So they will—from here on at any given moment.

What they will not do is tend to show up unequally in order to rectify whatever has already happened.

Unless they are defective, coins and dice and roulette wheels and other instruments of chance have no memories. (No consciences either, in respect to such matters as whether your baby needs new shoes.)

What has set me off on this is reading an article in a national magazine. Its author, having explained all this, takes off from it to a conclusion that could cost you money if you went with him. Since you may have encountered this thesis there or elsewhere, or even come to the conclusion yourself, it is worth exploring in the interests of protecting your pocketbook.

Says the author: "Psychologists and mathematicians have discovered that seven out of ten times, a person predicting the flip of a coin will call "heads."

I think that's very likely true. But now let's follow the author to his misguided conclusion.

"Therefore, make sure that you are the tosser, and your opponent the caller. Scientifically, you see, the chances are even for heads or tails to come up. But if he calls heads 70 percent of the time, you have a 3 to 2 chance of coming out ahead."

Somehow it sounds convincing, doesn't it? But the fact is that if the pigeon calls heads and tails with equal frequency he'll tend to be right half the time. If he calls heads on every flip he'll be right half the time, in the long run. Just remember that each flip is independent of the past and of the future—and for that matter of the call.

The Triangle of Chance

You MIGHT read that somewhere a woman has given birth to her eighth child, and for the eighth time it's a boy. How remarkable is this? In mathematical terms, what are the odds against it?

"When you toss a coin…" is a phrase that turns up in all sorts of scientific writing, on subjects as diverse as card playing and genetics. The toss of an "honest" coin is a perfect model for calculating probabilities.

Flip a coin once and the possible outcomes are two, with an equal chance for each.

Toss two coins (or one twice) and you may get a head and a tail or two heads or two tails. But not with equal probability, since there are two paths to head-and-tail: head then tail, tail then head. In somewhat more complex instances, this distinction caused a lot of confusion among learned mathematical people at one time, as it still does among amateur and professional gamblers—often to their financial woe.

If you flip a quarter 10 times, how probable is it that you'll get precisely four heads and six tails? Assuming boys and girls appear with equal frequency, which is close enough to true for purposes like ours, how often will a brood of eight children be equally divided?

You can list all the possible paths and add up the number that meet the requirement. But as the numbers dealt with become larger, you'll find it an increasingly tedious chore.

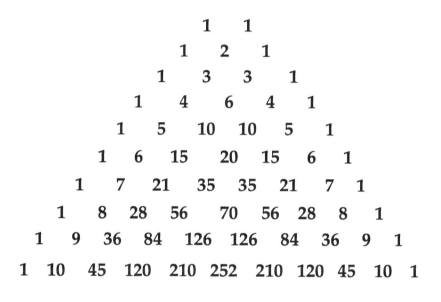

That's what makes a triangle devised by Blaise Pascal, more than 300 years ago, both fun and useful. As you will discover, the basic form displayed here can be expanded indefinitely to calculate the distributions of even longer runs of heads or tails—or girl or boy babies.

You'll quickly detect the pattern that produces the triangle. Each added number is the sum of the two numbers most nearly above it. You can go on adding lines indefinitely, if you wish to consider cases of more than ten coin tosses or children.

Now what does the triangle say about that family of eight kids— or any other number of youngsters?

After the zero line, which says in effect that all groups of no children are the same, line 1 says that with 1 child there are 2 possibilities (1 plus 1), which we hardly needed to be told.

Line 2 (for 2 children or 2 coin tosses) totals 4 when added across. Of these 4 possibilities, the first and last numbers represent the single possibility that both will be boys or both will be girls. The 2 in the middle is for the two chances in 4 that there will be one child of each sex.

Line 3, for 3 kids, totals 8 for the 8 possibilities. There is 1 chance in 8 of all boys, 3 of 2 boys and a girl, 3 of 1 boy and 2 girls, and 1 of 3 girls.

Jumping down to line 8 we find the answer to the question about a brood of 8. The total across is 256. So there is only 1 chance in 256 that all 8 will be boys and a similar probability—represented by the 1 at the right-hand end of the line—that all will be girls. Similarly, a 1-boy 7-girl mix will occur 8 times in 256 for a 1-in-32 chance, same being true for 7 boys and 1 girl.

The next number in the line says there'll be 2 boys and 6 girls 28 times out of 256. The 56 that follows is how many chances there are in 256 of getting precisely 3 boys and 5 girls, and the 70 says there are that many chances in 256 that the 4-and-4 mix that many parents might prefer can be expected that many times in 256.

Note that, as bridge players already know, while the equal distribution is the single most likely one it will occur far less often than the other possibilities combined.

The tenth line is, naturally, the one to examine to find the likelihood of heads coming up precisely 4 times in 10 tosses of a quarter. Adding across gives 1024. The numbers in line 10 say there is the usual 1 chance of all heads, 10 chances of 2, 45 of 3, and 120 of 4 heads. A quick exercise in division says that this probability of 120 in 1024 means about 1 chance in 8 or 9.

You don't need much math to use this wonderful triangle. But those who remember—probably not favorably—encountering the binomial theorem in elementary algebra will find that here it is again.

Think of the simple binomial a + b as one a plus one b, that is 1a + 1b. In the triangle, those 1s, the coefficients of a and b, appear as Line 1.

Now consider the second power, or square of a + b. Multiplying a + b by itself gives a result of which the coefficients are 1, 2, and 1. Sure enough, Line 2 of that remarkable triangle.

The coefficients of the third power of a + b prove to be 1, 3, 3, 1, and so it goes.

But you won't need to continue this expansion to find out how likely 8 boys in a row would be. The eighth line of the triangle says this is to be expected once in 256 runs of eight births. And, since a similar run of girls is equally likely, the chances that a brood of 8 will be same sex are 2 in 256 or 1 in 128.

Probability by Subtraction

W_{HETHER} we're buying, selling, investing—or merely crossing a busy street—life has us constantly calculating probabilities. Unfortunately, this is a often a tricky thing to do.

How maddeningly true this can be was nicely demonstrated by a newspaper columnist. Marilyn Vos Savant, in the weekly Parade magazine section, handed her readers this killer one Sunday.

You're on a game show. The host offers you the choice of three doors. One hides a car, the others hide goats. After you've chosen the left door—let's say—the host asks, "Aren't you glad you didn't choose the middle one?" (or it could be the right-hand one). And opens it to reveal a goat, as he knew it would.

"Now," he says, "would you like to change your choice?"

Should you?

When the "Ask Marilyn" columnist gave the answer Yes, 92 percent of the readers who contributed to the bushels of mail that followed disagreed, This included almost two-thirds of the replies that came from university campuses.

Perhaps despairing of proving her conclusion by mere explanation, the columnist eventually convinced many of the skeptics by challenging them to model the game with three paper cups and a coin. More sophisticated readers came around after writing computer simulations—by no means overkill when the problem is one dealing with probability, the branch of math that the great mathematician Charles Sanders Peirce has described as "the only one, I believe, in which good writers frequently get results entirely erroneous."

By both kinds of test—cup and computer—changing was proved to be the successful strategy almost exactly 67 percent of the time.

At the end of the controversy, all that was lacking was a direct, simple verbal explanation of why changing the guess was the way to improve the 33 percent chance—not merely to 50 percent, as a great many contended, but all the way to two thirds.

As I see it, at this point the plot demands a final scene in which none other than everybody's favorite master of logic intervenes.

Buried in the heaps of mail from readers is found an odd-shaped envelope, weathered with age, its postmark indecipherable.

As I have had occasion to remark to the sometimes obtuse Watson [it began, in a regrettably patronizing way] when you have eliminated the impossible, whatever remains, however improbable, must be the truth.

Could the signature be? It was! Sherlock Holmes!

In a trice, the great man had made it all clear. Given a 100 percent chance that a car crouched behind one of the doors and a 33 percent chance that this was Door 1, the remaining 67 percent chance belonged to the second and third doors jointly. One of these having been eliminated by the sight of a pair of hairy ears, giving it a probability of zero, the contestant merely need opt for the other to win 67 percent of the time.

Knowing that the chances are 2 in 3 that his first choice is a goat, our contestant should stand pat only if he prefers a Toggenberg to a Toyota.

Flip a Coin

How good are the chances of getting heads exactly half the time in a series of, let's say, 10 flips of a coin? Not so good as you might think. Not even 1 in 4.

Of the ways of figuring that out, one is especially interesting because both the method as a whole and its initial step are adaptable to so many other calculations.

You can begin by finding how many ways 10 objects (or coin flips) can be combined into sets of 5. This is the same as discovering how many different sets of 5 volumes each can be made from a 10-volume encyclopedia.

You figure this by dividing the factorial of the number of objects by the factorial of the number in the set you're assembling and by the factorial of the difference between these two numbers. Since the factorial of a number is simply the product of that number and all the smaller positive whole numbers, you'll find yourself dividing factorial 10 (3,628,800) by factorial 5 (120) twice, to get 252. (By the way, if you've forgotten, factorial 10, or 10 factorial, is commonly written 10!)

For your second step, take the probability of getting heads on any single flip and raise it to the power of the number of heads the problem calls for. Since heads is always 1 chance in 2, this means raising 1/2 (or 0.5) to the fifth power, which is 0.03125.

Third step: Raise the probability of NOT getting heads (in this example it is also 1/2) to the power of the number of times this is to happen (again 5) to get the same 0.03125.

Multiplying together the results of Steps 1, 2, and 3 gives the probability that the event in question will occur. It is 0.246, or not quite a 25 percent chance.

You can extend this routine to finding answers to questions of other probabilities.

If you use it to find the chance of precisely 6 heads in 10 flips and then of precisely 4 (which will be the same, naturally) and then add both these results to the previous one for 5 heads, you'll discover that the probability of getting between 4 and 6 heads in 10 flips of an honest coin is very close to 2 out of 3.

Zips and Plates

THE INVENTION by the U.S. Postal Service of the nine-digit zip code may just possibly approach the ultimate human limit. Anyway, the ability to recall—briefly—a nine-digit number is found at the "superior adult" level of a widely used intelligence test.

Designing such things as zip codes and license-plate numbers calls for a calculation of permutations. All this has to do with is the number of ways a set of things can be arranged and rearranged.

How this works is especially easy to follow when the things are digits. In our decimal system there are 10 of these, 0 through 9. You can use each followed by the same or another digit to form a two-digit code, producing all the numbers from 0 through 99, giving 10 times 10 possibilities. You can call that 10 squared and call the possibilities with three digits 10 cubed.

The five-digit postal zip code thus permits 10 to the fifth power, or 100,000, different zips. With the introduction of a 9-digit zip, this became a billion, which is several times as many zips as there are Americans. Even if the population triples, each citizen can still have his own personal postal delivery code.

With license plates that use letters as well as numbers, the possibilities multiply. If each letter and each number can occupy any position, a single character may have 34 variations (10 plus 26 minus 2—to exclude the letters O and I which might be confused with digits).

Such a 5-character zip or plate could come in the number of variations equal to 34 to the fifth power. This is upwards of 45 million and should serve any likely purpose. Such mixtures might be easier to remember than all-digit codes.

When letters and digits are not made fully interchangeable, as with California's license plates and the zip codes used in many countries, the calculation changes slightly. A rule that has been used for the 7-character California plates is 4 digits followed by 3 letters. So the number of variations available is 10 times 10 times 10 times 10 times 26 times 26 times 26, more simply expressed as 10 to the fourth power times 26 cubed. Either way, you'll find it comes out to more than 175 million.

Learning from a Sample

A GREAT PART of the figures you encounter in your reading about incomes, inflation, living costs, unemployment, health, and innumerable other matters are based on samples.

Although conclusions drawn from samples can be highly misleading and should be taken with several grains of salt, there are many things that cannot be found out except from a sample. As a maker of electric irons, you might want to test acceptance of a new and radically different model before investing large sums in producing it. You can't have it tried by everybody in the country; that would cost too much. So you test it on a sample of perhaps a thousand.

Or suppose you make light bulbs. You want to know how many hours they will last before burning out. Again you must test a sample, because if you tested them all to destruction you wouldn't have anything left to sell.

But you must be careful to choose a sample that is typical of the whole. Otherwise you may do something equivalent to what the Kinsey people did in their first famous study of sex: They issued figures based largely on college graduates and prison inmates, the two groups they found most accessible and willing to answer questions. How well these represent the nation as a whole is open to question—and statisticians have indeed been questioning it ever since.

A Gallup poll on adoption of the metric system produced the conclusion that 33 percent of people who had been to college had never heard of it. A large majority of those with less education were also in the dark.

So what would you make of the findings of a California newspaper that 98 percent of its readers knew about the metric system? The paper then boasted "how much more knowledgeable" its readers were than Americans generally.

But the discrepancy is much more easily explained: a badly biased sample. While Gallup had taken a cross-section of the public, the newspaper had relied on coupons clipped out, filled in, and mailed by readers. It is fair to suppose that most folks who had never heard of the metric system had simply selected themselves out of the poll by not bothering to participate.

Even such popular conclusions as the one that criminals are subnormal in intelligence are dangerous to believe, based as they are on biased samples. All studies "proving" this have been made on prison populations. What they prove, of course, is that criminals who get caught are not terribly bright. But we must accept the fact that we remain in ignorance about the IQs of those who contrive to stay at large.

The undersized sample must share guilt along with the biased one. Most of us use it informally, coming to conclusions about the intelligence of other races or the combativeness of Irishmen or the dumbness of blondes on the basis of the very small sample of each we have personally known—and possibly observed with something less than scientific objectivity.

This is foolish, true—but scientific teams have done the same thing many, many times. This is attested to by the horde of medical miracles, including cold cures, that have come and gone over the years. As a great 19th century physician once wryly advised, "Make haste to use a new remedy before it is too late."

How big must a sample be? There is no exact answer. A handful of cases may be indicative in some matters, while hundreds tell little in others. In an early test of polio vaccine, 450 children were vaccinated, while 680 were left unvaccinated as controls. Nothing was learned, because no cases developed in either group. Even at its worst, polio was a disease of such low incidence that groups of 7,500 to 25,000 would have been needed to produce significant conclusions.

In that word significant lies the answer for us when we want to judge a conclusion from a report of a medical experiment or other sampling study. A good report will give us what is called the significance of the result. If we find such words as "statistically significant at the five percent level," we can conclude that the odds are 95 to 5, or 19 to 1, that the result found is meaningful. That is, there are 95 chances in 100 that the treatment really did the work. And 5 in 100 that the result was pure chance and could not be relied upon to occur again.

The Rabbit-Sandwich Average

THE IMPORTANT thing when calculating an average is to avoid coming up with a rabbit sandwich. As we shall see.

You're not a snob and I'm not in the real estate business. But let's pretend for the moment that you are and I am. And that I want to sell you some property along a rural road I happen to be familiar with.

You'd like to know the average income of your prospective neighbors. I run my fingers over my calculator, entering some fairly accurate educated guesses. Not less than $100,000, I tell you, what with the presence of several doctors and lawyers and substantial business people.

Having bought on the strength of this snob-appeal, you are amazed a few months later when I ask you to join a campaign against a school-bond issue. The basis of our fight, I tell you, will be that people of this area just can't afford it. After all, their incomes average only about $25,000 a year.

Which figure is the phony?

Actually, I've told you the truth both times. But not the whole truth.

My newer figure is the kind of average called a median. It tells you that half the people in the neighborhood receive less than $25,000 a year, and half of them receive more. Since ours is an area of about ninety percent wage earners and retired people, this figure does represent their economic level rather accurately. And it is generally true that the median is the type of average that gives the most useful information about income distribution that can be crowded into a single figure.

As for my first figure, it too is an average correctly arrived at. It is the more familiar type, the arithmetic average or mean, reached by adding everybody's income together and dividing by the number of incomes. The weakness of this kind of figure for dealing with incomes is that a few extreme cases influence it to a highly disproportionate degree. This kind of average works out to that impressive $100,000 because at one end of our little country road are clustered the week-end retreats of three millionaires. City folk, you might say.

The way a few large figures can distort a mean might remind you of an old joke about the operator of a diner who offered rabbit sandwiches at a remarkably low price. When questioned about it, he admitted that he used some horse meat to keep his costs down. "But I mix 'em fifty-fifty," he avowed. "One horse to one rabbit."

It should be noted in passing that a third kind of average also crops up occasionally. It is called the mode, and it means the most common case. It is this kind of average we have in mind when we say, or did at one time, that the "average advertising man" wears a gray flannel suit. (It's not a subject with which I have been keeping up.) Or when we pass along the truism that the "average high school graduate" can't spell.

Standard Deviation Two Ways

Iꜰ ᴛʜᴇ ᴍᴏꜱᴛ common of all statistical calculations is the simple average (or mean) the second most often wanted is the standard deviation. Here is a choice of two routines for finding it.

Along the way, each method produces another result that is often wanted—the variance.

In the first method, you begin by finding out how much each of the scores or values you're working with differs from the mean (which no doubt you've already calculated by adding the values together and dividing by the number of them).

Now square each difference. Add up these squares. Divide by the number of scores or values. (This result is the variance.)

Now find the square root. That's the standard deviation.

To use the second method, begin by adding up the values. Square this total. Divide by the number of values. Subtract this result from a quantity obtained by squaring each value and adding all these squares together. Divide by the number of values (noting that this step gives you the variance). Once again, you can now obtain the standard deviation, or sigma, by deriving the square root.

If you'd like to test either routine, try these seven figures: 2, 2, 3, 5, 7, 9, 14. You should find a variance of 16.57 and a standard deviation of 4.07.

How to Get Facts from Figures

ALL OF US are reaching conclusions and taking action from reports of a statistical nature—newspaper and magazine articles, business presentations, advertising... That's why it's important to develop critical reading methods that spot such symptoms of unreliability as inconsistent reporting, unwarranted precision, and figures not quite attached to what they purport to talk about.

Inconsistent reporting methods lie behind many things we read. Newspaper stories once brought out the surprising "fact" that in Connecticut 2 percent of traffic accidents are fatal, while the figure for Georgia was 17 percent. This seems to suggest something important about the driving laws or habits of these states. But all it really meant, when examined more carefully, was that reporting methods differ. Connecticut had listed practically all accidents, however minor, and the fatal ones were a tiny number compared with mere fender bendings. In Georgia, only major crashes got on the list. Naturally a larger proportion of these involved fatalities.

Better reporting is one of the reasons for the alarming apparent increase in lung cancer in recent decades, when the true incidence has been shocking enough. More of the cases get into the published figures as diagnostic and reporting methods improve.

At one time, annual deaths recorded in New York State from coronary artery disease jumped from 8,119 to 36,390 in just six years. When the raw figures were published, a great wave of alarm quite naturally followed. But what had really occurred was a sudden change in the methods of reporting causes of death. Only after the earlier of the years was coronary artery disease regularly listed as such on death certificates.

The overly precise figure is another warning that what you are encountering is less than meets the eye.

A news report once said that the average Yale graduate of the class of '24 was now earning $25,111 a year—a princely income at the time. If you wondered how anyone could have uncovered such private information right down to the dollar, you might have read on and learned that the figure came from replies to a mail questionnaire. Since it is a fair guess that the greater part of those sons of Eli who had wound up in poverty were unreachable or unenthusiastic about telling their incomes, it is unlikely that the average was close to truth. And, anyway, few of us can calculate our own incomes down to the dollar.

Yet note how effectively those last three digits manage to imply that exact information is being imparted.

In the same category comes the sociological study that long ago found that a working woman in New York needed $40.13 a week to get by. The amount has changed as the years have passed—but using over-precision to lend respectability to a wild generalization has not.

Some figures are foolishly precise without at first seeming to be. Note the National Board of Fire Underwriters' announcement that fire losses in one recent year totaled $1,107,824,000. Impressively precise but not suspiciously so—isn't it?

In fact, it is being given down to less than one part in a million—

an implied accuracy of one ten-thousandth of one percent. Yet the figure was reached by adding to reported claims (which in themselves may be presumed to contain many exaggerations of actual losses) the Board's estimate of unreported losses. Surely such a total must be thousands of times as imprecise as it has been made to seem.

All overprecise figures, and especially this last one, belong in the same bag as a fine old joke. A farmer shows a tourist some bones found on his land and remarks that they are one million and three years old.

How can he know so precisely?

"Because," the farmer says, "an archaeologist told me they were a million years old—and that was three years ago in May."

The semi-attached figure is a handy device for those who would deceive. It's a source of a great many innocent, but costly, misunderstandings as well.

Thus you might have read about an overwhelmingly great increase in the populations of metropolitan areas of the United States in a mere ten years. Official census figures were cited by a news writer to back up the conclusions.

But what had really happened was that the Bureau of the Census had changed its definition of a metropolitan area. The new use of the term took in much more territory. What was probably a real but modest increase became all too striking. It should be noted that many a businessman makes potentially costly decisions on the basis of such population figures.

A San Francisco political figure, a Naval Reserve admiral, polled his constituents about attitudes toward the war in Vietnam. He merrily concluded that the people were predominantly hawks, because 42 percent voted for more use of military power, 27 for the current stand, and only 28 percent for immediate withdrawal.

The gimmick, as you can see, is that the politician's choice of questions forced the answers to an unknown degree. There was no resting place for the people who favored a position less aggressive than the administration's but short of simple withdrawal.

Moral: In reading the results of any poll, it is as important to study the questions as to study the answers.

Both unions and employers have been known to use semi-attached figures against each other. Thus an employer "makes a survey" by asking each workman if he has a complaint against the union. He probably has. By adding up all the assorted irritations, minor or otherwise, the employer is able to announce that "87 percent of the employees in this plant are opposed to the union."

Semi-attached figures have been with us for a long time. The death rate recorded in the Navy during the Spanish-American War was 9 per thousand. For civilians in New York City during the same period it was 16. Navy recruiters used these figures to prove that it was safer to be in the Navy than out of it.

These figures, of course, were not comparable. Those for civilians included old people and infants and the ailing, all of whom have relatively high death rates. Those for the Navy were based on a population from which all such people had been excluded.

Looking a Correlation in the Eye

I CAN'T THINK of anything to beat *post hoc* at leading people up the well-known garden path.

When two things are found to occur together regularly, they are, in statisticians' terms, correlated. The danger comes in figuring that this proves one has caused the other.

Post hoc, ergo propter hoc was the ancient logicians' way of reminding us that "after this, therefore because of it" is a piece of dangerously faulty reasoning.

It was once demonstrated that college students who smoke make somewhat lower grades on the average than those who don't. Aha! Smoking dulls the mind! For all I know, or anybody does, that may be the truth. But the correlation in question doesn't prove it. There are too many equally plausible explanations. Perhaps being smart has some tendency to keep you from smoking. Perhaps the sociable sort of fellow who takes his books less seriously than he might is also inclined to smoke. Who knows?

These three possible explanations of a relationship between smoking and low grades illustrate a useful rule. When two things, A and B, are associated, there are three possibilities: A causes B; B causes A; or both are caused by some third factor.

A leading medical journal reported that, according to a careful large-scale study, radiologists died 5.2 years earlier than did other physicians. This was immediately attributed to the effects of radiation, and it naturally produced a heavy dose of terror in radiology offices. Only sometime later did the voice of reason point out that radiology was such a new specialty that any of its practitioners who had died by that time were bound to have died young. There were no old radiologists.

What can you make of the conclusion of the Fair Campaign

Practices Committee that dirty politics doesn't pay off? On the basis of a national election the committee's chairman declared: "In 68 percent of the cases the candidate involved in unfair campaign tactics lost the election."

What the committee overlooked is the likelihood that the dirty tactics were often desperation moves on the part of a candidate who was already trailing badly. It is not unlikely that here B has caused A.

As J. Edgar Hoover (with growth of his FBI and his personal power at stake) and others since have pointed out many times over several decades, our national crime rate is growing. Something must have gone wrong in our schools or our law enforcement system or somewhere to produce this distressing change? Not necessarily. There are more young people in our population every year, and young people have always had a higher crime rate than adults.

The rising crime rate, then, must be seen at least in part as another symptom of the population explosion. As one expert has pointed out, the only sure way we know of fighting crime is birth control.

You can keep your guard up against unwarranted assumptions of cause and effect by remembering some of the amusing but spurious correlations that have been discovered.

One is a close association between the cost of rum in Havana and the salary level of ministers in Massachusetts. No one is going seriously to suggest that either of these things is a cause of the other. They are, of course, simply two products of a third factor, the historic increase over the years in the price of practically everything.

And then there is the correlation between the number of stork nests on Dutch or Danish houses and the number of children in the household. You can accept this as scientific verification of an ancient myth. Or you can note the presence of a third factor. Big families choose big houses; big houses have more of those toasty chimney pots in which storks love to nest.

How trickily confusing cause-and-effect relationships can be is nicely illustrated by what I hope you will forgive me for calling a lousy example.

The natives of the New Hebrides long believed that body lice bring good health—so no one should be without them. The basis was centuries of observation that had taught them that people in good health usually had lice, while sick people often did not. It took medical investigators a while to get this one straightened out. Almost everyone in the islands, it was discovered, had lice as a matter of course. When a man took a fever and became too warm for comfortable habitation, his lice (which may have brought him his feverish ailment in the first place) left him.

So here was a case in which cause and effect were not only the reverse of what observation indicated but also most confusingly intermingled.

"It ain't so much the things we don't know that get us in trouble," remarked the American humorist Artemus Ward long ago. "It's the things we know that ain't so."

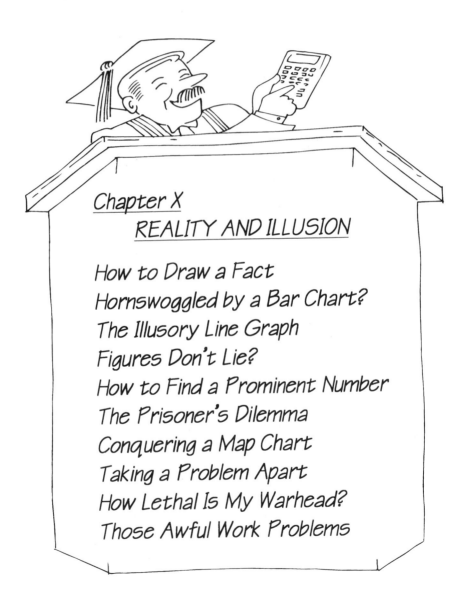

Chapter X
REALITY AND ILLUSION

How to Draw a Fact
Hornswoggled by a Bar Chart?
The Illusory Line Graph
Figures Don't Lie?
How to Find a Prominent Number
The Prisoner's Dilemma
Conquering a Map Chart
Taking a Problem Apart
How Lethal Is My Warhead?
Those Awful Work Problems

How to Draw a Fact

Pictographs constitute an especially useful device for transmitting complicated information simply. But when used carelessly or dishonestly, can they ever distort and misinform!

All too common is an illegitimate version of the pictorial graph in which your eye is forced to compare dissimilar shapes. A large utility company once presented a comparison of its annual tax bill with its expenditures for salaries and wages, indicating their relative size by pictures of moneybags. By drawing one fat bag and one sadly slack one, the artist managed to make a two-to-one ratio look like far, far more than that. If these were actual bags, the one at the left would hold eight or ten times as much as the slack one.

The artist should not be blamed too much. He was up against a problem knotty enough to have produced a great many misleading pictographs.

How would you solve it? How, that is, can you draw two moneybags so they represent 98 and 49 million dollars respectively? Actually, you can't. The ordinary beholder has difficulty in comparing bulks precisely. If you draw a true representation of a pair of bags of such relative size that one would indeed contain twice as much as the other, they will appear so nearly the same size that they will understate your case.

Draw them so that one occupies precisely twice the area on the page that the other does? Again no—because some eyes will see them as the solid objects they represent, and the bag covering twice the area would have much more than twice the volume.

Make one bag twice the height of the other? This is a common practice, and has been done with everything from moneybags to steel mills. But the answer is decidedly no. A bag that is twice as high will be twice as wide and twice as thick. The volume of one such bag will be to that of the other as the cubes of any dimension. One would actually contain not twice as many coins as the other, but eight times as many.

There is just one way out. Keep the bags uniform in size and vary their numbers. Draw two bags for the larger sum, one for the smaller. Or let each bag represent, say, 10 million dollars, using for the first quantity ten bags and for the second, five. For precision, you can omit a small part of the last of the bags in each set.

Whether your story involves populations or tools or telephones, you can tell it by using an appropriate number of little symbols.

Hornswoggled by a Bar Chart?

Bar charts can be deceptive in their simplicity.

A very tall and narrow one will tend to exaggerate, a short and squatty one to minimize.

A bar chart from which the zero line—and consequently some part of the area of the chart itself—is missing will exaggerate even more. Hence, with a chopped-off bar chart, as with a truncated line graph, you are entitled to a visual clue that something is missing. When a bar chart must be amputated, a bread-knife-shaped wavy line at the bottom is at least better than no warning that this unfortunate operation has been performed.

How chopping exaggerates can be seen by comparing a proper bar chart with its truncated brother. (These figures have to do with lines of advertising carried by a magazine.) As a careful look at the numbers shows, both of these represent the same data, but you'd never guess it at first glance. One of them makes the trend seem much more marked than it is. And that, of course, is the one that appeared in an ad.

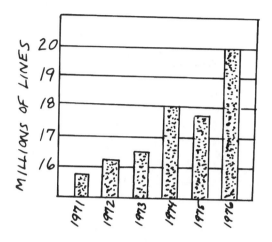

The trend or comparison offered by a bar chart can also be minimized by an unfair trick of geometry. The bars of a proper chart vary in only one dimension. The eye can easily compare these bars. But charts are sometimes drawn with all bars being of the same shape and proportion. The eye is forced to compare areas, which it cannot do well.

The Illusory Line Graph

LINE GRAPHS are widely used for pictorial presentation of everything from production and sales trends to medical and scientific data. Proper ones include numbers or times along both axes. True information can always be gleaned from them by careful attention to these numbers.

But the impression these graphs give is subject to distortion through use of a bagful of tricks, both intended and otherwise. How true this is can be understood by a glance at the set of graphs here. Since all four of them are based on precisely the same data, they should in all honesty create the same impression.

But, as you can see, they do not.

The first of the four is all that a line graph should be. It represents a substantial but not overwhelming trend, and that is just the impression it produces.

The second one employs horizontal expansion to make exactly the same trend look like something less than it is.

The third has been expanded vertically to do precisely the opposite. It makes the trend appear much more marked than it is. The shape of these last two charts should be taken as a warning signal. Whenever you meet a graph that is very far from square, it is likely to produce an impression different from what you would have obtained from the data themselves. So examine those figures and take your impression from them, not from the misshapen chart.

Even more deceptive is Trick 4, in which the same information is cast once more. Chopping a little off the top of the graph gives an impression of a trend that is all but bursting the top. More important, slicing off all the bottom below the curve itself leads the eye of the observer to believe that this trend goes from nothing to maximum.

It is not unreasonable to call the biggy on the opposite page a gee-whiz chart.

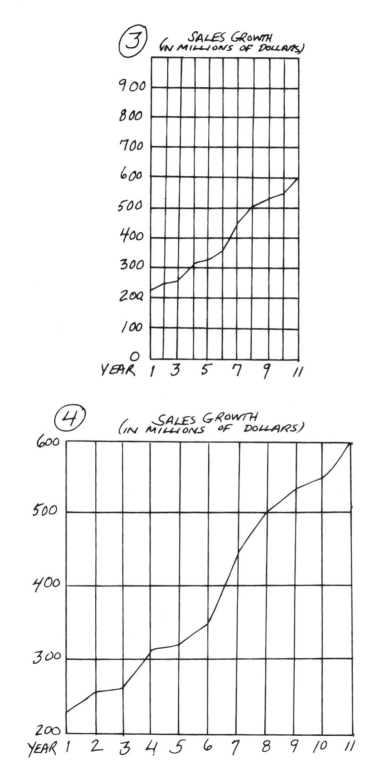

To guard against its misleading effect, it is well to remember that almost any graph in which the vertical scale does not begin with zero is deceptive.

If it should become necessary for reasons of space to violate this rule, the producer of the graph should feel obligated to offer a visual clue.

The most effective clue he can give is a break in the chart—such as a couple of wavy lines that might be easily drawn with the help of a bread knife—to give instant warning that some of the chart is missing.

Figures Don't Lie?

WHAT FOLLOWS the impact of statistics, or indeed any mass of figures, upon the unprepared human mind is likely to be a big buzzing blur.

That's why graphic methods were born. It's also why they offer a useful adjunct to any consideration of how-to-figure-anything.

A graph or chart or pictograph is a splendid device for converting a complex trend into a picture the eye can grasp almost instantly. Unfortunately, it is also a device by which figures can be distorted, trends exaggerated, differences minimized, and the eye fooled as effectively as by a practitioner of sleight of hand. When it comes to matters as serious as those about which we receive so much of our information from graphs today, it is not fun to be fooled. A little sophistication in how to draw an honest graph or, when reading, to distinguish one from its disreputable brothers is well worth gaining.

The distorted curve makes the simplest example. Let us suppose you are engaged in the manufacture of widgets of a prescribed weight. To discover how closely they hew to this weight, you ask that a number of widgets be snatched at random from production and individually weighed. A graph of these weights will probably fall into what is called the normal curve. It should look pretty much like this (A).

It will look like that instead of like either of the other two drawings because that is the proportion in which statisticians have agreed to agree that the normal curve should be drawn—the top of the curve being about three-fourths as high as the base line is long. How well this accepted shape is hewed to in

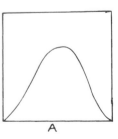

A

the graphing process is determined by how distances are selected on the horizontal and vertical scales.

Let us now compress the horizontal scale or expand the vertical one. We'll produce a curve that is tall and narrow, like this (B). The new curve seems to say that our widget production is very well controlled, with minimum variation from unit to unit. If you are Big Brass, this is the kind of chart a slightly shifty production manager might bring to impress you.

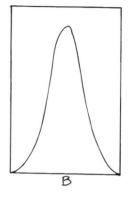

B

By contrary treatment, expansion of the horizontal scale, a curve with a flatter look is produced. What this one seems to say is that our widgets vary a great deal (C).

The data behind these three curves are the same. The information they transmit should be identical, and it is—if you take care to examine the numbers along the side and bottom that ordinarily accompany such curves. But where one of these graphs

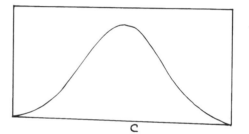

C

speaks to you in normal tones, the second whispers—and the third shouts.

Can you imagine a curve that reveals a quirk in human nature? Here is one that does this by catching your eye with a significant anomaly. It's a normal curve (D) with an odd little bump. Since the curve was produced by graphing the scores of a large number of students on a test, its normal shape was to be expected.

But why that bump?

Well, it occurs so often that it has been given a name, peak of charity. It is a product of the tendency on the part of most test graders to give the benefit

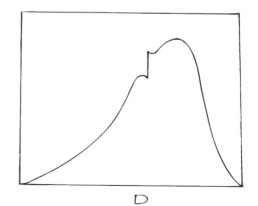

D

of any doubt to a student who is almost but not quite passing. This naturally pushes a significant number of grades from the area just below passing to that just above.

So the graphing process has revealed a phenomenon that might otherwise slip by unnoticed.

How to Find a Prominent Number

It's not likely to happen to you. I don't suppose it has ever happened to anyone, except as part of an experiment done by a psychologist. But suppose you were one of a small group of people promised a large reward if, being shown some numbers such as 7, 13, 99, 100, 261, 555, they could all independently hit upon the same one.

You would have to say to yourself: Which of these numbers has a special prominence that will catch the eye of the other persons in the group? Assuming they're all much like yourself in background and intelligence, it may well be the same one that catches your fancy—if you make your choice by a thoroughly logical process the others are likely to duplicate.

You can put up an argument for almost any of these numbers but if you think about it long enough you'll probably agree that only one of them really is unique. Two of them are of superstitious significance, two are symmetrical, one is basic in our decimal system. But only one is negatively prominent; 261 is meaningless to almost everyone.

Perhaps the only thing you can do with this idea is to turn it into a parlor game, but it is not without importance. It illustrates the idea of interdependence in thinking—the process involved when you lose your husband or your wife in a department store and try to handle the problem by using your head.

Let's say you're the husband. You begin by looking in the places your wife might logically go. Then you go where you should logically go, on the assumption that it will occur to her to look for you there. Then you ask yourself where she would expect you to expect her to go. And so on.

As Thomas Schelling, the experimenter responsible for the list of numbers, puts it: "What would I do if I were she wondering what she would do if she were I wondering what I would do if I were she...?"

It helps if the couple share much the same sense of humor, which could send them both promptly to the lost-and-found office.

Now suppose, to make it more fun, you are a secret agent

supposed to meet a colleague on a certain day in a certain city, but neither of you has been told place or hour. What do you have to pin your hopes on beyond sheer luck or coincidence?

Begin with place. Some cities might be an exception, but in most the city hall and the police station and the post office would all come quickly to mind. The first two having some drawbacks from the point of view of a spy, surely a spot in front of the main entrance to the central post office would be a good bet.

Time is possibly easier. Noon and midnight are the most distinctive hours. Unless at least one of you has reason to avoid being seen in daylight, the first of these is probably the choice which you and your accomplice would make.

See you in front of the post office at high noon, 007.

The Prisoner's Dilemma

Y OU DON'T have to be behind bars to face the prisoner's dilemma. It is the very model of many marriages and business affairs and relationships between governments.

It usefully illustrates why it is difficult for a couple of sensible people, or two great nations, to come to agreements even when they obviously offer a lot of benefit to both parties. A sad thing.

As formulated by Professor Albert W. Tucker, a Princeton mathematician, it goes roughly like this.

Two men charged with jointly committing an armed robbery are brought to the district attorney's office. He confirms what they already know. For lack of evidence, there's no way of convicting either of them of robbery unless at least one of them confesses.

"But," says the D.A., "I can get you both sent up for six months for illegal possession of firearms. And that's just what I'm going to do if I don't get a confession.

"If one of you confesses, he goes scot-free as a state's witness, and I'll see to it that the other gets the maximum of 20 years. If both of you confess, however, I won't need a witness. I won't need an elaborate trial either. So I'm prepared to offer you both the minimum sentence of two years for armed robbery."

The prisoners are returned to their cells.

If you were one of them what would you decide to do?

(We're going on the assumption that you have no opportunity to talk the thing over with your accomplice and have made no prior agreements covering an eventuality like this. But this doesn't really make very much difference, since then it would come down to whether you could trust him to keep his word—in short, to whether you are dealing with an honest crook.)

The simple, sensible answer is for both of you to keep mum and get off with six-month sentences. You see that right away and surely so does your partner.

But if your partner concludes that you will remain silent, will he take advantage of this to confess and get off free? If so, you'd better confess too. After all, two years is far better than twenty.

Also, will your partner—having realized that you will have followed this course of reasoning—dare remain silent even if his first impulse was to accept the six-months alternative?

So it goes, around and around in your head, with no real solution. What you may in the end decide to do will depend not only upon how much you trust your accomplice but also upon how much you think he trusts you. To say nothing of how much you think he thinks you think he trusts you...and so ad infinitum. It's like the picture on the oatmeal box that portrays an oatmeal box with a picture of an oatmeal box on it containing a picture....

By no means frivolous, or limited to criminal or to unreal conditions, this model problem takes on added significance every time disarmament talks resume. For decades, while the United States was spending itself into potentially crippling public debt on military preparedness, it was facing a U.S.S.R. that was producing weapons at the expense of tolerable living conditions for its people. Why in the world couldn't the leaders of these countries agree at least to slack off a bit for mutual benefit? Some of the reasons that the only sane solution, mutual disarmament, long proved impossible are explainable by the prisoner's dilemma.

In figuring out the best answer, if any, to a problem involving trust, whether it is between man and wife or between business partners or even between two world powers desperately seeking a safe way to reduce armaments, a handy device is a four-cell matrix.

Since this approach is all part of what is called the theory of games, it is usual to refer to the parties as players. Thus Player A represents you (or your country). You have your choice of strategies a1 and a2. The other fellow chooses between strategies b1 and b2.

In graphic, or matrix, form it looks like the accompanying illustration.

You might enjoy using such a matrix to outline other dilemmas and to play with this one and others of its kind with a real opponent.

See if the results differ, depending upon whether the opponent is a friend (or spouse) or not.

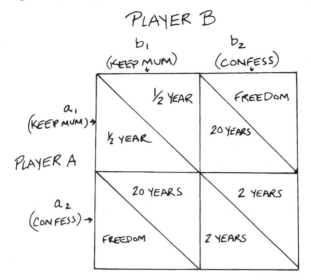

Conquering a Map Chart

W ANT TO show some fact clearly by illustrating it with a map chart?

Or are you just trying to read one of the things and understand what it truly means?

Either way, some discretion is required.

Take the classic example of The Darkening Shadow.

It was distributed by a large bank and widely reproduced by taxpayers' groups and newspapers and magazines. It was intended to impress us all with what proportion of our national income was being taken and spent by the federal government. By shading almost

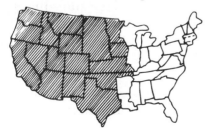

the entire areas of the states west of the Mississippi, it told us that federal spending had become equal to the combined incomes of the people of those states. And so it had.

But the trick is as unfair as it is old.

By choosing states of large

area and generally sparse population, the map maker has been able to cast his shadow over more than half the nation.

Had he chosen to start at the East Coast instead, his shadow would have covered such a tiny area that no one would have been impressed.

Yet this is a story that is worth telling, and by no means unimpressive even when presented objectively. All a map maker need do to cast an honest shadow is choose a block of states having total income in the same ratio to the national that their combined areas bear to that of the whole country.

Taking a Problem Apart

STARTING with a situation like the one in this drawing of a belt passing around a couple of pulleys, how do you figure out the length of the belt you'd need?

A good approach is to break up the complex problem into two or three simple ones you probably already know how to solve.

The length of the belt you see here is approximately equal to a couple of half-circles plus the hypotenuses of a pair of right triangles.

Suppose these pulleys have radii of 3 inches and 5 inches and their centers are 10 inches apart.

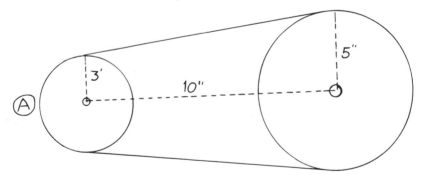

Since the circumference of a circle is equal to its radius times pi times two, half its circumference (what you're after here) is found by multiplying its radius by pi. So you can find the length of the two curved segments combined by adding 3 times pi to 5 times pi—that is, 8 times pi.

The straight segments are the hypotenuses of right triangles. Each triangle has a long side that is 10 inches and a short side that is

2 inches. Squaring each of these sides and finding the square root of the sum (10 squared plus 2 squared is 104) gives the length of each straight segment. Add 'em all up for the answer.

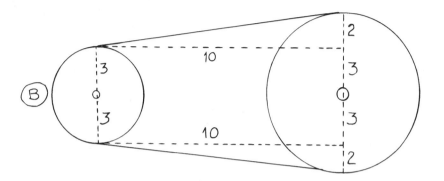

How Lethal Is My Warhead?

I<small>N AN ERA</small> in which we all must live with decisions about nuclear armaments and overkill, the calculation that follows is not without its grim interest—Cold War or no Cold War. It comes from Robert C. Aldridge, an aerospace engineer who helped design weapons for the Trident submarine of the '80s.

It seems that you can find the comparative deadliness, or lethality of warhead, of a submarine armed with ballistic missiles by this modest operation: Square the warhead yield in megatons and take the cube root; twice divide this result by the accuracy, expressed in nautical miles, and multiply by the number of missiles.

(If your accuracy figure is given in feet, divide it by 6080.2 to change it to nautical miles.)

It works like this for the older Poseidons, the subs of the '60s and '70s. Figure each as having ten 40-kiloton MIRVs (Multiple Individually Targeted Re-Entry Vehicles) with accuracy of 0.3 nautical miles. Since it takes a thousand kilotons to make a megaton, each Poseidon weapon rates 0.04 megatons.

Squaring that and taking the cube root (that is, raising 0.04 to the 2/3 power, which is the quick approach when working with computer or scientific calculator) and twice dividing by 0.3 gives a figure of 1.3. For ten of these MIRVs, the total lethality (and isn't that a great word to contemplate?) then becomes 13.

If you'd find it educational to compare various seagoing weapons, present and future, in this way, here are some figures to toy with.

Russian SS-N-8 (an earlier weapon): one 1-megaton warhead; accuracy, 0.8 nautical miles.

Soviet SS-NX-18 (a later entry): three 500-kiloton MIRVs; 0.25 nautical miles.

United States Trident 1: eight 100-kiloton MIRVs; 0.25 nautical miles.

Second-generation U.S. Trident 2 may be armed with fourteen 150-kiloton or seven 300-kiloton MIRVs (accuracy figure, 300 feet) or with seventeen 75-kiloton MARVs (Maneuvering Re-entry Vehicles), accuracy figure 90 feet.

If working with these last numbers gives you a calculated total lethality of more than 13,000, you've got the hang of it. It's sensible to round off your results since you're working with data that are only estimates of what the future could hold.

Those Awful Work Problems

Iᴛ ᴄᴀɴ ʙᴇ a bit on the nasty side. It's called a work problem, and most of us have encountered its kind, in school if nowhere else. It can go something like this.

You're an old hand with a lathe, on which you can make a little wooden bowl in 20 minutes. Your shop partner needs 40 minutes to do the same job. How many man-minutes should you charge against each bowl in your system of cost accounting?

You might guess it is the average—the mean—of the two figures, or 30, but it is not. It's a less familiar average, the harmonic mean. You find it by adding together the reciprocals of the time periods, then dividing into the number of these periods. In this instance, you divide 1 by 20 and 1 by 40 and add together those two results, which you then divide into 2, to get the result of 26.67. Which is quite a bit different from 30. It is so much different, in fact, that if you're serious about this lathe thing, it could tell you why the business is going broke.

What you don't know can't hurt you? Well, it can dangerously mislead you when the subject is averages.

With numbers as simple as these, you could have an exercise in mental arithmetic. The sum of the reciprocals is seen to be 3/40. To divide that into 2, invert and multiply—as we learned to do in school—so the final step is dividing 80 by 3.

Or, for a slight variation, try dividing the sum of the reciprocals by 2 instead of t'other way around. Then find the reciprocal (that is, complete the division of 3 by 80).

And that's still a harmonic mean.

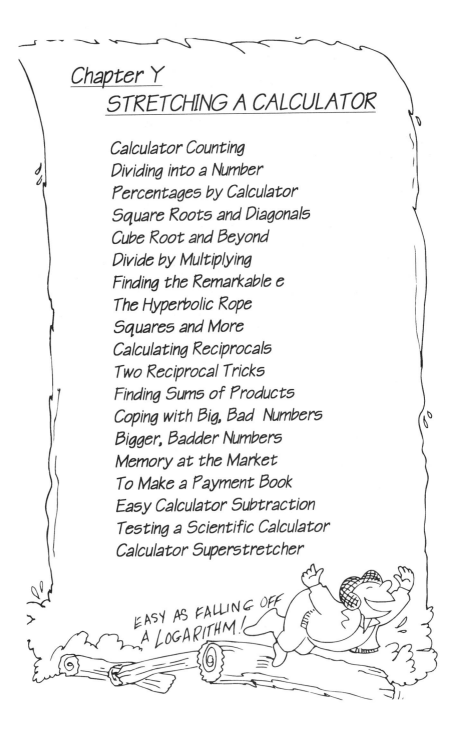

Chapter Y
STRETCHING A CALCULATOR

Calculator Counting
Dividing into a Number
Percentages by Calculator
Square Roots and Diagonals
Cube Root and Beyond
Divide by Multiplying
Finding the Remarkable e
The Hyperbolic Rope
Squares and More
Calculating Reciprocals
Two Reciprocal Tricks
Finding Sums of Products
Coping with Big, Bad Numbers
Bigger, Badder Numbers
Memory at the Market
To Make a Payment Book
Easy Calculator Subtraction
Testing a Scientific Calculator
Calculator Superstretcher

EASY AS FALLING OFF
A LOGARITHM!

Calculator Counting

You won't get any help on this from the booklet that comes with your calculator. More's the pity, since this is a task each machine does in its own way.

It's a useful little trick for counting supermarket items, or traffic, or the box office. Or keeping track of Scouts you're trying to get onto the bus without leaving one or two behind.

With some calculators, generally those with an automatic constant that works with all functions, you can count by just hitting the one and then the plus repeatedly.

If your calculator is an old timer with a constant (K) switch, turn that on first. If it has a K key, the procedure will probably be to hit successively the 1, equals, K, and plus keys.

Often you punch 1, plus, and then equals repeatedly.

To count by 2s, just use the 2 key instead of the 1. To count backwards or correct a mistake, touch the minus.

If these methods fail, it may be that the constant on your machine works only with multiplication and division. This situation demands a crafty trick. Load up the entry with zeros and ones in such a way that each time you multiply, the product increases by slightly more than 1—but so slightly more that the difference is at the tail end of so many decimals that the machine will drop it overboard.

You do this, simply enough, by multiplying 1.0000001 by 10000000. (You needn't count those zeros; just put in enough to fill the machine.) Now each time you touch the equals key, the digit at far right will increase by 1.

That's how you do it for a model designed so the first number you enter becomes the constant. For machines with which it's the second number, you do the same multiplication but in reverse order. If instead, you divide 10,000,000 by 1.0000001 you will be able to perform the opposite operation—count down from 9,999,999.

Dividing into a Number

A common bugaboo of calculator users is finding they need to divide a number already on the display of the machine into another number.

Suppose, for example, you've added up a long series of numbers and now want to divide the total into 365. This will stump the ordinary pocket calculator.

How can you do it without copying down the total and then re-entering it after feeding in 365 and the division sign?

Answer: Divide the total by itself twice, accomplished on most calculators by pressing divide and then equals. Then multiply by the 365. When you multiply by the reciprocal (that's what you got by the double division of a number) it's the same as dividing by the number.

Percentages by Calculator

Many personal calculators—those just above the bottom price range—boast percentage keys.

For regular business use, you may find that additional key well worth its extra cost. But for occasional use, you can get along just fine without it. In effect, you teach the simple electronic helper you already own how to handle percentages without recourse to a special key.

Basic rule: Instead of hitting the number and then the percent key, just hit the decimal point and then the number, so that you've moved the point two places to the left.

Thus 25 percent becomes 0.25; 12-1/2 percent becomes 0.125; and 4 percent becomes 0.04.

To complete the conversion of your mini calculator into an efficient percentager, keep in mind that a discounted price can be found directly by first subtracting the discount percentage from 1. For example:

You're thinking of ordering a $127.50 bicycle on which you've been offered a 22-1/2 percent discount. Sounds complicated, but.... Just enter 1 into your machine, subtract 0.225, multiply by 127.50— and read off that you'll be writing a check for $98.81. (And if that price is subject to 6 percent sales tax, multiply by 1.06 to find the grand total the easiest way.)

Square Roots and Diagonals

Y\ou're laying out a cabin and want its corners square. Well, how do you find the length of the diagonal of a rectangle? You find it, as you probably recall, by squaring the lengths of two adjacent sides, adding these squares together, and finding the square root.

That's great unless there's no square root button on your calculator and you're not one of those precious few who can recall from school days how to paper-and-pencil it.

Happily, there's a neat way to approximate a square root. It's said to go back to Sir Isaac Newton, who would have given an arm and a leg for your little electronic marvel.

You begin by guessing what the square root might be. Any reasonable number will do, but a careful guess will speed things up a little.

Then, on your calculator, you divide the original number by the guessed root, add the amount of the guess, then divide by 2. What is the square root of 139, for example?

Recalling that 12 times 12 is 144, you might take 12 for your guess. You divide 139 by 12, add 12, divide by 2, to get 11.791666. Check by squaring that on your machine to get 139.04338, surely close enough for most purposes.

If you're really fussy, repeat, using the new root instead of your guess. Divide 139 by 11.791666, add 11.791666, divide by 2, and you'll get a square root that squares to 138.9999. Closer than that you can't get, as you'll prove if you try taking it another step.

Since square root is an important addition to the partnership between your mini and you, try another example: square root of 150 this time. Divide 150 by the same guess, 12, add 12, divide by 2, to get 12.25. Squaring that to check it yields 150.0625. If that's not close enough to suit you, divide 150 by 12.25 this time, add 12.25, divide by 2, and you'll get an approximate square root that squares to almost precisely 150, as you will see.

Cube Root and Beyond

T\here is a rather complicated formula you can use for finding cube roots, fourth roots, and so on. But it's often easier to go the route of plain guess-and-try if you have a calculator handy. That way you don't have to remember a formula.

It goes like this.

To cut-and-fit your way to the cube root of 1,234, for instance, you might reasonably begin by guessing 10.5 since you can see that 10 is the cube root of 1,000 and so is a bit small.

When you cube 10.5, you find that's too small too, so you could try 10.7, which cubes to 1225 plus. You're close, so you might try 10.72. When that proves too small but 10.73 proves too big, you know the answer's between them. Within a couple more tries, you'll find that the cube of 10.726 comes to 1233.9948, giving you a result that's as on-the-nose as an answer can be unless you want to get into a ridiculous number of decimal places. You've gone as far as your calculator can go, and quickly too.

For a really extreme example of how well this somewhat random walk works, I decided to go for the 10th root of 98,765 (a number with no significance, just some digits strung in descending order). In nineteen minutes I had the answer to the seventh decimal.

A reference to this in a magazine column I was conducting brought me a stack of mail, mostly from readers who were pleased to note that they had beat me all hollow.

Divide by Multiplying

F ACED BY the chore of dividing quite a few different quantities by the same many-digit number, you'll probably reach for your pocket calculator.

DIVIDE OR MULTIPLY — IT'S ALL THE SAME TO ME!

OH, PETER! YOU MAKE ME BLUSH!

In the absence of such a device, however, you may find the following trick worth knowing, and interesting as well.

It is based on a couple of facts. One is that it is easier to multiply than to divide. The other is that you get the same result when you divide by a number as when you multiply by its reciprocal. You can see why this is when you recall that a reciprocal is just 1 divided by the number.

Say you have many numbers to be divided by 1.20671. Find its reciprocal by dividing 1 by 1.20671. The answer, rounded to 0.8287 (assuming that's far enough to carry it for your purpose) becomes the basis for a table you construct by multiplying it by each number from 2 to 9.

Now suppose the first number you need to divide by 1.20671 is 92. Set it up as the product of 92 and the reciprocal of 1.20671, like this:

```
     .8287
   x  92
   ---------
    1 .6574
   74 .583
   ---------
   76 .2404
```

Since those two partial products came right out of your table, all you've had to do is add them together.

Finding the Remarkable e

THE REMARKABLE constant called e that is used in so many scientific, financial, and engineering calculations can be defined in many ways. It is, for instance, the number of dollars you'll have at the end of a year if you deposit one dollar at 100 percent interest compounded continuously.

You can approximate its value in quite a short operation. Just increase 1 by an extremely small amount and then raise this total to the power of the reciprocal of that small amount. Easy IF you have a calculator.

If the "small amount" you use is 0.0001, your result will be correct to three decimal places, the form in which the value of e is usually given: 2.718. Having added 0.0001 to 1 you can complete the calculation (raising that total to the power of the reciprocal of 0.0001) in an instant with a scientific calculator.

Using a simple calculator, you must first find that reciprocal by dividing 1 by 0.0001 to get 10,000. To find that 10,000th power, you first find the tenth power (with most calculators, by pressing times once and equals nine times) then repeat three more times. If the result your calculator gives is off in the third decimal place, an error from internal rounding, you may want to try 0.001 as the "small amount" to see if the result of this shorter procedure isn't just as good.

The Hyperbolic Rope

I<small>F YOU</small> should ever wish to graph the shape taken by a heavy rope or chain, you'll find a need for something called hyperbolic functions. Only a few of the more elaborate scientific calculators have these built in, but you can calculate them quite readily on your own. Even a simple calculator will do, although the process is a little faster with a scientific model.

If your calculator has that useful constant e on it, begin by raising e to the power of the angle you're working with and also to the minus power of that angle. Subtracting the second of these from the first and dividing by 2 gives you the hyperbolic sine (commonly abbreviated sinh). Adding the two together instead of subtracting produces the hyperbolic cosine (cosh) after the division by 2.

To find tanh, divide the difference between the two quantities by their sum. Dividing the sum by the difference produces the hyperbolic cotangent, or coth.

The gimmick in all this is that you may be working with a calculator that lacks an e key or a means of finding negative powers (to say nothing of the powers that are not whole numbers, called for if the angle you're working with is not a whole number of degrees).

Handle the absence of an e key by entering its value, which is 2.71828. To get a negative power, take the reciprocal of the positive power, either with the special 1/x key or by dividing 1 by the positive power. To work with angles that are not whole numbers of degrees, you'll need the help of logarithms.

Squares and More

M<small>OST CALCULATORS</small> can be taught to square and cube and all that, without your having to enter the number more than once. That's important if it's a number like 3.1415927, as it not infrequently is. When you've taught your mini the procedure, you'll find it makes life almost as simple as if you had a special squaring button.

You'll need squaring, and the higher powers, too, for interest calculations—both the kind you pay when you buy a car or a house and the nicer kind that the bank or savings and loan pays you.

You'll need squaring for geometrical figuring that you do in laying out shop and building projects and for almost anything having to do with light and optics. For instance the amount of exposure your camera lens gives is calculated by comparing the squares of various "f" stops. At f/1.2 you're getting 14 times as much exposure at a given shutter speed as at f/4.5, and squaring the two stop numbers and dividing the bigger one by the littler is a way to find that out.

With the automatic constant found on most calculators, all you need do is hit the number, the times, and then the equals (repeatedly for higher powers). If the constant is governed by a switch, turn that on first.

Calculating Reciprocals

A RECIPROCAL may not sound like anything you'd ever want to calculate. But, as using this book will demonstrate, a reciprocal has its uses. The 1/x button isn't on scientific calculators for nothing—but the truth is you don't really need it. As we shall see.

You can employ reciprocals to figure electrical resistance easily and quickly, since the joint conductance of several conductors is found by adding together the reciprocals of the resistances.

And there are those wonderfully pesky tank problems too: If one hose could fill the tank in two hours and the second hose.... And work problems: Jim could do the job alone in 8 days; John could do it alone in 5. How long will it take them working together? (Please, no jokes.)

Since the reciprocal of a number is the fraction or decimal you get when you divide 1 by that number, there's a special trick for getting it. You could, of course, just ring up 1 and then divide by the number. But often the number of which you want the reciprocal is the one already on your machine and you don't want to have to copy it off on a piece of paper (humiliating, anyway) and then put it back into the machine after punching the 1.

The quick trick is this: Just divide the number by itself twice. The first step automatically produces the 1 you need, and the second step gives you the reciprocal.

Carry this operation one step further. Punch the key you're using one more time. You'll get the reciprocal of the square of the number you started with. And so on.

Two Reciprocal Tricks

THERE ARE two sly ways to tackle one whole family of problems that cannot be solved by direct and obvious use of the ordinary small calculator. Problems in this family are represented by the example that follows.

You've dug out the information that in a certain small county, total personal income for last year was $98,765,432. You know the population consisted of two cities having populations respectively of 12,345 and 4321 and of the remaining rural areas with 12,321 persons.

What's the average (mean) income of the inhabitants of this county? Obviously it is the dollar amount divided by the sum of the three population figures.

If your calculator has a memory, you can use it to solve this type of problem. Or you can resort to pencil and paper.

But how can you solve it quickly, without paper and pencil, without memory feature, using a typical mini calculator that has a constant and 8-digit capacity , as most do?

Two ways. Take your choice.

One is to add up the population, find the reciprocal of this total, then multiply that by the dollar figure. (Dividing by a number and multiplying by its reciprocal come to the same thing.) Remember that you get the reciprocal of a number by dividing it by itself, which gives 1, then by itself again. The constant lets you do this, usually by hitting the divide key three times or the divide key once and equals key twice.

The other method also makes use of reciprocals.

All you do is turn the problem upside down to work it. Then you do the same with the answer to restore it to propriety.

In our example you add the populations together and divide by the dollars (instead of the reverse) and then restore things by getting the reciprocal. As in the other method, you do this by hitting the divide key three times (or, depending upon how your tiny miracle machine operates, the divide key once and the equals twice).

You may be dismayed to discover that these methods will give you results differing very slightly from each other and from the precise result obtainable by other approaches. This is the result of some digits having dropped off by the wayside as the calculation proceeded. They are probably no loss.

If you need extreme precision, test the method first on the kind of figures you're working with. On approximate data of the sort we're talking about—population counts and income estimates—the final result cannot be exact anyway and should be rounded off. Whether the method selected gives $3397, $3407, or $3408 (and some odd cents, in each case) rounding off to $3400 is advisable. Otherwise you're guilty of offering, or using, figures laden with false precision.

Finding Sums of Products

ONE TIME you are likely to encounter sums-of-products is as assistant shopper in charge of numbers at the supermarket. You might want to know the total cost of several items for which you have unit prices. Two pounds of eggplant at so much a pound, plus 3 eggs at some price per dozen...and so on. The complete eggplant dish is going to cost you a sum of products.

You can solve this kind of problem by rearranging it into a straight-line task.

Suppose you have these three products to add up: 5888 times 4; 6981 times 166; 59.4487 times 5. How do you go about it?

With any calculator, but especially with the common types having a constant that works on addition, there's a terrific time saver. It works best when at least one number in all but one of the pairs is small.

In the example given, begin by multiplying together the pair in the middle, because they're both big. Then just add 5888 four times and 59.4487 five times.

There's another approach. It may be a little more trouble, at least the first time, but it will work with large numbers as well as small and the simplest electronic calculator can handle it.

The idea is that a problem like 2 times 3 plus 4 times 5, though it may lead to utter confusion, is at least well arranged for working with paper and pencil or with an advanced calculator. But a simple calculator finds it devilish until it is rearranged by dividing both parts by the last number, 5, and then multiplying the whole thing by 5. As in the first picture. Now you can proceed straightforwardly to multiply, divide, add, multiply, an easy routine for any calculator.

$$\left[\frac{2 \times 3}{5} + 4 \right] \times 5$$

When you handle that earlier problem by this system, the second drawing shows the form it takes. The answer you should get is 1,182,695. This method can produce a very slight inaccuracy from early rounding.

$$\left[\left(\frac{5888 \times 4}{166} + 6981 \right) \times \frac{166}{5} + 59.4487 \right] \times 5 =$$

Use a similar rearrangement to make a sum of quotients feed in smoothly. To do 32 divided by 8 plus 39 divided by 3 plus 21 divided by 7, rewrite the problem as you see it in the third picture.

$$\left[\left(\frac{32 \times 3}{8} + 39 \right) \times \frac{7}{3} + 21 \right] - 7$$

That may look formidable until you've become used to it, and possibly not worth while for this simple computation. But for big numbers, it's a trick that can ease life considerably.

Coping with Big, *Bad* Numbers

Wʜᴀᴛ ᴄᴀɴ you do when there's need to multiply numbers too big for your calculator?

Suppose you were to ask almost any calculator to multiply 456,789 by 987,654. It would give you an answer with zeros shown in place of the last digits.

Despair not. There is a trick by which you can hogtie numbers of any size, as long as the smaller of them has no more than seven digits.

Using this method, you multiply just as you were taught in grade school: multiplying by one digit at a time, then adding up the partial products. But you do this with a calculator, thus saving a lot of effort and avoiding mistakes.

(You can do the adding step by machine, too, if you like. But you probably can do this part with pencil and paper faster than by re-entering into the machine. Try it both ways.)

Ordinarily, if your shorter number has seven digits, you should multiply it by the other number one digit at a time. If, however, it has six digits, you can speed up things greatly by multiplying by pairs of digits.

Multiplying by one digit at a time, you'll get the result you see in the first calculation.

```
        456789
    x   987654
      1827156
     2283945
     2740734
    3197523
    3654312
   4111101
   451149483006
```

That's simple enough. It's the routine you've known since grade school, except that a calculator is doing most of the work. Let's look at the special case mentioned, where one of the numbers has exactly six digits. Speed the work by doubling up, taking the digits in pairs, as in the calculation below.

$$456789 \ \times \ 54 : \qquad 24666606$$

$$\times \ 76 : \qquad 34715964$$

$$\times \ 98 : \qquad \underline{44765322}$$
$$451149483006$$

The calculator can handle paired digits as easily as single ones, cutting your work in half.

Be sure to take advantage of the constant on your calculator, or the memory feature, to save finger work. In doing the problem you see here, you'll have to enter 456789 only once.

Bigger, Badder Numbers

A POCKET CALCULATOR is great on big numbers but not on big, big ones. Past eight (on some machines, six) digits, something funny happens. Work stops and a warning sign pops on. With some calculators, there's no warning and digits begin falling off at the right like lemmings off a cliff.

Expensive machines solve this problem with something called scientific notation, and so can you. What the scientific machine does is show the first eight digits—all it has room for—and tell you how many zeros you must tack on to make your answer the right size.

To get the same result, you can reduce the problem to manageable size and keep track of the decimal places.

For a problem involving the number of days in a year, you might want to raise 365 to the 6th power, a number with far too many digits for ordinary calculators. Move the decimal point two places to the left and operate with 3.65 instead. The answer you get will be 2364.5972.

But you know you've swiped two places off the original number and then used the number six times. Add those 12 places you owe and you have the result: 2,364,597,200,000,000. (In scientific notation, which is often the easiest way to record and use such a number, you could call it 2.3645972 times 10 to the 15th power.)

Sometimes you can bypass the problem altogether. The formula for figuring out how many different poker hands can be dealt calls for dividing the product of all the numbers from 52 to 48 by the product of all the numbers from 5 to 1. If you write that down, you'll discover you can reduce the size of the numerator and get rid entirely of the denominator, by some simple canceling. This process brings the chore down to the capacity of the calculator and tells you there are 2,598,960 things that could happen to you on the first deal in your next draw session.

Would you like to try both canceling and mental scientific notation on a single problem? Calculate the number of possible hands at bridge—all the possible groups of 13 cards that can be dealt from a deck of 52.

Following the rule for combinations, the denominator will consist of the digits from 1 through 13 multiplied together. And above the line you put the numerator, consisting of the 13 digits from 52 down through 40.

The number you come up with will tell you that the chance of getting 13 spades in your next bridge hand is 1 in 635,013,550,000.

Memory at the Market

If your pocket calculator has a memory built in, you can make it act like two machines at once. And two is how many you really need for some chores, such as grocery shopping cannily done.

Way to go is this. Having found, with the help of your calculator, that the little can of sausage is running one third more per ounce than the big one, and welcoming a chance to save, you flip the bigger can into your basket and its price into the memory bank of your calculator.

Now you can clear the machine for price-checking the next item, while keeping a running total in the memory. It will be all ready to compare with what you're told you owe at the cash register. Which is known as checking the checker.

To Make a Payment Book

W ITH A POCKET calculator to do the hard work, it's a simple chore to set up a payment book. You can do one for any kind of installment loan or purchase, including a home mortgage. The routine described here includes alternate approaches to discovering the amount of each payment, to use if you don't have that figure to begin with.

Including this item led to warm discussions with some of the people who helped edit these pages. They found enough repetition of earlier installment-payment ideas to suggest the word redundant. I won on the basis that this word for a worker who must be fired because he isn't needed also applies to a fail-safe machine in which if one part breaks down a spare takes over at once.

Step 1. To find the payment amount, express the interest rate as a decimal and divide by the number of payments a year. (Save this result separately for later use.) Add 1. Raise this sum to the power of the total number of payments to be made over the period of the loan.

Step 2. Divide this number by itself less 1. Multiply by the amount of the loan and by the amount you saved in Step 1. (If you'd like to check yourself, try this computation for a mortgage loan of $45,500 at 12.5 percent for 30 years. The monthly payments should come out $485.60.)

For an alternate route to the same result, follow Step 1 above. But instead of Step 2, find the reciprocal (that is, divide 1 by the number you have). Then subtract 1 and change the sign. (If your calculator lacks a change-sign key, probably marked CHS or +/-, just ignore the minus sign from here on.) Next, divide by the interest rate in decimal form, which would make 12.5 percent become 0.125. Multiply by the number of payments per year. Find the reciprocal. Multiply by the amount of the debt.

Use these headings: PAYMENT, INTEREST, PRINCIPAL, BALANCE.

The amount calculated by either of the methods just given will go in the first column.

For the second column, multiply the amount of the debt by the annual interest rate expressed as a decimal and divide by the number of payments per year.

Subtract the second figure from the first and enter it in the third column.

Column 3 subtracted from the previous balance gives you the new balance for Column 4.

Although the simplest of pocket calculators will serve, a change-sign key is useful and a memory even more so. A programmable calculator is a great help. If you've never used one and want to get started, this is an excellent introductory exercise. That's also true of the BASIC program in a personal computer.

Easy Calculator Subtraction

Y ou've added up a set of figures and you want to subtract their total from 365. How can you do this without copying down the total on a scrap of paper and then re-entering it after entering the 365?

Use the change-sign key, if your calculator has one, then add to 365.

On some mini calculators, the minus key can be used to make this sign change.

But if your calculator has no special features at all, just work backwards: subtract the 365 from the total you have. Then mentally change the sign of the result.

Testing a Scientific Calculator

S cientific or "slide rule" pocket calculators are the ones that can come up with a logarithm or a cosine as blithely as the ordinary machine can multiply your income by your tax rate.

They used to cost an arm and a leg, meaning a three-digit sum. Then they came down, down, down until anyone with the price of an ordinary calculator could own one by looking around a bit for a bargain offer.

Choosing among the models offered is a little tougher.

You can begin by running through the same routines you'd use to compare ordinary electronic calculators. Test for convenience in key operation, visibility of results, presence of the keys you want most, type of memory, and so on.

Less obvious are these three specific tests you might want to try.

1. To test for speed, punch in sine 45 degrees. Some machines will dredge this up in about half a second, while others can take as long as three seconds. The reason speed is important is that with one of the slower calculators you may enter the next step in a problem too soon and be misled by a wrong answer as a result.

2. To test for capability in one field where some calculators are

lacking, try this. Ask for the sines of 45, 135, 225, and 315 degrees. You should get the same answer each time, except that two will have minus signs. If, instead, the machine you're testing gives zero answers in two cases, that means it's not capable of dealing with negative functions of angles.

3. It may surprise you to find how inaccurate a scientific calculator can be. Enter some easily remembered number that fills the display—eight digits with most calculators. Then hit successively either the InX and e to the power of X keys or those for log X and 10 to the power of X.

Theoretically, each pair of demands will give you back the number you started with. In practice, there'll be some error, and naturally you'll prefer as little of that as possible.

For example a slightly antiquated "slide rule" calculator that I own changed 88,888,888 into 88,889,050 with one of these routines and into 88,888,980 when I used the other.

Only one of six calculators I tried returned the last four digits of my telephone number perfectly. Four converted 8052 into 8051.98 and the sixth came up with a remarkably unprecise 7578.

Calculator Superstretcher

Sooner or later, every calculating device runs out of working space. Most, even the expensive scientific types, are limited to eight or ten digits.

As one of the books on using electronic calculators says, "if you wanted to multiply something like 55,612.31 by 58.412 and wanted to know the answer to more than one decimal place, you would be out of luck, because with that calculation the machine would run out of display space after 3,248,426.2. Back to the pencil-and-paper method you go."

Well, yes—and no.

The author is right, in a way. You can't do the problem directly on your eight-digit calculator. But you can do it with the help of your calculator and only limited recourse to paper and pencil. The method will save you some tedium and a good deal of time, while reducing the chances of making a mistake.

Think of the problem cited as being (55,612 plus 0.31) times (58 plus 0.412).

Now multiply each of the first pair of numbers by each of the second pair, doing the chore on your calculator and writing down each product, like this:

$$
\begin{array}{llr}
55{,}612 \text{ times } 58 & = & 3{,}225{,}496. \\
55{,}612 \text{ times } 0.412 & = & 22{,}912.144 \\
0.31 \text{ times } 58 & = & 17.98 \\
\underline{0.31 \text{ times } 0.412} & = & \underline{0.12772} \\
& & 3{,}248{,}426.25172
\end{array}
$$

The calculator has done the hard work. All you've had to do with paper, pencil, and brain is the comparatively simple addition.

You may have noticed that the last two multiplication steps in this problem could have been done in one. I've taken the longer way here to demonstrate the routine for even bigger numbers than these.

Something else may occur to you. Given numbers with as many digits as these but no decimal points, the principle still works. You could divide a number like 5,561,231 by 100 as a preliminary step. Just don't forget that the final result will then have to be multiplied by 100.

For even longer numbers, use 1,000 instead. And so on.

I guess the important principle here is that what your calculator can't do alone, it and your brain can do together.

Chapter Z

COMPUTER DOES NUMBERS

Your 50-Pound Pocket Calculator?
BASIC Tips for Math-by-Computer
Table of Powers
Watch Your Money Grow
Anatomy of a Program
From Interest to Yield
How Big a Beam?
Rent or Buy?
Loan Payments Figured Fast
Print a Payment Table
The Mortgage-Payoff Miracle

Your 50-Pound Pocket Calculator?

T HE TIMES they are achangin' for sure.

When I started to amass this compendium, the only tools I dared assume a reader would have besides the book itself were a pencil and a piece of paper.

Those are still all you'll need for most of the stuff covered herein.

But in a few cases—including some pretty useful calculations— this assumes that you're willing to embrace such sticky buddies as tables of logarithms and trigonometric functions, requiring you to relearn many things since schooldays happily forgot.

I believe that most people in the closing decade of our high-tech century will prefer to find an electronic route.

For most of the figuring with which this book deals, you can leap from data to conclusions with a simple *pocket calculator* at two or three dollars (a miracle). For the rest of the stuff, a *scientific calculator* at anything from ten dollars up (an even greater miracle) will facilitate things immensely.

Still better, for those willing to learn to use it, is a calculator that is not only loaded with scientific functions but is programmable as well.

Programmability is especially desirable—almost essential—for otherwise repetitious chores, such as comparing results for various angles, dimensions, sums of money, or interest rates. But a versatile *programmable calculator* can cost more than many people are eager to pay for a device likely to be used only now and then.

And the outlay will be doubled or more if a page-size calculator printer is added so that detailed tables can be created.

That's where a *personal computer* comes in. Not, perhaps, as something to buy for our limited purpose but as something to turn to an additional important use if you already own it.

Turn your hand to it and you will quickly discover that any computer will do plain arithmetic in a trice, even when faced with pretty staggering numbers. It will also do complex scientific and financial figuring of a kind that, if approached by the calculator route, would demand an expensive and specialized variety.

Combine computer with the printer that most owners already have and it can hand you your results spread out on paper in whatever format you fancy. This is especially nice if you're creating a bulky table of financial calculations—a schedule of loan payments, for instance.

One particularly nice aspect of doing your calculating by computer is that you already own the darn thing and won't have to lay out one single additional dime to add it to your calculating arsenal.

In making use of this fabulous capacity you're just using a little of the potential for which the idea of a computer was dreamed up in a previous era and brought to practicability in this century—number crunching. It's a talent too easily overlooked in the maze of spread-sheeting, data-basing, and word-processing, not to mention game-playing, that were all most of us were thinking of when we acquired our computers.

I'm mostly talking about the IBMs and their innumerable clones that predominate. Tucked into their operating systems, alongside DOS, is something called BASIC. It can be used for all sorts of programming chores, including the simple and straightforward ones we're concerned with. It sits there in DOS computers, waiting patiently, and hardly anybody pays any attention to it.

As I write this, Macintosh computers come with an application called Calculator that's fine for ordinary arithmetic. For more demanding applications, a Mac user group or a dealer can steer you to low-cost or free software that sits on your screen and looks and works just like a sophisticated scientific calculator.

But although I work on a Mac myself, I regret to say I have yet to uncover a simple and affordable way to use it for math programming. I find this, like other aspects of our national allergy to quantitative thinking, deplorable. But this Mac deficiency is belatedly beginning to heal itself with the arrival of an operating system compatible with DOS software. Or so I'm told.

There's one further way out.

If you use a spread sheet in any breed of computer, my friends assure me that there lies your best answer. You may own such a program, stand-alone or as part of an integrated "works" program. Some cost little, as shareware or freeware.

Once called The Universal Machine (I understand that was long before our forebears actually got around to inventing it) the computer has a few of its own ground rules for doing arithmetic. Some are in the established language of math. A couple are pure computerese.

• If a statement contains parentheses, you're supposed to deal first with what's inside them. If there's more than one set of parentheses or brackets, work from the inner ones out.

• If there's anything to be raised to a power, that operation comes ahead of ordinary arithmetic. Then the school kid's memory aid, My Dear Aunt Sally, takes over to remind you that—working from the left—you Multiply and Divide, then Add and Subtract, using the familiar plus and minus signs on the keyboard.

• But note that confusion with the letter X is avoided by using the asterisk or star (*) instead of the usual sign for multiplying.

• The right slash (/) says divide.

• The tiny tent (^) we know as a caret (or occasionally the upward-pointing arrow or a pair of asterisks) says raise-to-the-power-of and is used to get squares, cubes, cube roots, and all that kind of stuff. Thus the cube, or 3rd power, of 10 is written 10^3 and the cube root is 10^1/3.

If you're new at the game, you'll probably want to start by practicing some typical maneuvers right from the keyboard—without programming.

How much is 3 times 75? Type PRINT 3*75 and a return. You'll be rewarded by the answer.

Try using a question mark instead of PRINT. Works the same in most BASICs. If you use this time-saver in programs and play them back with the command LIST, note that the question mark has undergone a sea change.

As you will have noticed, the PRINT command means print on the screen. With your printer on-line and a send-to-printer command, such as LPRINT, substituted for just plain PRINT, the answer will be typed out for you.

Sometimes you'll want to record the problem as well as the answer. Use quotation marks to do this. To demonstrate that what follows quotes doesn't function but merely prints, type PRINT (or LPRINT) followed by "3 X 75 =" 3*75. With most programs, your computer will print out the stuff in quotes and then perform the quoteless instruction, giving you 3 x 75 = 225.

You've probably noted that until you type a return, you can correct errors—in either direct mode or a program. But when you've finished a line, there's a new problem. You can't get back to change it. What to do in that case is just type in the whole line all over again, including the original line number, wherever you happen to be.

The erroneous line will still show on the screen but it will be as dead as a dodo and the corrected line will function in its place.

If you then view the program by typing LIST and a return you'll find things nicely in order.

BASIC Tips for Math-by-Computer

P<small>ENCIL</small> and paper are the only tools you need to perform most math chores you meet in the course of spending money, running a business, designing or building a house. But each further step—adding a simple calculator, a scientific calculator, then a programmable one—speeds up the figuring. A personal computer, desk top or lap top, running the program called BASIC, wraps up the virtues of all these approaches into one. Some form of it comes with most computers.

For most calculations, you can use BASIC to turn a set of instructions, or a formula, into a program that will run through a calculation for you. As you progress, you'll eventually find use for such commands as these:

GOTO (to jump the program—which is just a series of steps—directly to whatever line you specify);

GOSUB (to send the program through a set of commands ending with RETURN);

FOR NEXT, a valuable tool when you encounter certain equations you wouldn't care to tangle with barehanded.

IF THEN, a command taking a form like IF A=1 THEN 90, that jumps execution provisionally to a new place, Line 90 in this instance.

There are also a couple of housekeeping commands you may find especially handy:

NEW clears memory and gets you off to a fresh start. Your clue that you should have used it is often the appearance of an asterisk after a line number, indicating that the number is already in use.

To clean up a messy screen so you can start fresh, type CLR.

Oddly missing from indexes of any books on this subject I've looked into is the word roots. I guess we should all be flattered by the assumption that we remember, from an algebra class or somewhere, that to find the cube root of a number, we need merely raise it to the 1/3 power (fourth root, 1/4 power, and so on).

It's not easy to extract information about such fundamentals as these from manuals. Most of the time, their authors are all too eager to rush you on into the complex stuff at which they shine.

With this modest boost, you should be able to employ many others of the math functions built into BASIC.

Some examples:

Square root is wanted so often it rates a command of its own, as in SQR(99) for finding the square root of 99.

Used similarly, LOG finds natural logarithms, from which common logs can be obtained by multiplying the natural log by 0.434295. (To go the other way, from common logs to natural, divide by that same decimal fraction. Or multiply by its reciprocal.)

SIN, COS, TAN, and ATN (that last is arc tangent) are built into at least some forms of BASIC in radians, for those who want to know the angle at which to saw a rafter—or to use trigonometry in any other way. The rest of the trig functions can be derived from these.

Take it from there.

Table of Powers

ONE OF THE THINGS a personal computer and its BASIC program can do that you may not have thought of is write mathematical tables.

Try this one.

First bring up BASIC.

When your computer responds (most likely you'll be greeted by an "OK" as a prompt) type the following. End each line with the usual return.

```
10 FOR N=1 TO 15
20 PRINT N,N^2,N^3,N^4,N^5
30 NEXT N
RUN
```

This should give you a lovely screenful of table. For the numbers 1 through 15 you have all but instantaneously compiled a chart of squares, cubes, and fourth and fifth powers. Think what would have happened if, instead of 15 you'd typed 150! (And stop a moment for respectful contemplation in honor of the fellow who first ground out any halfway comparable table with pencil and paper or—for all I know—sharp stick and clay tablet.)

Now, if you have a printer hooked up to your computer, write that tiny program all over again. But this time instead of PRINT in the second line type LPRINT.

Oh, by the way, don't be put off by what may seem the eccentricity of numbering program lines not consecutively but by tens. That's merely so you'll have unused numbers galore if you want to insert more lines later. As for LPRINT, that's just BASIC'S odd way of telling the computer not to put stuff on the screen but to feed it to the printer instead.

Watch Your Money Grow

YOU MAY BE relieved to find that this whole dread subject of interest compounding comes down to a single short, if nasty, expression.

The sum of money you start with is called PV, for present value. How much you'll end up with is called FV (naturally) for future value.

That nasty expression (Line 130 in the program) is the quantity by which the known PV must be multiplied to tell you the unknown FV.

Interest can be compounded at any interval and any rate that you, or your lender or your borrower, happens to choose—and the market will bear and the law will allow. Over any period of time.

In seeing how this works, let's make our example a $1749.45 savings deposit that is to be compounded daily at 8.75 percent. What will that amount to after 27.5 years?

And, what is 8.75 percent interest when expressed as annual yield? This is valuable to know for comparison with any other opportunities described to you in terms of annual yield.

To begin with, examine that nasty-but-short multiplier mentioned a few paragraphs back. Its formula tells you to express the interest rate per period as a decimal, add 1, and then raise this sum to the power of the total number of periods.

Sounds like a chore? Have no fear. Your computer is here.

If you're interested in making this kind of calculation only once you can do it directly and forget programming.

But you're likely to want to do this kind of calculation many times, so you'll save time in the long run if you'll type in a neat program you can save and use repeatedly. By tucking in new figures as you run it, you can get practically instant answers whenever you need them.

Type the following.

(If you're working with MBASIC and you first type AUTO and a return, you won't have to remember to supply the line numbers; the little devils will appear automatically, thus filling you with delight the first time or two it happens. To cancel AUTO when you're done with it, when Line 160 shows, hit control-C and a return.)

```
10 REM (FILE NAME: DY360INT)
20 PRINT "RATE=?"
30 INPUT R
40 PRINT R
50 PRINT "AMOUNT=?"
60 INPUT A
70 PRINT A
80 PRINT "YEARS=?"
90 INPUT Y
100 PRINT Y
110 LET N=360*Y
120 LET I=R/36000
130 LET B=A*(I+1)^N
140 PRINT "BALANCE=$";B
```

After the return following the last line, type RUN and then another return. Answer the queries that appear, using the rate, time, and amount suggested earlier.

(By the way, don't use commas to set off trios of digits when you plug in amounts such as the $1749.45. Commas having a special meaning in BASIC, your computer will be confused and will be inclined to speak sharply to you.)

If you've done everything right, you should find your for-example nest egg swollen to a nice $19,389.40. And you'll know you have a working program you can apply to your own fiscal life.

Anatomy of a Program

SOMETHING for which you may well wish to use your daily-interest program "Watch Your Money Grow" is to compare alternative rates or periods, when making a choice among investments or loans.

For this purpose, a printed record is helpful. In order to make one, LIST the BASIC program (that is, type LIST and a return) and then change PRINT to LPRINT wherever it appears.

Take the time to examine the daily-interest program, called

DY360INT, line by line. You'll see how each one of those lines does its work. The 360 in its name, by the way, means that it follows the convenient, if slightly weird, banking convention of calling 360 days a year.

The first line has been used for a remark. Whenever a line starts with REM (or a single quotation mark, which your computer will have turned into REM by the next time you see it) the program will retain that line and LIST it as a sort of memo but won't RUN it. Use such lines as reminders to yourself—or any other user—of what the program is up to.

The next three lines ask you for the interest rate; stow it away in a specified place you are calling R, and display it.

The two sets of three lines each that follow do the same thing for the amount of money involved and for the number of years.

Line 110 multiplies the number of years by 360, to change it to the number of days, stowing this result in a place called N.

In a single step, Line 120 does two things. It divides the interest rate you put into R by 100 to change it from percentage (the form in which you entered it) to the decimal form called for by the formula. At the same time it divides by 360, to change the rate from yearly to daily.

Now all the data are in and converted to the form the standard compounding formula asks for.

That formula, as you can see from Line 130, says that you start by adding 1 to the daily interest rate in decimal form.

Then you raise this sum to the power of the number of compounding periods—which for daily interest is the total number of days the money will be working.

Next you multiply by the number of dollars you started with, designated as A for amount. Line 140 instructs the computer to print that quantity, called B, as your new balance.

From Interest to Yield

A COMPUTER program for calculating compound interest has another use. It can convert an interest figure into an annual yield. You'll recall yield as that puzzling second figure that crops up alongside interest in such places as bank ads touting certificates of deposit.

It's a useful number for making different investments directly comparable. It has the virtue of combining the effects of rate of interest and frequency of compounding into a single figure.

To find it, using a compound-interest formula, enter the interest rate as usual when asked. But for amount and for number of years, simply enter 1.

The "balance" the program will print will be the amount your single dollar will have grown to in a year of daily compounding at the interest rate you have named.

Now drop the 1 from the left, and move the decimal point two places to the right. For 6 percent interest, you will discover the annual yield, with daily compounding, to be 6.1824 percent.

(You might find it interesting to try a series of guesses to determine, by trial and error, the level of interest rate at which yield becomes a whole percentage point higher than rate.)

Would you like your compound-interest program ("Watch Your Money Grow") to show yield directly? Add these two lines:

150 LET L=(B-1)*100
160 PRINT "YIELD=;L;"PERCENT"

When you're using your daily-interest program to compare alternative rates or periods (perhaps for making a choice among investments or loans) a printed record is helpful. In order to make one, LIST the BASIC program. That is, type LIST and a return and then change PRINT to LPRINT wherever it appears.

How Big a Beam?

Hᴇʀᴇ, in the form of an easy-to-use computer program, is a fundamental building formula that has saved me time and money.

I've used it many times over the years, whenever the urge overtook me to build a new house for my family or to put up a cabin or other smaller structure. It's a calculation that has done even more for me since I enlisted the aid of a computer and turned it into a program that does the job in a trice or two.

Yes, it is a chore to type in the rather bulky set of instructions—unless you enjoy doing it. But you could think of it as an investment; it has to be done only once.

This figure-a-beam program comes from formulas that calculate how much weight a beam can carry without danger of collapsing and without bending more than 1/360 of the length of the span.

(Bending, or deflection, doesn't have much of anything to do with safety. It deals with the less urgent matter of avoiding bounciness that might be unpleasant in a floor or unattractive in a roof, or cause cracking in the plaster of a ceiling.)

Notice that this program uses LPRINT commands. In the strange language of BASIC, that means your results will be printed out for you on paper—assuming that your computer is hooked up to a printer. If not, or if you'd rather read the figures off the screen, just change each LPRINT command to PRINT as you copy the program.

Before getting into the program for finding how much a beam can hold, you may want to calculate how much weight you're going to ask it to hold.

To do this, you should first figure out how many square feet of floor or roof the beam—or girder or rafter or joist—is to hold up.

First, you need to know the span—the distance in feet between the points of support. (This is the same span you will use in the program, but note that there you have to multiply the feet by 12 since the program calls for the span in inches.)

And you need to know the width of the area to be supported. Normally this is the same as the center-to-center distance, less the width of a beam, since the beam or joist is supporting half of each of two spans.

However, at the edge of an area, you'll have a different deal. There's only half of one interval.

(If the floor or roof to be supported extends to form a cantilever, you have to count in that whole distance, since it has no other support. An edge joist or the last beam in a series may also have another chore you must allow for. It may be supporting a wall, and possibly a second floor and a roof, and you must add in all the weights involved.)

Having found the area (the span in feet multiplied by the interval in feet) you multiply by the bearing capacity required. Building codes commonly call for 40 pounds to the square foot for house floors. Less may be acceptable for second floors or roofs.

This result is the total load that you are going to ask each span of the beam to carry.

There's still another factor that comes up occasionally. It becomes significant only when the depth of a beam is about one tenth or more of the span. It's called horizontal shear and refers to the loss of load-bearing capacity as the beam tends to split apart internally.

Shear is rarely anything to be concerned about. Generally a safe limit for this is 120 pounds per square foot. To see if your simple beam (not a cantilever) meets this test, first divide the calculated load by 4 times the product of the breadth of the beam and its height; and then multiply by 3. This applies to any simple beam, whether fixed at the ends or freely supported and whether evenly loaded or with the load concentrated at the center of the span.

For the cantilevers covered by the program, the numbers are slightly different. Whether the cantilever has its load evenly distributed over the span or concentrated near the free end, divide the weight to be carried by 2 (instead of 4) times the product of the breadth and depth of the beam. Then multiply by 3.

If the result in either case is no more than 120, there should be no problem with horizontal shear.

After you've worked with the program that follows you probably could write one for shear to use with it, if you need it. Base it on the information in the preceding three paragraphs. If that gives you any problems, you'll find elsewhere in this section a dual-capacity program for calculating beam load and horizontal shear.

```
10'PRINTS STRENGTH & DEFLECTION OF WOOD BEAM
   (BEAMCALC)
20 INPUT "WIDTH IN INCHES";B
30 INPUT "DEPTH IN INCHES";H
40 INPUT "f=(USE 1100 FOR COAST DF #2)";F
50 INPUT "SPAN IN INCHES";L
60 Z=B*H*H*F/L:LPRINT CHR$(26)
70 LPRINT    "For    SIMPLE    BEAM    or    CANTILEVER
   BEAM"B"by"H"inches,
80 LPRINT "f= "F", span "L inches, allowable bending loading
90 LPRINT "in pounds and (given in parentheses) loading that
100 LPRINT "would cause deflection of 1/360 of the span are:
110 LPRINT:LPRINT "Evenly loaded, freely supported:
120 LPRINT INT(Z*4/3) "( DEF "INT(31290*B*H*H*H/L/L) "）
130 LPRINT:LPRINT "Evenly loaded, fixed bearings:
140 LPRINT INT(Z*2) "(DEF "INT(156440!*B*H*H*H/L/L) ");LPRINT
150 LPRINT "Load concentrated at span center, freely supported:
160 LPRINT INT(Z*2/3) "(DEF "INT(19555*B*H*H*H/L/L) ");LPRINT
170 LPRINT "Load concentrated at center of span, fixed bearings
180 LPRINT INT(Z*4/3) "(DEF "INT(78222!*B*H*H*H/L/L) "）
190 LPRINT:LPRINT "CANTILEVER, evenly loaded:
200 LPRINT INT(Z/3) "(DEF "INT(3258*B*H*H*H/L/L) ");LPRINT
210 LPRINT "CANTILEVER, load concentrated at free end:
220 LPRINT INT(Z/6) "(DEF "INT(1222*B*H*H*H/L/L) "）
230 END
```

When you start this useful program running, it will ask you four questions. If yours is, for example, a 6-inch-by-8-inch beam placed on edge, your first two answers will be 6 and 8. For a surfaced beam, enter actual dimensions in decimal form, 5.5 and 7.25.

Lumber grades for which you would enter 1100 for f include Coast region Douglas fir and No.2 southern pine joists and planks. Select and structural grades may run as high as 1500 to 1950.

Span is the actual unsupported distance in inches. If you were placing joists across a series of beams 6 inches wide that were on 8-foot centers, the span would be entered as 90 (inches) since the actual distance between the beams would be 8 feet minus 6 inches.

When you run this program for a beam of actual width 6 inches and depth 8 inches, stress grade 1400, for a 90-inch span, you should get these useful results:

Such a beam, when it is evenly loaded with its ends not fastened in place, will safely support 7,964 pounds. Fixed in place, it will handle 11,946 pounds.

It will support a 3,982-pound load concentrated at the center of the span if freely supported, 7,964 pounds if ends are held.

In each instance, you'll also be given a DEF figure that is larger than the one for load, showing deflection is not a problem.

If the beam is a cantilever, you'll learn that it can handle 1,991 evenly loaded pounds—but only 995 pounds if that load is concentrated at the free end.

Rent or Buy?

M<small>OST</small> of the time, when it comes to a choice between remaining a renter and becoming a homeowner, the bulk of the long-run considerations favor buying.

But we live in the short run, too. For a real-world decision, it is often necessary to compare the payments about to be undertaken with the amount now going for rent. Or with the maximum you could lay out each month if you had to. What follows is a comparative calculation. It estimates the monthly mortgage payment that will put no greater strain on your budget than the rent you now pay.

```
10 'RENTBUY
20 INPUT  "RENT= ";R
30 INPUT  "ANNUAL INCOME= ";M
40 INPUT  "INCOME TAXES= ";T
50 V=T/M
60 X=R*12*(1+V)
70 Y=X/10000*.043
80 N=V-Y
90 A=(M-X)*N
100 P=R+(T-A)/12
110 PRINT  "PAYMENT=$ ";P
```

In running this program, if you enter instead of your actual rent the heftiest rent you could handle, the calculation takes on a slightly different usefulness. Now it gives you an educated estimate of the largest mortgage payment you dare sign up for.

Naturally there have to be some assumptions embedded in any crystal-ball calculation like this. One is that your income is not about to change drastically. Another is that income-tax rates and the allowable deductions for mortgage interest and other itemized expenses are not drastically changing.

A further assumption built into this program is that you will benefit substantially from the mortgage-interest deduction. If your income, interest payments, and deductible expenses are low, this is not so. Taking the interest deduction at the cost of the standard deduction may gain you little or nothing.

At higher levels of income and mortgage payments, however, this luxury-housing subsidy in the law functions very well, creating enormous tax breaks.

Considering all these imponderables, it would be well to confirm estimates from this program by using totally independent calculations before taking any irrevocable action.

The program asks for three bits of information. In addition to the rent you're handling now (or could manage if put to it) it wants to know your annual income and income taxes.

Note that the .043 figure in Line 70 is for a couple. To run the program for a single person, enter .052 instead.

Replying to the program's questions by successively entering present rent of $900, income $50,000, taxes $9573, a couple should find that mortgage payments of about $1277 would entail no added budget stress. For a single person, it's $1362.

Loan Payments Figured Fast

A<small>LL</small> <small>THIS</small> useful schedule of time payments—for car loans, home mortgages, or whatever—asks you to do is fill in three variables.

Since you're dealing with monthly payments, the formula then expresses (in Steps 120 and 130) the period of years and the rate of interest in monthly terms.

It applies a standard formula, one that has in its day generated whole books of bankers' tables.

Written in BASIC language as I present it here, with a separate line for each tiny step in the interest of clarity, it's less compact than it could be.

But I think you'll find it easy to follow and to type in.

```
10 REM AMORTIZATION
20 REM (FILE NAME 'PAYMENTS')
30 PRINT "RATE=?"
40 INPUT A
50 PRINT A
60 PRINT "AMOUNT=?"
70 INPUT B
80 PRINT B
90 PRINT "YEARS=?"
100 INPUT C
110 PRINT C
120 LET D=C*12
130 LET I=A/1200
140 LET R=B*I/(1-(1+I)^(-D))
150 PRINT "PAYMENTS=$";R
160 PRINT "TOTAL=$";R*D
```

Line 140, the formula that does the work, calls for—take a deep breath now—raising 1 plus the monthly interest rate (in decimal form) to the negative power of the number of payments, subtracting this from 1, then using this mishmash as a quantity by which to divide a figure you get by multiplying that monthly interest rate by the amount of the loan.

Don't be alarmed even if you're not used to math by computer. You don't have to do anything with the program or that formula. Just type it in. Ah, the beauty of computer calculating!

Print a Payment Table

```
10 LPRINT "AMORTIZATION TABLE"
15 LPRINT
20 INPUT "INTEREST PERCENT(TYPE 0 TO QUIT)";A
30 IF A=0 THEN STOP
40 INPUT "PRINCIPAL AMOUNT";B
50 INPUT "NUMBER OF YEARS";C
60 INPUT "PAYMENTS PER YEAR";D
65 INPUT "PAYMENT AMOUNT";R
70 LPRINT A"% " "$"B  C"YRS "D"PER  YR " "PAYMENTS
    $"R
80 I=0
85 LPRINT
90 LPRINT "PMT","INTEREST","APPL TO PRIN","BALANCE"
100 A1=A/100
105 A2=A1/D:B1=B*A2:B2=R-B1:B=B-B2
110 I=I+1
120 LPRINT I,B1,B2,B
130 IF B>.1 THEN 100
140 LPRINT
150 LPRINT
160 GOTO 20
```

What this table shows is amount of payment, how much of it goes for interest, how much for principal, and balance owing.

You'll probably find most of this program easy enough to follow. However, you might wonder why Line 130 doesn't use B>0 as its cue to stop running. Reason is that if the payments, because rounded to the nearest cent, wind up a penny or two short, the table will call for a tiny additional payment. This being silly, it seems sensible to alter the command to say that when the balance drops to less than a dime, let's just forget it.

After saving this table-printing program and using it, you may become critical of one aspect. You may find it a nuisance to have to dig out a separate program to calculate the payment amount you'll need for Step 65. So why not combine the programs?

All you need do is LIST the program you already have and then type below it the supplementary steps that follow.

Now when you LIST or SAVE the program, the new steps will have inserted themselves between Steps 60 and 70.

Since one of these new steps uses its line number, Step 65 finds itself automatically canceled—an easy way to get rid of it now that it is no longer needed.

```
62 LET I=A/(100*D)
64 LET R=B*I/(1-(1+I)^(-D*C))
65 LPRINT "TOTAL PRINCIPAL AND INTEREST=$";R*D*C
67 LPRINT
68 LPRINT "TOTAL INTEREST=$ ";R*D*C-B
69 LPRINT
```

The Mortgage-Payoff Miracle

THERE IS no free lunch. But for millions of home buyers, there is a plan that can produce results that verge on the miraculous. You may view it as a computer-assisted bit of self-hypnosis, but all the same it could save you $80,000 on a $100,000 mortgage. And that ain't hay.

Here's the idea: By making your loan payments every two weeks, in an amount less than half your present or contemplated monthly ones, you can pay off a 30-year mortgage in less than 21 years.

For anyone who gets paid every week or two weeks, this can enormously simplify budgeting. Your paycheck has become smaller but on a consistent basis that's not hard to live with—since you're compensated by never having to think about your home-loan payments. If your bank frowns on payments other than monthly, there's still a way to get this big saving. Make the payments into an interest-earning savings account and pay the mortgage holder from that. Make an additional payment on the principal whenever enough accumulates. Work out with the bank how this is to be done.

In setting up the loan, try to find a bank that will give you a reduced rate of interest because it's a 19-year loan rather than a

30-year one. (Some will, some won't.) Then you'll have an even earlier payoff and an even greater saving.

You can see why many people, not quite sure how they can arrange a payment program like this for themselves, have paid promoters substantial sums to do it for them.

You can test it for yourself by using one of the mortgage-payment (amortization) routines in this book. The easiest way, given the aid of a home computer, is to type in and use the following program, written in BASIC, that will tell all. Unlike most programs you might come across—and most tables of payment schedules—this one can process payments made at any interval.

```
10 REM FILENAME AMORTANY
20 INPUT "INTEREST PERCENT";A
30 INPUT "PRINCIPAL AMOUNT";B
40 INPUT "NUMBER OF YEARS";C
50 INPUT "NUMBER OF PAYMENTS PER YEAR";D
60 LET I=A/(100*D)
70 LET R=B*I/(1-(1+I)^(-D*C))
80 PRINT "TOTAL=$ ";R*D*C
90 PRINT "TOTAL INTEREST=$";R*D*C-B
100 PRINT A"% " "$"B C"YRS "D"PER YEAR" "PAYMENTS $"R
```

Try it first for 30 years with 12 payments a year, and then for 21 years with payments every two weeks—26 payments a year.

Try other combinations till you find the one that fits you best.

Think of the great trips you can make to Paris and Tahiti during those years of no mortgage payments.

INDEX